£9

DEMOCRATIC SOCIALISM AND THE COST OF DEFENCE

DEMOCRATIC SOCIALISM AND THE COST OF DEFENCE

THE REPORT AND PAPERS OF
THE LABOUR PARTY DEFENCE STUDY GROUP

Edited by
MARY KALDOR, DAN SMITH and
STEVE VINES

CROOM HELM LONDON

© 1979 The Labour Party
Croom Helm Ltd, 2-10 St John's Road, London SW11.

British Library Cataloguing in Publication Data

Kaldor, Mary
    Democratic socialism and the cost of defence.
    1. Great Britain — Military policy
    2.'Disarmament
    I. Title   II. Smith, Dan   III. Vines, Steve
    355.03'35'41   UA647

    ISBN 0-85664-886-8

The report of the Labour Party Study
Group on Defence Expenditure, The Arms
Trade and Alternative Employment was
originally published by Quartet Books
as Sense about Defence, London, 1977.

Reproduced from copy supplied,
printed and bound in Great Britain
by Billing and Sons Limited,
Guildford, London, Oxford and Worcester.

CONTENTS

Acknowledgements
Foreword <u>Olof Palme</u>   1
Introduction <u>Mary Kaldor, Dan Smith and Steve Vines</u>   4

PART ONE: THE STUDY GROUP REPORT
Foreword <u>Ron Hayward</u>   18
Introduction <u>Ian Mikardo</u>   19
<u>Background to the Study</u>   23
1. Why Cut Defence Spending? A Summary of the Arguments 25
2. Current Defence Effort and the Necessary Level of
     Cuts   31
3. Defence Expenditure and the Economy   49
4. The Strategic and Political Implications of Defence
     Cuts   51
5. Creating New Jobs   93
6. The Arms Trade   115

PART TWO: THE STUDY GROUP PAPERS
<u>A. Political and Strategic Issues</u>   129
1. Observations on the Proposal to Align UK Defence
     Expenditure with the Average Percentage Spent on
     Defence by the FRG, Italy and France <u>Paul Cockle</u>   130
2. Defence Programme Options to 1980-1 (and Postscript)
     <u>David Greenwood</u>   137
3. Note on the Budget Approach and a Programme Which
     Retains the UK Nuclear Strategic Force <u>Paul Cockle</u>   196
4. Defence Programme Options to 1981: Politico-
     Strategic and Economic Implications <u>Alan Lee
     Williams</u>   203
5. Strategic and Political Implications of Reduced
     Defence Programmes <u>Dan Smith</u>   209
<u>B. Military Expenditure and the Economy</u>   251
6. Note on the Military Expenditure and National
     Product: UK and Certain Other Countries <u>Frank
     Blackaby</u>   252
7. The Resource Cost of Military Expenditure <u>Ron Smith</u>   262
8. Defence Costs and the Defence Industry <u>Mary Kaldor</u>   281
9. The Employment and Other Economic Consequences of
     Reduced Defence Spending (and Postscript) <u>David
     Greenwood</u>   315
10.Note on the Employment Consequences of a £1,000
     Million Cut (at 1974 Prices) in Military
     Expenditure Over Five Years <u>Frank Blackaby</u>   355

C. Industrial Conversion and Alternative Employment   360
11. Military Expenditure Cuts: Note on the Transfer
     of Resources Frank Blackaby   361
12. Aspects of Conversion of Arms Industries Dan Smith   366
13. Alternative Employment for Naval Shipbuilding
     Workers: A Case Study of the Resources Devoted to
     the Production of the ASW Cruiser Mary Kaldor and
     Albert Booth   393
14. Tornado: Cancellation, Conversion and Diversifica-
     tion in the Aerospace Industry Dan Smith   416
15. Community Planning and Base Conversion Dan Smith   445
16. The Lucas Aerospace Corporate Plan Steve Vines   458
17. The Vickers Proposals: Building a Chieftain Tank
     and the Alternative Use of Resources The Vickers
     National Combine Committee of Shop Stewards   479

PART THREE: THE MINISTERIAL RESPONSE
Study into Defence Spending — Summary of Conclusions
John Gilbert, John Tomlinson and James Wellbeloved   505

Appendix One: A List of Papers Submitted to the NEC
Study Group on Defence Expenditure, the Arms Trade   549
and Alternative Employment
Appendix Two: A List of Products Identified as Being
Suitable for Manufacture in the Converted Defence
Industries   551
Notes on Contributors   556
Index   558

ACKNOWLEDGMENTS

The early stages of the Labour Party Defence Study Group's
activity greatly benefited from the work of the Group's
first joint secretaries, Keith Hill and the late Chris
Ralph.

We are also indebted to Dennis J. Nisbet, Member of the
Society of Indexers, for compiling the index.

FOREWORD

by Olof Palme

World military expenditure today equals almost 400 billion dollars per year.   This is more than twenty-five times the amount of money spent on development aid in a world where 570 million people are undernourished, one billion people lack adequate housing and where 70 per cent of the population is still without safe water.

The arms race has now reached such proportions that political leaders everywhere are beginning to realize that, instead of strengthening security, armaments, nuclear as well as conventional, are becoming a threat to the peace and security of the nations involved as well as to the world as a whole.   That unbridled bilateral arms races between nations or alliances can create instabilities and tensions and increase the risk of local conflicts escalating up to general war is now commonly recognized.

Not only is the arms build-up a growing threat to world peace, it also represents a tremendous waste of human, material and technological resources in a world plagued by poverty and hunger.   At a time when the gap between the rich and the poor countries is widening this situation is intolerable from the point of view of solidarity and justice. Nobody can defend a world order based on such grotesque priorities as those which appear when the amounts spent on armaments are compared with those allocated for development. The special session on disarmament which was held by the General Assembly of the United Nations in May/June 1978 may be seen as an expression of the concern felt by the international community about the arms race and its consequences for security and development.

But in spite of a growing awareness of the dangers inherent in the present situation the obstacles to disarmament remain formidable.   Thus the current state of international relations with its faltering spirit of detente is hardly conducive to disarmament efforts.   Distrust is a powerful promoter of armaments.   And we democratic socialists also refuse to acknowledge partition of the world between two immovable and opposing blocs that produce tension and run the risk of dangerous confrontation as a permanent fact of international relations.

In the existing situation, lasting security for the world cannot be achieved merely through equilibrium between the power-blocs on the basis of shared spheres of influence.

Nor can it be brought about by a balance negotiated between the superpowers alone.

The extension of détente must lead to greater co-operation between nations, thus progressively reducing the sources of conflict and grounds for interventions by the great powers in fields that encroach upon the sovereignty and independence of states.

1

But there are also the self-generating forces at work within each nation, political expediency, pressures from interest groups and progress within weapons technology.

We all have an instinctive tendency to be conservative in matters of national security. And political leaders are not always encouraged to become more open-minded and far-sighted by the representatives of the bureaucracies, the military-industrial complexes and other organizations connected with national defence, who – perhaps not unexpectedly – tend to see matters in the light of the more immediate interests of their own organizations and to indulge in 'worst-case-analysis' of the adversary's intentions and capabilities. Under such circumstances the qualitative arms race acquires a momentum of its own and its pace is ultimately only set by the potential for advance within weapons technology. New weapons then see the light of the day, not because they are needed, but because it is possible to develop and produce them. And if there is no strategy for their deployment, a strategy is invented.

Many approaches and technical solutions to disarmament problems can be suggested, but the crux of the matter is that of political will. And the political will to restrain the arms race can and must be supported and reinforced by an en-lightened public opinion. I believe that the general public must now be engaged if there is ever to be any progress in arms limitation and disarmament negotiations. And I am certain that such involvement could also make itself felt in social systems where there is no freedom of expression.

In order to mobilize public opinion both at the inter-national level and the national level we have to make it aware of the magnitude of the resources that are now being devoted to armaments and the tremendous potential for development that these resources represent. In particular we have to deal with assertions that defence cuts will inevitably lead to unemploy-ment and economic stagnation. At the international level a fresh attempt in this direction has just been made by the special disarmament session of the UN General Assembly, which – at the initiative of the Nordic countries – has requested a study from the Secretary General containing an objective analysis of the economic consequences of armaments and disarm-ament respectively.

Against the general background I have just indicated, the importance of the present Report of the Labour Party Defence Study Group stands out clearly. I cannot think of a better and more timely contribution to the international discussion of the conversion problems and possibilities than this Report. It contains a penetrating and lucid analysis of those very issues that need clarification internationally as well as nationally. Common sense may suggest that it should be possible to manage the conversion of resources liberated through defence cuts, whether unilateral or negotiated, in such a way that they further economic advancement, but a well-documented in-depth analysis will certainly strengthen the case of common sense.

Since so much of the discussion about the economic
implications of disarmament is focused on the unemployment
aspect I believe that labour movements and their members have
a special responsibility for examining this issue and for
working out plans and suggestions for alternative production
and employment.   It takes a good deal of hard work to examine
all the relevant aspects of the conversion problems.   And it
also takes courage to question and scrutinize matters which,
because they pertain to national security, are often regarded
as being almost sacrosanct.   The Labour Party Study Group
therefore deserves our respect and admiration for this
pioneering effort, which I think will stimulate thinking
everywhere and serve as an example to other labour movements
all over the world.

In this way this book becomes an important contribution to
the work of the Socialist International in the disarmament
field, which was high-lighted by the Helsinki Disarmament
Conference in April 1978, and which will be continued in the
SI Study Group on Disarmament.

INTRODUCTION

by Mary Kaldor, Dan Smith and Steve Vines

In this volume we present not only the report of the Labour
Party Defence Study Group, which has already been published,(1)
but also the main discussion papers upon which the study
group's work was based. The work encapsulated here was
pioneering in many respects and represents an exercise which
is still unique. This is so not just in the sense that the
questions the study group asked itself demanded original
research employing fresh approaches, but also in another more
important sense; the study group sought solutions to a
variety of problems, seeking to fuse the search for a more
economic defence policy with the concern to revitalize arms
control, seeking to harmonize a policy for international
relations with an industrial strategy for the domestic
British economy. The study group was not asked to create
new policy for the Labour Party, rather, to define existing
Party policy and to assess how and with what consequences it
might be implemented; yet within those confines, there was a
great deal of scope for imaginative approaches to these inter-
locking problems. The purpose of this introductory essay is
to attempt to situate the report and the discussion papers
within their proper political context, to define, in effect,
what it was about the political experience and perceptions of
the Labour Party in the mid-1970s which made this exercise
both necessary and possible. Our hope is that these remarks
will show that although the study group's report was the
product of a particular political situation at a particular
time, it will be of lasting interest especially because of the
principles and underlying methodology of its approach, and
that, although this work was done in Britain referring to
specifically British conditions, problems and possibilities,
it has a wider international relevance making it worthy of
close attention abroad as well as in Britain.

   These days, no essay on matters relating to arms and disarmament is complete without a reference to the apparent
paradox of a climate of international détente accompanied by
generally rising military expenditures by the main military
powers, and by a continuing and even intensifying super power
'arms race'. Despite the evident continuation of conflict
between east and west, there can be little doubt that the
international atmosphere shows a remarkable improvement in
east-west relations compared to the 1950s and most of the
1960s. Whatever détente has not provided, it has at least
diminished the likelihood of a major nuclear conflagration and
removed, at least for the present, some of the major bones of
contention, especially in Europe. These are no mean achievements.

   Yet those changes in international relations summarized in
the term 'détente' have shown little capacity to affect the

4

arms programmes of either NATO or the Warsaw Pact. Since the early 1960s an impressive number of bilateral and multilateral arms control agreements has been signed, resulting from negotiations in a variety of ad hoc and permanent fora, yet for only one category of weaponry, bacteriological, has an actual disarmament measure been agreed. Most of the agreements have been cosmetic, barring states from what they had no intention of doing, or banning actions (such as atmospheric nuclear tests) the need for which had been super-seded by the main participants. Meanwhile, weapons develop-ment has continued apace, providing for both conventional and nuclear weapons greater accuracy, or destructiveness, or lethality, or all three. Since the Strategic Arms Limitation Talks between the USA and USSR began in 1969, the number of American strategic nuclear warheads has more than doubled (from 4,200 to 8,500 in 1977) while for the USSR the figure has almost quadrupled (from 1,100 to 4,000). (2) Since NATO and the Warsaw Pact began negotiations on force reductions in central Europe, both sides have increased their forces in key categories. (3)

The stubborn unresponsiveness of arms programmes to changes in international relations has induced scepticism about détente in many quarters. Much of this may spring from the dashing of hopes which were in any case unrealistic; détente has brought change and improvement, but it should never have been expected to bring a sudden fundamental reform to inter-state relations which were entrenched in hostility, suspicion, and conflicting interests. Détente is not a phenomenon introduced into inter-national relations from 'outside'; it is rather a process which is itself, and always has been, a part of international relations. It cannot outgrow the limitations of the inter-national system by itself; what is required is determined and consistent political effort on all sides, based on the recognition that, whatever the differences, certain interests are shared, not least an interest in survival. Thus the appropriate response to both the apparent unresponsiveness of arms programmes to this process and the set-backs it undergoes is not to condemn détente, whether as a Soviet con-trick, super power collusion, or any other slogan, but to address ourselves to ensuring that its genuine gains can be made less precarious and developed further. Crucially, this means, in terms which now feature regularly in post-summit meeting communiqués, finding a military corollary to détente, and therefore con-fronting the obstacles to that corollary.

Validation for arms programmes is traditionally provided through the actions or perceived intentions of the internation-al adversary. It is no surprise, when relations between NATO and the Warsaw Pact have thawed, to find this form of valida-tion treated with increasing scepticism, while theories which explain the 'arms race' in terms of domestic pressure gain wider currency. Such theories exist in many forms, variously stressing the role of the profit motive, the entrenched power of military industrial corporations, the influence of bureaucratic politics and inertia, technological momentum, the pathological condition of the societies producing large quantities of arms, and so on. For many of them, the target of their analysis is the military industrial complex of any or all of the major arms producing countries, a group of institutions whose constitution and role are also defined in a variety of ways. It is not necessary to subscribe to one rather than any other of these contending theories in order to accept that, taken together, they have demonstrated that a strong input to arms programmes is domestically determined and has little if any relevance to the international scene.

Neither a crude depiction of the malevolent influence of a shadowy military industrial complex, nor a monocausal explanation of domestic factors as the sole determinant of arms production is required to agree that some parts of the process of producing armaments simply are not susceptible to changes in the international environment. From this it is but a short step to acknowledge, first, that international negotiations alone are inadequate to the task of controlling and reducing armaments, and, second, that any serious effort to develop an international programme of arms reduction or disarmament must include unilateral actions by governments. Such actions must include not only the reduction of certain armed forces but also, perhaps more importantly, change in the institutions now involved in the development, production, procurement and deployment of armaments.

Unilateral actions along these lines could be conceived of in three ways. First, almost irrespective of a desire to contribute to worldwide arms reductions, they could simply amount to a realignment of the provision of armed force to a level which is economically more feasible given other social priorities and calls on the national income. This approach played no small part in the work of the study group; the report argues that Britain overspends on its military establishment and needs to reduce the burden, if for no other reason, as part of a sounder allocation of economic resources. As the study group's work proceeded, especially in assessing the prospects for converting military industry to other work, it was realized that such a reallocation could create opportunities holding great potential benefit. Second, it can be argued that Britain's armed forces have tasks and capabilities which are of decreasing relevance in the modern strategic and political environments, but which remain because of the inherent conservatism of military institutions; abandoning these tasks and capabilities need have no ill effects on security and could be seen simply as a way of enhancing the efficiency of the forces. It may be somewhat ironic if the occasion for effecting such changes is a proposal to reduce military spending, but it is no bad thing of itself, and may well be far harder to achieve in the absence of firm budgetary pressure. Third, reducing military spending and reorienting the military posture could be consciously directed towards promoting international arms reductions and furthering détente. The advantage of unilateral initiatives of this kind is their potential for breaking through the log-jam at international negotiations where the participants' unwillingness to make the first move provides dim prospects of making any progress. Evidently, this aim would place certain constraints on the options for military restructuring - for example, a cheaper but more aggressive posture would necessarily be ruled out, as would greater reliance on nuclear weapons. Whether or not the desired effect would be achieved is, of course, an imponderable, but there could be some impact, creating opportunities for similar action by other states.

If unilateral action is to be conceived of in this third way, however, unilateralism alone is probably not enough. If we avoid monocausal diagnoses of the problem, one-track solutions make little sense. Unless unilateral action is accompanied by active participation in the available international fora of arms control and disarmament, it stands no chance of being anything more than a matter of economic and military management; what is important is that it should also be an integral part of a new foreign policy. The past fifteen years are eloquent evidence of the inadequacy of international fora alone to the task of slowing the pace of arms programmes and reducing weaponry in which the military of any side is strongly interested; therefore some emphasis must be switched to tackling the task within the domestic arena, but quitting the international arena altogether is no answer. If defence policy is to be consciously designed to, among other things, contribute to making genuine disarmament measures possible, overall foreign policy cannot adopt an isolationist attitude.

Thus, the work of the study group represents an attempt by a major political party to re-examine defence policy from the perspective of disarmament; that it was in Britain that this task was undertaken must be explained in terms of Britain's historical situation. In brief, Britain can no longer afford, and probably does not need, the range of military capabilities, including a large and advanced arms industry, that was inherited from an earlier imperial era, and this has given rise to a debate about British defence policy that is unprecedented in post-1945 history.

Now that Britain has disentangled itself from all but a few of its extra-European military commitments, the contribution to NATO accounts for about 90 per cent of the current defence budget. Despite Britain's poor economic performance, it continues to play a leading role in NATO, fulfilling major responsibilities on land, at sea, and in the air, as well as maintaining a small nuclear force. Although the military budget is lower than those of both France and the Federal Republic of Germany, Britain has the most fully rounded military capability of any European NATO member. This relatively ambitious posture is supported by an extensive arms industry, probably the largest in Europe. Apart from the Swedish arms industry, it was the only one to survive uninterrupted through-

out World War II, since when it has been perennially subject
to problems of excess capacity. The cost of maintaining
Britain's current defence posture and the concomitant military
industrial base, reflected in the high share of Gross National
Product devoted to defence compared with other European
countries, (4) is out of proportion to Britain's economic and
political status, and may, as the study group's report suggests,
have contributed to the decline of both. Indeed, it is argued
that since as far back as the 1880s military spending has
imposed a burden on the British domestic economy which helps to
explain both Britain's economic decline and consequent loss of
political power.

Among its other effects, this continuing decline in turn
limits the resources available for military spending, producing
a mismatch between the size of the potential output of the arms
industries and the size of the industries' major market, the
defence budget. It is this, together with changing production
processes, which puts pressure on employment in those industries, felt most severely in the declining regions of Britain
– such as the North East of England, Northern Ireland, the
West of Scotland – which are the areas where the older heavy
industries are located and where, in particular localities,
dependence on the defence budget has long been high.

In the sphere of military operations, this basic mismatch
goes far to explaining some of the anomalies of British
defence policy. While the effort is made to sustain a
rounded military industry and military posture, unavoidable
pressures on the budget produce a concentration of military
effort into western Europe and the eastern Atlantic, matched
by the concentration of resources onto a few major weapon
systems which are both extremely expensive and extremely
complex. In their attempts to stem the rising cost of weapons
procurement, successive governments have narrowed the range of
military equipment produced without equally narrowing the range
of military missions; the producers of equipment have been
amalgamated, and the military have been persuaded to carry out
different tasks with the same types of equipment. The result
has been the emergence of multi-role weapon systems, such as
the Tornado Multi-Role Combat Aircraft and the three Invincible
class Anti-Submarine Warfare Cruisers, the two most important
examples of this tendency, both of which are described in detail
in the study group papers. These projects, and a few others
like them, are extremely costly, imposing a disproportionate
burden on the defence budget which detracts from the investment
that should be made in other less glamorous but no less
important military capabilities; for example, given the way
in which budgetary resources are shared out between the three
services, it can be shown that the commitment to Tornado has
directly contributed to the decline of RAF transport capacity.
For a relatively poor country like Britain, some tough
decisions are necessary in order to reassert a balance
between the various military roles it wishes to sustain, or
else it is likely to end up doing everything, but doing none

of it properly; narrowing down the range of missions undertaken may need to be among the decisions

These problems coincide with a general questioning of the
role of the British armed forces. War in Europe has come to
seem a more remote prospect, and actual military engagements
have been less grand and often closer to home - Northern
Ireland, the 'Cod War', the firemen's strike. One school of
thought, associated with the writings of Brigadier Kitson,(5)
has argued that the concentration upon an unlikely East-West
war has diverted attention away from the threat from the
'enemy within'. This questioning of the general role of
armed force, which takes place both inside and outside the
military is accompanied by a more detailed questioning of
how its role should be carried out.

The success of small accurate guided missiles in Vietnam
and in the 1973 Middle East war suggests that weapon systems
like Tornado and the Invincible cruisers may no longer prove
to be the appropriate form of warfare, because of their
vulnerability and the expense of replacing them. These
technological advances suggest the need for a major re-
ordering of the military postures of NATO states. And in
the British forces, these doubts are compounded by dis-
satisfaction with pay and equipment; penny-pinching in
manpower and support costs has been the corollary of attempting
to maintain a wide range of military capabilities based on
sophisticated weapon systems. Piecemeal adjustments of the
defence budget in the 1970s have barely touched the major
projects, and the effects have thus been felt elsewhere in the
budget, apparently resulting in a serious decline in morale
and difficulties in recruiting and retaining military personnel.

A further consequence of the gulf between the international
status to which Britain still aspires and the resources it can
afford to devote to defence is the increasing dependence on
the arms trade as a way of sustaining military industry. Heavy
sales to repressive regimes abroad have given rise to moral
indignation, combined with a more pragmatic concern at the
possible political and economic effects. On the one hand,
many of Britain's best customers in this field are inherently
unstable, and the trade ties Britain into a network of agreements and commitments which is both discreditable and potentially dangerous; on the other, Britain's enthusiastic
participation in the trade contributes to a process whereby a
scarce economic surplus worldwide is frittered away on
purchase of military equipment instead of being used to tackle
some of the major problems of underdevelopment. At a verbal
level, British government ministers have committed themselves
to seeking and supporting restraints upon the arms trade, but
these words will not be translated into action without, as a
precondition, a serious effort to resolve the contradictions
of defence policy.

These various manifestations of the longstanding dilemmas
within British defence policy have emerged in a period of
changing international relationships, reflected in the growing

multipolarity of world affairs — the evolution of west Europe
as a political and economic entity, the relative decline of the
USA, the intensification of the Sino-Soviet conflict, the
growing power of parts of the Third World, and so on.   One
outcome of these general changes is the parallelism of the
recognition of the need to establish a secure   modus vivendi
with the USSR, with a new alarmism about Soviet strength in
Europe and worldwide.   This has generated renewed emphasis
upon the need for NATO solidarity and increased national
contributions to the alliance, which has both exacerbated
Britain's own defence problems and, by threatening to erode
detente, added to the urgency of rethinking orthodox
approaches to defence and national security.

The study group's proposals for reducing the scale of the
military effort in a way that would serve the goals of disarmament and social well-being are not just of insular
interest, for many of these problems are affecting other
countries as well.   The cost of carrying out given military
tasks is rising everywhere as a consequence of technological
change, and while Britain's economic ills are particular to
its situation, restraints on public spending and the problem
of excess capacity are widespread.   It is not only military
doctrines in the British forces that are challenged by new
technologies, and all countries with advanced arms industries
have increased their dependence on the arms trade.   Within
NATO, questions of equipment standardization, military
integration and industrial specialization have come to the
fore because of shared economic and political problems while,
outside NATO, the Swedish government must now decide whether
it wishes to have a home-produced frontline combat aircraft
and sizeable aircraft industry in the future.

It is therefore to be expected that other countries will
have to face similar issues to those which in Britain have
prompted a new defence debate, marked by an upsurge in the
number of newspaper column-inches and radio and television
minutes devoted to military issues.   Around one central
point the debate is sharply polarized: while the Conservative
Party seems set to resolve the mismatch between resources and
defence policy by increasing the budget, the Labour Party, as
spelled out in this report, would reduce the tasks to bring
the defence budget down and convert some military industrial
capacity to other work.

A number of features mark out the distinctiveness of the
report's proposals, whose substance will be missed by attempting to approach them as 'just another' set of defence cuts.
While the report recognizes the inter-relationship between
defence policy and arms control, it also recognizes the interdependence of defence and industrial policy.   Arguing the
need for a unilateralist ingredient in arms control policy is
based on an awareness of strong institutional pressures for
armament, including, but not confined to, pressures from
military industry.   Any attempt to secure lasting and
effective measures of disarmament must tackle these institutional pressures directly.   Equally, a positive industrial

strategy is ultimately fettered by defence since it absorbs
research and development resources and skills and distorts
the whole direction of industrial innovation, introducing a
tendency to concentrate in civil fields on projects which
most closely resemble military projects, such as the nuclear
power industry and Concorde .

Post-1945 British defence policy contains numerous examples
of government attempts to change defence policy without
tackling the underlying institutional structures - the 1957
White Paper, the 1966 White Paper, and the 1974 Defence
Review; contrariwise, governments have tackled the institutional structure, as in the cancellation of TSR-2 during the
1960s, without attempting to alter the policy - the TSR-2
requirement was allowed to stand, and will now be fulfilled in
part by Tornado. Either way, previous attempts have failed
to achieve lasting solutions to the problems, and the 1974
Defence Review is certain to be added to the list of imperm-
anent answers because of the institutional pressures. Policy
change and institutional change must go together if the
results are to be lasting.

Much of the report is devoted to assessing and affirming
the possibility of institutional change within the arms
industries through the mechanism of converting some of their
capacity to other work. But the proposals for change in the
armed forces - maintaining only one service in its present
state while the other two are shrunk to less ambitious
dimensions - should be seen within the same context, as
proposals for institutional reform as part of a programme of
reducing military spending. These proposals are slanted
towards taking advantage of recent technological developments
in light missiles to deploy comparable defensive strength at
reduced cost. This will require not just shrinking the
services, but carrying through changes in force structure and
doctrine, weaning them away from their present emphasis on the
major weapon systems and this, it must be stressed, denotes
fundamental change with the likelihood of strong institutional
resistance.

Industrial conversion can thus be seen as one part of a
two-pronged programme of institutional reform. On the other
hand, understood as a question of shifting economic resources
from one use to another, it can also be seen as an aspect of
economic development. Although some who favour this re-
allocation of resources argue that government planning is not
necessary, the report insists upon the opposite view.
Resources, it argues, will not necessarily move by themselves,
or not to the right destination; without planning, and in
the context of domestic and international economic crisis and
the poor investment record of British industry, there would
be a danger of collapse especially in the least efficient
sectors dependent on defence. It should be noted at this
point that some critics of the report appear not to have
read, or to have misunderstood, the section in which these
issues are discussed.

The industrial strategy set out in Labour's 1974 election manifestos and 1976 Programme goes some way towards answering the problem, providing key mechanisms in the proposed National Planning Commission, National Enterprise Board, and planning agreements. But even as presented there the strategy has problems in coping with questions of industrial innovation. There is a danger that the industrial strategy might simply reinforce misdirections in technological policy, concentrating upon sustaining declining industries, or tilted towards capital intensive sectors. This problem is highlighted when considering conversion of military industries because important parts of those industries are in long-term decline and themselves represent a misdirection of technological policy.

Part of the answer lies in 'grass roots' initiatives, of the kind discussed in the study group papers, and most notably the plan for alternative production advanced by the shop stewards at Lucas Aerospace. These proposals, and similar approaches developed elsewhere, have grown out of the concrete conditions in which these workers find themselves, conditions of deskilling and steadily diminishing employment even in the absence of defence cuts. Jobs are lost in these industries for precisely the same reason as in other industries − the development of new technologies, mechanizing production, reducing the human presence in production processes − but in the defence industries the pattern has been more pronounced than in most others, because they are generally more capital intensive and technologically advanced.

It is therefore understandable that the first signs of a break in this chain of new technology and unemployment have come from workers in the defence industry who, situated at the sharp end of technological progress, have begun to challenge the assumption that they must foot the bill for technological advance. They have used their skills and detailed knowledge of their industries to develop alternatives to the waste of resources. In doing so, they have offered a profound challenge to orthodox assumptions about industry and planning.

Proposals such as those at Lucas Aerospace and Vickers do not offer a fully fledged plan for taking over management functions. That would be inappropriate: important experience is lacking; workforce skills may in some cases be unsuitable or obsolete (a problem at Vickers, and one also recognized in the Lucas shop stewards' proposals for improved employee development schemes) or may lead to imposing inappropriate technocratic solutions to social problems − only a limited range of the National Health Service's problems can be solved through the medical machinery proposed in the Lucas plan. Instead, the essence of the proposals, and the mainspring of their challenge to orthodoxy, is to extend collective bargaining into the choice of products and production process-es, embracing a perspective of rehumanizing industry by virtue both of what is produced and how it is produced.

The Lucas plan in particular raises the critical question of technological priorities. How can it be right to devote such sophisticated skills and resources to the task of flying a

businessman to New York in time for lunch, while old age
pensioners in Britain die from hypothermia and almost half the
world's population lives in poverty? Why are Lucas workers'
skills devoted to making components for advanced weaponry,
when there are four potential users for every one kidney
dialysis machine in Britain? In seeking to assert a different
order of priorities, the Lucas Plan follows two complementary
paths. The emphasis is on socially useful and socially
responsible production – making things which people really
need, which do not damage the environment, and which eliminate
health hazards in production processes. This is paralleled
by the proposals in the plan which would reduce the alienation
of worker from product – the plan rejects the premise that
those who sell their labour power should have no control over
its use. Thus the Lucas plan affirms the right of workers to
have a say in what they produce and how, and to choose to
produce goods with a genuine social value according to their
scale of priorities.

This does not just involve identifying alternative products
and production methods, but also establishing direct links
between the workforce and the consumers (for example, the NHS
for kidney machines, the Tanzanian government for the hybrid
road/rail vehicle), in order both to strengthen the bargaining
position and to ensure directly that what is produced is what
is really wanted – a socialist alternative to the mediation of
the market mechanism between producer and consumer.

The combination of proposals for socially useful production
and changed production processes has met with a stubbornly
negative response from the Lucas management which, apparently
resting on management prerogatives, seems to prefer industrial
strife and redundancies to new products. Together with the
inability or unwillingness of the government to intervene
decisively, the response to the proposals is a sad commentary
on present attitudes and institutional inflexibility. It
indicates the need for a sustained programme to overcome such
obstacles, and the possibility that new statutory powers would
be required by government.

While such proposals, stemming from the immediate problems
faced by the workforce, must primarily be seen as products of
current industrial conditions, they also lay the basis for
establishing an alternative institutional structure. As the
Lucas management's response shows, this alternative will not
come about of its own accord, which is one reason for government involvement in planning. Here, a careful balance must
be struck between the need for government to be involved as
co-ordinator and provider of needed resources (possibly
including statutory instruments), and the need for this
involvement not to swamp the process. It is important not
to stifle the genuinely democratic planning of new production
and the communication between producer and consumer upon
which the whole enterprise rests.

A central consideration here is the politics of technology
– the question of who controls the use of technology and to

what ends. The perspective on industrial conversion outlined here includes a democratization of the politics of technology, through the mechanism of collective bargaining, both so that people may have greater control over decisions affecting their lives, and because this will ensure important inputs to the planning process. It would not be counted a particular gain if the workforce were still excluded from planning decisions, but by a government bureaucracy instead of a business bureaucracy. Government involvement in planning must therefore permit and encourage the emergence of alternative institutional structures in the industries, which could generate a new set of institutional interests vested in peace and disarmament rather than in the production of arms. This could both ensure the permanence of military reductions, and begin a steady advance towards disarmament, recognizing that disarmament is a process and not a single act.

Of course, these are great ambitions, but it is exactly such ambitions and such new institutions which are needed to reach solutions to the problems of world armaments. Clearly, whatever the specific differences between each country's industries and military posture, such proposals and perspectives for industrial conversion have more than purely British interest.

We have thus come full circle in showing how the study group's approach was based on the links between international relations, defence policy, and industrial policy, and how the perceptions on which the approach was based grew from a mixture of problems, some specific to Britain's situation and history, others relevant throughout the international community. It is not argued that achieving these changes in policies and institutional structures will be an easy task; one of the merits of the report is its awareness of the complex interdependence of the problems, and the consequently necessary variety of measures needed to achieve solutions. What the report argues is that this task, though complex, is possible given the right combination of political will and planning, and, more important, that it is necessary.

One further comment on the report is required. When it was prepared for presentation to the Labour Party National Executive Committee and for original publication, great care was taken to ensure that all the figures in it were as up to date as possible. For publication in this volume, the figures could once again have been updated, but were not for the simple reason that they would almost immediately be outdated once more. What is of lasting interest is not the figures, but the way in which the general problem is identified and defined, and the principles and methodology underlying the proposed solutions. It is to these features, in both the report and the papers, that we especially direct the reader's attention.

14

Notes

(1) The Labour Party Defence Study Group, _Sense About Defence_ (London, 1977).

(2) The Stockholm International Peace Research Institute, _World Armaments and Disarmament, SIPRI Yearbook 1977_ (London, 1977) (the 1978 _Yearbook_ did not update these data).

(3) _The Military Balance 1973-1974_ and _The Military Balance 1977-1978_ (London, 1973 and 1977).

(4) Whatever the other discrepancies between their data, both SIPRI and the IISS show Britain spending a larger proportion of its GDP (in the case of SIPRI data) and GNP (IISS) than other west European members of NATO; see the 1978 SIPRI _Yearbook_ and _The Military Balance 1977-1978_ .

(5) See F. Kitson, _Low Intensity Operations_ (London, 1971), and his more recent _Bunch of Five_ (London, 1977).

PART ONE: THE STUDY GROUP REPORT

FOREWORD

The Labour Party is publishing this document both as a contribution to the widespread debate about defence as well as a demonstration of ways in which the massive defence burden can be reduced. Successive Labour Party Conferences and the 1974 Manifestos have addressed themselves to this problem and expressed concern about the high levels of defence spending but this will be the first time that we have presented a detailed study which examines the practical consequences of taking action to reduce military expenditure.

This document presents the findings of a three and a half year study by a group of defence specialists who have been discussing and contributing written evidence to the National Executive Committee (NEC).

The National Executive Committee would like to record its gratitude to the members of the Study Group on Defence Expenditure, Alternative Employment and the Arms Trade which met under the Chairmanship of Mr Ian Mikardo MP. Members of this group and staff at Transport House have devoted long hours of labour to the production of a Report for the NEC on which this book is based and we believe that their efforts have been well rewarded by the quality of the outcome of their work.

Ron Hayward

(The General Secretary)

INTRODUCTION

by Ian Mikardo, MP

When the Interim Report of the Labour Party
National Executive Committee's (NEC) Study Group on
Defence Expenditure, the Arms Trade and Alternative Employment was published in 1976, it was seen in the press as representing a radical departure from previous policy. The truth of the
matter is that Labour's interest in reducing defence spending has
a long and consistent history. The discussion was brought to
prominence in the early 1930s when Labour Party Conferences
were already expressing concern about the waste of resources
devoted to the production of arms, as well as the danger that
this posed for world peace.

At the 1931 Annual Conference of the Labour Party, Hugh
Dalton (later to become the Chancellor of the Exchequer) moved
a resolution which read in part:

'This Conference reaffirms its belief that the present expenditure on armaments by the nations of the world is a danger to
peace and to the security of the peoples, and represents a wasteful
and unproductive use of a large part of the world's resources.'

The resolution was carried almost unanimously. The Labour
Party's concern, however, didn't stop here. It is a tribute to the
foresight of our predecessors to record that, as long ago as 1929,
the Transport and General Workers' Union took the initiative
in calling for the establishment of a committee to examine the
economic consequences of disarmament. A tripartite committee

was established in the same year, consisting of representatives from the General Council of the TUC, the National Executive of the Labour Party and delegates from the Parliamentary Labour Party. In 1931 this committee made the following recommendations:

'(a) That alternative work of suitable character not at present available to industry should be provided.

(b) That specific Government action to increase the employment in the industries chiefly affected should be taken; and in the event of neither (a) nor (b) having the desired effect –

(c) There should be adequate monetary compensation to the work people displaced.'

In 1977, after a period of some five decades, we are still facing the same problems in the defence sector; in many ways, these problems are more serious. The consumption of valuable resources by the defence industries has caused serious structural distortions in the British economy, and the very nature of modern-day warfare is more terrifying and promises even greater devastation than could have been contemplated by our predecessors.

Nevertheless, our approach to these problems remains much the same as before. In practice, this means that we seek to retain a level of defence capability appropriate to our needs, and that any cuts consequently required should not cause unemployment but be transformed into opportunities for greater economic prosperity and a more rational apportionment of the taxpayers' money. Indeed, it could even be said that the aim of our present proposals is relatively moderate. The Labour Party has translated the deeply rooted feeling that current levels of defence spending are too high and simply transformed this view into the proposal that an economically weak country like Britain should not devote a greater part of her Gross National Product to defence purposes than our European allies like Germany, France and Italy. The remit was merely to quantify the effects of the policy decided by the party and to analyse, in depth, the consequences of carrying out a reduction in defence spending of this order. This book represents the product of the NEC Defence Study Group's research and is, to the best of my knowledge, the most detailed analysis of its kind yet to have been produced.

Much of this book's emphasis relates to the economic and industrial impact of defence expenditure, and there are sound reasons for this, Nevertheless, I think that it is worth stressing our very real concern about the dangers of a mounting military build-up leading to a possible Third World War. It is alarming that the tremendous skills and ingenuity of mankind have concentrated to such a great extent on the destruction of human life rather than its enrichment. While this book is being printed new weapons will be tested around the world to find better ways of killing people; today, we have a new bomb (the neutron bomb) that can kill people as a result of irreversible brain damage yet do no harm to buildings; tomorrow, another weapon might be developed which just causes some new and hideous form of physical or mental disability – and all this is done in the name of 'progress'. In 1955, when the nuclear arsenals were relatively sparse compared with today, Dr J. Robert Oppenheimer, the chief designer of the first atomic bomb, was asked on American television whether it was 'true that humans have already discovered a method of destroying humanity'. He replied: 'Not quite. You can certainly destroy enough of humanity so that only the greatest act of faith can persuade you that what is left will be human.' This, therefore, is the spectre of world war at present.

With this background in mind, I hope you will find this book worth reading and studying. What we seek to do in this book is to make a positive contribution to the wider debate about defence questions. Unfortunately, however, press comment to date has been not only misinformed, but clearly misleading. We hope that the publication of the full text of this report will provide an opportunity for a discussion on this subject based on its content rather than on emotive reactions.

BACKGROUND TO THE STUDY

The Labour Party has advocated a reduction in defence expenditure and cutbacks on the arms trade for many years. Major economic reasons for this attitude are:

1. Money spent on defence could be better used to provide public services such as housing, education and health.
2. The defence industries impede productive economic development by wasting valuable economic resources and diverting investment funds away from those industries which make socially desirable products.

This problem has been highlighted by the recent debate on public expenditure. The Labour Party cannot accept that Britain, with all her economic troubles, should go on spending a greater proportion of her Gross National Product on defence than most other Western European nations. For this reason, the 1974 General Election Manifesto promised to: 'reduce the proportion of the nation's resources devoted to defence so that the burden we bear will be brought into line with that carried by our main European allies'.

An important objection to such proposals is that they would lead to unemployment in the defence industries. We know this, and have therefore stressed methods of creating alternative employment as well as the need to use the considerable skills of workers in these industries to make a positive contribution to Britain's economic revival.

The National Executive Committee set up a study group to look into the implications of Labour's Manifesto commitment, and in particular to:

- assess the reduction in defence spending needed to meet the Manifesto commitment;
- provide a list of options of defence cutbacks which would achieve this projected reduction;
- examine the implications of these cutbacks for Britain's military and political strategy;
- calculate the impact of such cutbacks on our economic and employment situation;
- consider ways of ensuring that unemployment – even in the short term – was not a result of this policy; and
- look into possible criteria which might be used to control arms exports and examine both the impact on employment of a possible reduction in the arms trade and the opportunities for alternative employment.

The report which follows is based on a detailed, three-year study of the issues raised. The study group received evidence from a wide range of defence specialists, trade unionists from the defence industries, and politicians with an interest in this area.

The study group believes it necessary to underline the point that vast sums of taxpayers' money are spent under a cloud of quite unnecessary secrecy. No information was asked for which might be considered as undermining national security, nor would it have been thought appropriate to do so. In spite of this, in the earlier stages of the study, requests for information were frequently met by recourse of the Ministry of Defence to the claim that the disclosure of information would be prejudicial to national security, though this was not a claim that was repeated at later stages. Not only did this waste the group's time, but it illustrates that those responsible for a sector which currently consumes almost 11 per cent of the public purse are operating in an unhealthy atmosphere of distrust and secrecy which offers no guarantee that public responsibility for defence activity is not being evaded.

1

WHY CUT DEFENCE SPENDING?
A SUMMARY OF THE ARGUMENTS

> *'Just as it would be wrong to endanger national security in our concern for social justice, so it is no good having a defence policy which could bankrupt the society it is designed to defend.'* – Rt Hon. *Fred Mulley, MP, Secretary of State for Defence*[1]

Any discussions on defence spending usually centre on the alleged dangers of reducing expenditure. Yet it is equally valid to ask whether the real threat to our society does not in fact come from excessively high levels of expenditure which waste precious resources and so retard economic recovery. At the same time it is important to recognize that any reductions in defence spending must be accompanied by positive action to create alternative employment for those who work in the armed forces and defence industries. Such a programme of conversion would demand urgent governmental action, and the departments concerned should therefore initiate feasibility studies. And such work should be seen as an integral part of Labour's industrial strategy.

Although this study concentrates on the economic aspect of defence spending, this is to be seen as a by-product rather than the core of the subject. Defence expenditure has a clearly defined purpose: it is a form of insurance which we would not need in an ideal world, but which the present state of international relations obliges us to provide. Yet it would be absurd to find that the chosen form of insurance was itself contributing to the danger it sought to avert. In other words, does not a high level of defence spending represent a threat to world security, and may not any moves to reduce this commitment lessen the tension? Meanwhile the high level of spending and the associated wastage of resources contributes to maintaining inequalities between the industrialized and non-industrialized nations – itself a further source of tension. And, as it has been argued, 'Our nuclear missiles are already killing people now, because we waste on the missiles resources which could be invested in ensuring food supplies for the world.'[2]

The arguments against maintaining the present level of defence expenditure illustrate the background to Labour's policy.

The Economic Argument

Example 1: The new Tornado (Multi-Role Combat Aircraft) project will cost about £7,000 million[3] – more than the total cost of Britain's health and personal social services for 1976/77.

Example 2: The price of the Frigate *Ambuscade* at £16 million would provide a new 508-bed hospital in Bangor.

Example 3: The submarine *Superb* is more expensive than building 4,000 new homes.

These examples indicate the magnitude of the defence burden – a burden which overall consumes 11 per cent of the nation's total public expenditure (including capital spending). More alarmingly, our defence bill accounts for nearly a quarter of the current expenditure on goods and services provided by central and local government.[4]

These facts are already fairly well known, but may hide the even greater costs of wasted resources. The defence industries employ some of our most skilled workers, consume products appropriate for capital investment and have in part been responsible for the distortion of industrial development in Britain. To state it bluntly, the defence industries have swallowed an important part of the resources so urgently needed for Britain's industrial regeneration.

Certainly Britain has an impressive record of arms production, but it has been at the expense of the socially productive side of British industry. It is often said that Britain's role in the international arms industry is a key factor on the positive side of our balance of payments. This is an exaggeration: at their peak, arms exports account for less than 3 per cent of total export earnings. Moreover, profits from arms exports are qualified by the need for a large home market, both to inspire confidence in the product (nobody will buy defence products from a country not prepared to use them itself) and to cover the costs of research and development. Later we will see how the export record on the military side of industries which produce both civilian and military equipment is less impressive than the sales of civilian goods. Thus considerable potential exists for both import substitution and the sale of advanced technological merchandise which could follow from a reduction in arms production.

As we have said, crucial resources and key workers are being diverted into defence production. This might not be so damaging if it could be shown that they were being sensibly used. Chapter 3 demonstrates that this is hardly the case. First, the structure of the defence industries and military bureaucracy tend to force the Government to adopt the most expensive solutions to strategic problems. Secondly, these industries are less successful in creating jobs and generating exports than their civil counterparts. This certainly suggests that some conversion of the defence sector to civil use would be more likely to improve than to damage our prospects for economic advancement.

The conversion of the defence industries is not merely an abstract proposal since the kind of changes suggested have been made elsewhere and on a much larger scale. Chapter 5 gives an industry-by-industry breakdown of conversion potential. In Britain today, much thought is being given to ways of releasing the defence industries from exclusively military production. And much of the present initiative is coming from the shop floor itself, thus giving the plans a sound basis in practical experience.

The Arms Race and the Arms Trade
The dilemma of a country like Britain, whose present economic weakness is counterpoised by a relative military strength and a capacity for high arms production, is best understood in the global context. Certain protagonists of the *status quo* argue that Britain would be foolish to abandon her share of the world arms market for an uncertain future bereft of military might. This argument is then taken further, and the existing size and importance of the defence industries is seen as a main reason for continuing to develop their potential. Arms manufacturers and

the government defence bureaucracy are then in a position to argue for an expansion of the military machine almost as an end in itself. This process is a continuing one, not only in Britain but in most other industrialized countries, both East and West. As others step up their arms production, so the domestic hawks see their hand strengthened as they point to increasing military power elsewhere. This is the process commonly called the 'arms race' – a race which develops an ominous momentum of its own and which can be stopped only by enlightened political action. Yet, as our studies indicate, the pressure on the defence budget from rising costs of various kinds is such as to require defence cuts simply to stand still.

The British arms industry is dependent upon arms exports. It seems rash indeed to rely on the arms industry as a major factor in Britain's economic stability, such investment being based on the inherent political instability of the customers. In other words, Britain's economic future would be dependent on recurrent wars in places like the Middle East and the continuation in power of repressive régimes. Such an attitude is not only immoral and anti-socialist but also short-sighted in both political and commercial terms.

Since Britain is a leading manufacturer of arms, she has at least a limited influence in controlling the international arms trade. In Chapter 6 this point is discussed more fully and the following proposals are made:

1. The British Government should continue to work for a multilateral arms control agreement involving both importers and exporters; while recognizing that the problems of trade are inextricably linked to the problem of production.

2. The Government should consider refusing to supply arms to states involved in international disputes, or to those régimes with a proven record of using torture and repression against their subjects.

3. A Register of Arms Sales – possibly under the auspices of the UN – should be set up. Such a central pool of information is vital if the arms trade is to be controlled.

4. Even if an international register cannot be agreed, a British register should be established.

Undoubtedly there would be considerable problems in implementing these proposals, but it could certainly be done, given the right political atmosphere.

The Political and Strategic Argument

Our high level of defence spending is a serious economic handicap. Nevertheless, Britain must have a military capacity appropriate to her defence needs. The justification of the current level of military forces is based on an official assessment of the international military balance. Our studies show, however, that the military balance is open to widely differing interpretations. Furthermore, orthodox assessments of the military balance over-emphasize technical and quantitative criteria and do not take such factors as political will or economic need sufficiently into account. It therefore seemed appropriate to examine not only the capabilities but also the intentions and interests of the USSR.

In our view, the USSR is not restrained from attacking the West by feelings of benevolence. But she is prevented from doing so by the overwhelming dangers such a step would present to the Soviet leadership itself. It is with this fact in mind that we should look at the political and strategic consequences of cutting defence spending.

Britain contributes a disproportionately high percentage of her Gross National Product (GNP) to NATO, the Western military alliance. This would be less of a problem if the United Kingdom were economically strong and the other NATO members weak. But exactly the converse is true. This imbalance is exacerbated by the steady escalation of international military tension. If Britain could cut her military spending to a level comparable to that of her European NATO allies, it would make a positive contribution to lessening international tensions and therefore actually increase international security.

Defence cuts could be carried out in various ways with varying political and strategic consequences. Three different approaches – by no means mutually exclusive or exhaustive – are discussed in Chapter 4, and are briefly:

1. A reduced British contribution to NATO might induce some of our allies to fill the gap. This would not be the aim in proposing defence cuts, but it might be the response of some of our allies with stronger economies than our own.

2. It might be possible to maintain much the same strategic capability at reduced cost by adopting more cost-effective methods, for example, by replacing expensive sophisticated weapons like Tornado with cheaper weapons which make use of the new technology of precision guidance.

3. We could seize the opportunity of our own defence cuts to make a contribution towards arms control.

These proposals are being made in the political climate of détente and increased effort towards mutual disarmament. Britain has the opportunity to play an active and leading part in this process. There are many reasons why we should make sure that she does.

1. *Sunday Times*, 31 October 1976.
2. Robin Cook, quoted in 'Arms Jobs and the Crisis', *CND*, July 1975.
3. See Chapter 5, under 'Alternatives in the Aerospace Industry' (p.88–9), forTornado costs; other figures from Cmnd 6735, Cmnd 6721 (vol. 11) and *Labour Research*, July 1976.
4. Figures from Government Expenditure Plans, Cmnd 6721 (11), 1977–8.

2

CURRENT DEFENCE EFFORT AND
THE NECESSARY LEVEL OF CUTS

The starting point for any programme to reduce defence expenditure must be an examination of existing defence effort. The following section therefore broadly summarizes the current situation.

Expenditure
Despite the prevailing impression that the Government is making great cutbacks in defence expenditure, the following figures confirm that this is far from the case. Table 1 compares actual defence expenditure for the first half of the 1970s, current estimates and the Government's budgetary projections to 1980/81 on a common prices basis.

Only in the year 1977-8 has the Government managed to bring about an actual reduction in the defence budget, and this is only a temporary halt. Annual expenditure is expected to rise again to over £7,000 million in the early 1980s according to information given to the Select Committee on Expenditure.[1]

The 1977/8 figure in Table 1 is the sum actually voted in the Defence Estimates for the current financial year. It corresponds to about 5·5 per cent of the estimated Gross National Product (GNP) at factor cost for 1977/8. It is envisaged that the £6,329 million will be spent as shown in Table 2. A further breakdown of these figures is contained in Table 3, which shows the allocation of defence expenditure to the various defence 'mission' and 'support' programmes.

Employment

The military services and industry in Britain employs, directly or indirectly, about a million people. Normally each year at least 180,000 of these can be expected to change their jobs.[2] Government figures show that the Ministry of Defence is responsible for the employment of 337,100 servicemen, and of 200,000 workers in arms industries on work projects for the Ministry, while a further 70,000 to 80,000 workers are engaged on defence exports. In addition, some 250,000 workers are indirectly employed on Ministry contracts and about 100,000 on defence exports. About 956,000 civil servants are also employed by the Ministry of Defence.[3] (See pp. 79–80 for a further regional and occupational breakdown of these figures, as well as a look in detail at some case-studies of employment in large-scale military projects.)

Military Capability[4]

The total personnel strength of the armed forces of 337,100 is relatively small in relation to the size of the British population, but conceals the leading role which Britain plays in NATO. Indeed, it is commonly accepted that Britain is the main NATO European power, largely because of Britain's naval superiority and nuclear capability. The French are also a nuclear power, but are no longer an intrinsic part of the alliance. Other indications of Britain's military strength are the large number of tanks owned by this country as well as the high degree of sophistication of the RAF. It is also widely accepted that British military personnel are of a very high standard.

Britain is also one of the few European powers to deploy numbers of troops in other parts of the world like Hong Kong and Cyprus, not to mention the 55,800 troops in the British Army of the Rhine (BAOR) in Germany. The overseas British military capability will be discussed in more detail when we come to examine the various options for cuts in the armed forces.

TABLE 1. DEFENCE EXPENDITURE, 1971–81

		£ million at 1977/8 Estimates Prices		
1971–2	*1972–3*	*1973–4*	*1974–5*	*1975–6*
6,638	6,460	6,395	6,164	6,448
1976–7	*1977–8*	*1978–9*	*1979–80*	*1980–81*
6,544	6,329*	6,275†	6,550†	6,550†

* Incorporates reductions in previously planned expenditure announced on 22 July and 15 December 1976.
† Provisional figure.

SOURCE: *The Government's Expenditure Plans*, vol. II, Cmnd 6721 (II), converted to 1977/8 Estimates Prices.

TABLE 2. MAJOR CATEGORIES OF EXPENDITURE, 1977/8

	£ million
Civilian pay	1,077
Equipment	2,350
Buildings and miscellaneous stores and services	1,105
Forces pay and allowances	1,452
Forces pensions	345
Total	6,329

SOURCE: *Statement on the Defence Estimates 1977*, Cmnd 6735.

The Amount of the Reduction

While Britain spends considerably more of its Gross National Product on defence than almost any other Western European nation, the share of national resources devoted to domestic investment is among the lowest in Western Europe. Britain has the largest arms industry in Europe, and it is one that is still growing. At the same time, civilian production in defence-related sectors, such as automobiles, mechanical engineering and shipbuilding, faces the most severe international competition since the war.

By how much, then, can we afford to cut defence expenditure, and how can we compare our defence burden with those of our allies? The following method is one suggestion.

The standard basis for comparing the resource cost of the defence burdens of various nations is to express the expenditure on defence as a percentage of the Gross Domestic Product (GDP). This method is not perfect, but in the words of the 1975 White Paper, 'it is the best single readily available measure of the defence burden in relation to a country's resources'.[5] The gap between Britain and her allies is, in any case, too significant to be explained by any methodological fault. In 1976, Britain's defence expenditure on the basis of the standard NATO definition was estimated at 5·2 per cent of her GDP at market prices,[6] whereas the comparable figures were 3·9 per cent for France, 3·5 per cent for Germany, and 2·6 per cent for Italy.[7] The weighted average figure for our 'main European allies' is 3·4 per cent.[8] Therefore, if it were only a question of an overnight reduction in the UK defence budget, the Manifesto commitment would imply that the budget should be brought down from 5·2 to 3·4 per cent of GDP: a cut of one third. However, it seems more sensible to envisage a reduction over a reasonable and specified period of time so that industrial conversion can take place more easily; the assumption here is that it would be a time period of five years.

To quantify a precise reduction in British defence expenditure to be aimed at in five years' time, we need to make two further assumptions. The first concerns the rate of real increase in the British GNP over the next five years. Between 1954 and 1973, the average annual rate of growth in the economy was 2·8 per cent. However, we start, in 1976, from a year in which there was heavy unemployment, but when we are beginning to move into a period when the balance of payments constraint on economic growth is likely to be eased by the inflow of North Sea oil. A 'central' assumption has therefore been made that, between 1976 and 1981, the average annual rate of growth in the UK economy will be 3·5 per cent a year.[9]

The second assumption concerns the likely movement, over the next five years, of the share of GDP devoted to defence among the three allies – France, Germany and Italy. The long-term trend has been for the aggregate share of their defence expenditure in GDP to fall – that is, while their military expenditure has increased, it has not increased as fast as the GDP. Thus, to avoid the objection that the calculations are based on assumptions which magnify the figure of the defence cut required in the UK, we need to assume that in all three countries military expenditure will in real terms rise significantly faster than it has done in the past. We assume that in all three countries the increase in real military expenditure is 3 per cent a year, or about double the rate of increase of the last decade. Secondly, we assume that the growth rates of GDP in the three countries from 1976–81 are fractionally lower than the long-term trend. On these two assumptions, the weighted average figure for military expenditure as a share of GDP in the three countries taken together becomes 3·2 per cent in 1981. This is more or less in line with the decisions of the May 1977 NATO Council meeting.

TABLE 3. FUNCTIONAL ANALYSIS OF DEFENCE EXPENDITURE, 1977–8

Major Programmes (and Sub-programmes) £ *million*

MISSION PROGRAMMES	
Nuclear Strategic Force – (Polaris)	96
Navy General Purpose Combat Forces	843
(a) Amphibious forces	29
(b) Aircraft carriers	15
(c) Submarines	130
(d) Cruisers	63
(e) Destroyers and frigates	319
(f) Mine counter-measures	41
(g) Other ships	116
(h) Aircraft	92
(i) Overseas shore establishments	33
(j) Fleet HQs	5
European Theatre Ground Forces	1,091
(a) BAOR	779
(b) Berlin	22
(c) Home forces	290
Other Army Combat Forces	70
(a) Hong Kong and Far East	18
(b) Mediterranean	44
(c) Other areas	8
Air Force General Purpose Forces	1,034
(a) Air defence	160
(b) Offensive support	52
(c) Strike/attack reconnaissance	280
(d) Maritime aircraft	50
(e) Transport aircraft	65
(f) Tanker aircraft	17
(g) Operational stations	148
(h) Other aircraft/civil charter/HQ/general support	262
Total mission programmes	3,134

Table 3 continued overleaf.

Reserve and Auxiliary Formation	105
Research and Development*	823
(a) Military aircraft	319
(b) Guided weapons	98
(c) Other electronics	107
(d) Naval	79
(e) Ordnance and other army	62
(f) Other R & D	158
Training	565
Production, Repair and Associated Facilities in the UK	447
(a) Naval dockyards	87
(b) Repair and maintenance	122
(c) Storage and supply	179
(d) Quality assurance	59
War and Contingency Stocks	108
Other Support Functions	1,139
Miscellaneous Expenditure and Receipts	8
Total Support Programmes	3,195
Total	6,329

* Excludes £3 million of metereological R & D included under 'Other Support Functions'.
SOURCE: 'Statement on the Defence Estimates 1977', Annex B, *Defence Data 1977/8*, Cmnd 6735.

If we bring together these central assumptions about the UK growth rate with the growth rate in Germany, France and Italy and the likely trends of military expenditure in those three countries, we can see that, to match the percentage of the national product devoted to military expenditure in the UK with that of our main European allies by 1981 (or as soon as possible thereafter), the annual military budget in this country will need to be brought down by something like £*1,825 million at 1977/8*[10] *Estimates Prices:* a cut of around 28 per cent on the expenditure projected for 1980–81 and 1981–2.

We have looked into the consequences of varying the central assumptions on which this calculation is based. What, for example, would be the consequences of a faster growth rate in the UK of 4·25 per cent a year; or of a slower growth rate of

2·8 per cent? What would be the consequences of a higher and lower share of national product being devoted to military expenditure in the three continental countries in 1981, taking 3·4 as the high percentage (which would mean no drop at all from the 1976 figure), and 2·8 per cent as the low point. This would be more in line with the long-term tendency for a slower rate of growth. On every combination, a substantial cut in UK military expenditure is implied: the lowest figure means a cut of some £1,250 million, the highest a cut of over £2,500 million at 1977/8 Estimates Prices. These calculations simply illustrate the size of cuts that would be needed to fulfil the Manifesto commitment; obviously they should not be seen as a rigid framework for approaching the problem. The options discussed in the following sections are, in fact, based on a rather smaller cut in military expenditure than that implied by the central assumptions set out above: on a cut of 20 per cent rather than 28 per cent.

How the Cuts Could Be Made
There are many approaches to finding ways to cut defence expenditure. This section deals with the question of looking at ways in which the defence budget *could* be reduced over the next few years if alternative judgements and assumptions were made about the nation's basic security interests, and about what constitutes 'proper provision' for defence. The options outlined are not, however, in any way prescriptive or exhaustive. The argument is based on an analysis prepared in 1975. The calculations are therefore expressed in 1975/6 Estimates Prices and designed to expose ways of attaining savings on the defence budget of £1,000 million at these prices by 1980–81, implying a reduction in the defence effort of about 20 per cent. In an updating and revaluation of the exercise undertaken just before the completion of the present report, it was concluded that the programme changes discussed might yield savings of around £1,300 million at 1977/8 Estimates Prices by 1983–4.[11] Needless to say, neither these figures nor the details of the various options should be seen as rigid formulations; both are flexible and subject to adjustment. Yet it is clear that if a cut of this order were implemented, a major reshaping of the military effort would be essential. The options outlined below have been selected from the many available choices, and a workable outcome would need the adoption of a mixture of these possibilities. Although each option has been costed, it must be emphasized that all costings are estimates based on budgeting projections; that such pro-

jections are in themselves tentative and subject to change. Therefore the figures given should be seen as rough guidelines in the same way that the whole exercise should be viewed as an attempt to quantify the extent of budgetary changes rather than an attempt to put forward an alternative defence programme.

Unexpected pressures on the defence budget from 'over-runs' on important equipment programmes, increased personnel costs and overall financial requirements may, in any case, force us to adopt some of the proposed modifications. It would be preferable if these could be openly anticipated and publicly discussed before they become inevitable.

Estimated Savings Possibilities on Major Programmes/
Sub-Programmes
The strategic and political consequences of adopting the selected programmes are analysed in Chapter 4. The following summary of Appendix V therefore considers only budgetary questions. The assumptions that David Greenwood has identified about current defence priorities which would be brought into question if a programme of cuts were to be implemented are as follows:

(1) A continuing commitment of ground and air forces to Allied Command Europe is crucial. But neither existing force structures nor present force levels need be regarded as sacrosanct.

(2) The Eastern Atlantic and Channel areas – including the North Sea and the Norwegian Sea – are equally vital. But the share of the Alliance effort that the United Kingdom provides might be reduced; and the nature of that provision would bear scrutiny.

(3) Ensuring the security of the 'home base' is a third co-equal priority. With the assertion of limited sovereignty over offshore areas the scope of this responsibility will grow.

(4) Neither strategic nuclear forces nor garrisons in remaining dependencies can be classed as 'essential' elements in a national order of battle constructed on the *par inter pares* basis envisaged.

This assessment leads on to the following options.

(a) Nuclear Strategic Forces
An immediate decision to withdraw Polaris submarines from service and a rundown of all associated activity. This rundown should take no more than two to three years. A plausible guess of

likely savings over a four-year period would be (£m at 1975–6[12] Estimates Prices):

1976–7: nil
1977–8: 50
1978–9: 50
1979–80: 75
1980–81: 75

(b) *Navy General Purposes Combat Forces*
Possible savings in this sphere would include:

(1) Paying off *Ark Royal, Hermes, Fearless* and *Intrepid* progressively over three to four years.
(2) Rundown of patrol submarine activity.
(3) Deletion from the programme of provision for the second and third Anti-Submarine Warfare (ASW) cruisers and, possibly, some 'stretching' of the building programme for the new destroyer and frigate classes.
(4) Reduced provision for Afloat Support (Royal Fleet Auxiliaries).
(5) Cancellation of the plans to acquire the Sea Harrier for the ASW carrier and cruisers; reduced purchases of helicopters and reduction of other costs for naval air support.

In round figures, the year-by-year distribution of these 'savings' possibilities might be as in Table 4.

TABLE 4. £m AT 1975–6 ESTIMATES PRICES

	1976–7	1977–8	1978–9	1979–80	1980–81
(1)	5	10	15	20	20*
(2)	5	10	30	40	50
(3)	–	50	100	100	150*
(4)	–	—	—	—	10*
(5)	–	10	30	40	50*
(6)†	–	—	—	—	—

[say] 275

* Denotes related sub-programmes.
† The final line is included for comparison with Table 2; no 'savings' are credited here because even a smaller ship navy would justify the retention of some overseas facilities, e.g. Gibraltar. The total is rounded to the nearest £25 million: a range 'guesstimate' would be £250–300 million.

(c) *European Theatre Ground Forces*

The whole question of troop deployments in Europe is currently under discussion and the following reductions may be envisaged:

(1) A phased rundown of the BAOR to about 30,000 by 1980/81, with appropriate adjustments in equipment purchases.
(2) No change in the Berlin garrison.
(3) Cuts in Home Forces to reflect the reduced roulement requirement of a smaller Rhine Army.

The order of magnitude of budgetary 'savings' attainable on this programme would be as in Table 5.

TABLE 5. £m AT 1975–6 ESTIMATES PRICES

	1976–7	1977–8	1978–9	1979–80	1980–81
(1)	—	50	100	150	200*
(2)	—	—	—	—	—
(3)	—	—	25	50	75*
					275

* Denotes related items. 'Savings' possibilities under (1) would be lower if, as a counterpart to 'savings' on RAF Germany, it were decided that the Rhine Army should have enhanced missile air defences and/or improved artillery capabilities.

(d) *Other Army Combat Forces*

Relatively minor savings are foreseen for the reduction of army strengths outside Europe; they would, however, contribute to an existing rundown of forces in this sphere. A phased reduction or gradual increase in contributions towards maintaining these overseas troops might yield 'savings' in the intervening years as follows (£m at 1975–6 Estimates Prices):

1976–7: nil
1977–8: 5
1978–9: 10
1979–80: 20
1980–81: 25

(e) *Air Force General Purposes Forces*

Savings in the Royal Air Force would demand the reshaping of

40

the force to concentrate on the effective performance of a more limited range of roles in place of the present broader spectrum of mission capabilities. Possibilities for 'savings' might then include the following:

(1) Deletion from the programme of provision for the air defence variant (ADV) of the Multi-role Combat Aircraft (MRCA) Tornado. Phantoms would have to fulfil the air defence task throughout the 1980s if the Tornado interceptor were cancelled. Improvements in them and their Sparrow missiles would therefore be necessary. Increased provision of surface-to-air missiles (SAMs) is another possibility for reinforcing the army's ability to defend itself from air attack. Savings on the cancelled Tornado programme would therefore need to be assessed alongside these considerations.

(2) Additional expenditure on Jaguar, and perhaps on Harrier, would also need to be set against Tornado savings. These aircraft have a strike/attack capacity, but may not be as cost-effective as a plane like the Fairchild A10. So here is another possibility for replacement.

(3) Arising from a reassessment of mission priorities and the creation of a 'lower cost air force', a substantial reduction in strike/attack/reconnaissance could be envisaged by cancellation of the 'common' strike/reconnaissance Tornado version.

(4) Such a reassessment would also allow for reductions in the support programme; namely, reduced provision for tanker aircraft and transport squadrons.

(5) Further closures of operational stations, or reduced running costs of existing ones.

(6) Some reduction in the expense of general support and headquarters. These are largely administrative costs.

Table 6 gives a broad estimate for a plausible scenario for 'savings', these figures being subject to wider margins of error than those used previously. This is because the net 'savings' illustrated would be applicable only in the immediate future owing to the effect of cancellation charges and because much of this speculation relates to the provision of substitutes where the problem of costing is complex. Nevertheless, this is what the position in round figures might be.

TABLE 6. £m AT 1975–6 ESTIMATES PRICES

	1976–7	1977–8	1978–9	1979–80	1980–81
(1) ⎫					
(2) ⎬	—	—	180	200	220*
(3) ⎭					
(4)	—	—	—	5	10
(5)	—	—	5	10	15
(6)	5	10	15	20	30*
					275

* Denotes related sub-programmes. The attribution of early 'savings' to (6) presupposes immediate decisions on Tornado to permit 'support' economies; that of a minor and late 'saving' to (4) is explained by the fact that further transport squadron cuts would not be possible until the end of the period, and could not even then be extensive.

(f–l) Support Programmes
A further 'savings' potential with regard to shifts in defence programme options of this size lies in the reduction of support programmes. At present there is a ratio of 54:46 per cent between expenditure on mission and support programmes. It would therefore be reasonable to expect support function to yield at least a third of possible savings. The cumulative effect of this would be to produce a 'saving' of something like £325–350 million in the period to 1980/81. The source of these savings could therefore be as listed in Table 7.

TABLE 7. SUPPORT PROGRAMME
£m at 1975–6 Estimates Prices

(f) Reserve forces		15
(g) R & D not included elsewhere		50
(h) Training		75
(i) Production, repair and associated facilities		50
(j) Other support functions (miscellaneous heading for Whitehall organization, local administration, personal pensions and services for armed forces, etc.)		125
(k–l) Stocks/miscellaneous expenditure and receipts		35
		350

Defence Programme Options to 1980–81
Having identified areas where savings could be made, we now

need to suggest various ways in which these savings could be combined throughout the services to achieve the target level of cuts. Implicit in this exercise is the attempt to achieve a balance which will ensure that essential defence needs are met. This means that, while the exercise allows for a relaxation of assumptions on the right level of provision, it is also involved in a more stringent definition of security priorities.

By showing the scope for possible 'savings', three leading options have emerged as providing likely arrangements for a reduced defence effort. In this scenario, the rundown of the Polaris force and facilities, and the dropping from the current planned programme of expenditure on army combat forces outside Europe, are common to all three option programmes. Cuts in two of the three remaining mission programmes (b, c, or e), would also need to be implemented. And finally, it would be necessary to make appropriate cuts in support programmes if a reduction of around one sixth of present planned expenditure is to be achieved by 1980–81.

A fourth option would be to retain the Polaris programme and make larger cuts in other areas; or a further possibility is to tackle the question by attempting a mix of options within the various sub-headings of the major mission programmes.

On this basis, the following options emerge:

(1) Adopt the smaller ship fleet philosophy and make reductions in European Theatre Ground Forces (ETGF) while planning to maintain a 'full spectrum' Royal Air Force, complete (in time) with nearly 400 Tornados. *Or*
(2) elect to move to both a smaller ship fleet *and* a lower-cost air force, thus avoiding the necessity for reducing ETGF levels. *Or*
(3) preserve the existing naval programme, with its balanced fleet philosophy ('ships of high quality'), allowing the burden of adjustment to fall on the Royal Air Force, notably RAF Germany, and on the BAOR and the UK land forces.

Options 1 and 2 would be 'available', so to speak, with or without a somewhat diminished submarine force; Options 2 and 3 with a larger or smaller complement of maritime, transport and tanker aircraft.

The fourth option, which envisages the retention of Polaris, would involve further reductions in all three major conventional force programmes. In the naval sector, this could mean stretching the fleet submarine, destroyer and frigate building programmes. In the European theatre, it implies a marginally sharper rundown of men to around 27,500 with an appropriate adjustment in support functions and home forces. And it is assumed that the necessary air-force cuts will come from the Tornado and related programmes sector.

There are at least four other implicit options: those which follow from the remaining permutations possible on the 'heavy-weight' service programmes. They do not, however, hit the 1980–81 budget target: one of them would yield greater 'savings', the rest a good deal less. But they are none the less worth stating, giving, as they do, a general indication of the broader range of choice.

In principle, if all the 'rationales for options' were thought especially compelling, it would be possible to:

(5) Opt for the smaller ship fleet *and* reduced ETGF levels *and* a lower cost air force – entailing an all-round diminution of military stature.

But if, on the other hand, reconsideration of priorities and 'proper provision' were thought possible for only one service's roles, the options would be:

(6) The smaller ship fleet. *Or*
(7) the reduced ETGF levels. *Or*
(8) the lower cost air force.

In summary, the 'guesstimates' for costings of the options discussed would be as in Table 8.

TABLE 8. PROGRAMME OPTIONS TO 1980–81: OPTIONS FOR DEFENCE BUDGET TARGET (Target 'Savings': £1,000m at 1975-6 Estimates Prices. These figures would be around £1,800m in 1977-8 Estimates Prices)

Defence Budget Saving 1980–81 (£m)

Option 1

(a) Rundown Polaris and facilities	75	
(b) Smaller ship fleet	275	(225)
(c) Reduced ETGF levels	275	
(d) No cost to UK arrangements	25	
	—	
Mission programmes	650	
(f-l) Proportionate reductions	c. 350	
	—	
	c. 1,000	(925)

Option 2

(a) Rundown Polaris and facilities	75	(225)
(b) Smaller ship fleet	275	(225)
(d) No cost to UK arrangements	25	
(e) Lower cost air force	275	(265)
	650	
Mission programmes	650	
(f-l) Proportionate reductions	c. 350	
	c. 1,000	(900)

Option 3

(a) Rundown Polaris and facilities	75	
(c) Reduced ETGF levels	275	
(d) No cost to UK arrangements	25	
(e) Lower cost air force	275	(265)
	650	
Mission programmes	650	
(f-l) Proportionate reductions	c. 350	
	c. 1,000	(985)

Note: Figures in parentheses identify the effects of excluding 'non-related' items, i.e. submarine reductions in (b) and tankers and other aircraft reductions in (e).

Option 4 would demand greater reductions in two or more of the main mission programmes, plus the necessary proportionate reductions in support programme outlays to 'compensate' for retention of the Polaris force. The composition of savings attainable for Options 5–8 can be inferred from the information for Options 1–3.[13]

As explained at the outset, this detailed analysis was originally undertaken in 1975. It is clear, however, that the possibilities outlined as programme change options remain more or less valid today. The savings they would allow could, however, hardly be attained by 1980–81. Moreover, the value of the eventual reduction in the defence budget which they would make possible needs to be translated into 1977–8 Estimates Prices for comparison with the current and planned future defence expenditure levels set out in Tables 1–3. In fact, the transposition in timing is not at all troublesome if it is assumed (a) that not even early decisions on a reshaping of the defence effort could begin to yield significant savings before 1979/80, but that (b) the savings allotted in the above paragraphs to 1977/8 to 1980/81 would be attainable in the period 1979/80 to 1983/4, subject only to minor modification (to reflect, for instance, increased cancellation costs on projects now further advanced).

Conversion of the projected savings to 1977/8 Estimates Prices is similarly straightforward. The most succinct statement of the essential argument of this chapter is therefore the presentation in Table 9. It shows the budgetary projections for the current defence programme (from Table 1), and the time profiles of expenditure which would be associated with the three principal options discussed above (pp. 43–5), on the retiming and revaluation assumptions just stated.

TABLE 9. PROGRAMME OPTIONS TO 1983–4 (£m at 1977–8 Estimates/1977 Survey Prices)

Programme	1979–80	1980–81	1981–2	1982–3	1983–4
Current					
(From Table 1)	6,550	6,550	6,550	(6,900)	(7,200)
Option 1	6,543	6,250	6,000	6,000	5,900
'Savings'	7	300	550	900	1,300
Option 2	6,535	6,325	5,900	5,900	5,900
'Savings'	15	225	650	1,000	1,300
Option 3	6,550	6,400	6,000	5,900	5,900
'Savings'	—	150	550	1,000	1,300

SOURCE: See Appendix V to full report, Postscript.

Conclusion
The courses of action reviewed would allow an estimated reduction of around £1,300 million in the defence budget by 1983/4 – a 20 per cent reduction measured against the forecast expenditure level of the late 1970s. It must be re-emphasized that this exercise is not prescriptive; it does not provide a blueprint for a new defence strategy in the 1980s. The intention has been to give an impression of the kind of changes needed to achieve a re-shaping of the armed forces if savings of this kind are to be achieved.

Notes and References
1. See Second Report from the Expenditure Committee, Session 1976–7, HC 254 (76/77), Report, p. x.
2. F. Blackaby, 'Note on the Employment Consequences of a £1000m Cut (at 1974 Prices) in Military Expenditure Over 5 years'. See Study Group Paper (hereafter SGP) No. 10
3. *Hansard*, col. 42, 9 October 1976; and Cmnd 6735.
4. Information for this section from Cmnd 6735 and *The Military Balance 1976–1977*, IISS, 1977.

5. 'Statement on the Defence Estimates 1975', Cmnd 5976.
6. This estimate is taken from *Economic Trends*, April 1977, adjusted to the standard UN definition. The definition of GDP at market prices is approximately 3·5 per cent below the figure given in UK national accounts.
7. Estimate calculated from *UN Monthly Bulletin of Statistics*, March 1977. This is an estimate NIESR based on known movements in outputs and price changes.
8. The weights are derived from the estimates of GDP to US dollars in *UN Monthly Bulletin of Statistics*, March 1977.
9. 'World Armaments and Disarmament', *SIPRI Year Book*, 1974 p. 133.
10. 1977/8 Estimates Prices are the basis on which the government is planning its current expenditure programmes; these figures represent actual price levels in September 1976.
11. The material which follows is based on studies prepared for the NEC by David Greenwood: *Defence Programme Options to 1980–81* (the original analysis, completed September 1975); and a postscript to that paper (written May 1977). Both appear in SGP No 2.
12. These are the figures from the original analysis; for revised assessments, attributed to the period to 1983–4, see the Postscript in ibid.
13. For details, see **SGP No 3**.

3

DEFENCE EXPENDITURE AND
THE ECONOMY

Decisions about the level and nature of defence spending are not only political. They are also decisions about the social allocation of resources, about how to decide between competing claims on public expenditure and how to reconcile strategic requirements, as seen by defence planners, with economic and social needs. There seems, in fact, to be a lack of evidence for the notion that higher levels of defence spending or the maintenance of current levels bring economic advantages. Instead we have drawn two somewhat contrary and disturbing conclusions: first, that governments have tended to adopt the most expensive solutions to strategic problems, often in contradiction to declared policy objectives – which can be explained only in terms of the structure of the institutions responsible for the research, development, production and operation of military equipment; secondly, that these same expensive solutions have had harmful effects on the civilian economy. Defence spending actually affects the economy at a number of 'sore points':

1. It requires a relatively high skill-content, both in the production of weapons and in their operation; and a perennial problem of the British economy has been the shortage of skilled workers.
2. It absorbs a very high proportion of the research and development effort of the United Kingdom.

3. Because consumption and welfare spending represent a relatively stable proportion of national income, military spending is, in expenditure terms, directly competitive with investment.

4. It has a specifically heavy impact on the engineering sector of the economy, which is the sector also responsible for a relatively high proportion of our exports and investment goods.

5. Particularly because of the cost of maintaining troops in Germany, there is a high balance of payments cost; and again, the balance of payments has been one of the perennial weaknesses of the British economy.

6. Finally, the whole of defence expenditure falls in the public expenditure category; and it is widely held that the levels of taxation, particularly on the average wage-earner, associated with the present level of public expenditure are damagingly high.

All these factors suggest that the restructuring and redeployment of Britain's defence industry is an essential condition both for cutting defence expenditure and finding cheaper solutions to strategic problems, as well as for regenerating the British economy. The present chapter provides the background to these issues. Chapter 5 will detail specific proposals for conversion and diversification opportunities.

Defence Decision-making and the Economy: Choice of Projects
A major cause for concern is the cost of particular weapon programmes which have come to dominate the British defence budget. Current examples are the Multi-Role Combat Aircraft (Tornado) and the ASW Cruiser. The Expenditure Committee of the House of Commons recently concluded that:

> The Ministry (MOD) have occasionally appeared to aim for so high a standard that either projects have had to be abandoned or sacrifices have had to be made elsewhere in the defence programme in order to limit or accommodate escalating costs.[1]

The increase in the cost of new generations of military equipment has been remarkable. For example, it has been calculated[2] that the real cost of producing 385 Tornados will be slightly greater than the entire production costs of Spitfire before and during the Second World War. Taking account of inflation and other cost increases, it was shown that the cost per ton of warships has increased by anything from a factor of 10 (the difference between an early post-war 'Bay' class frigate and the last of a long series of 'Leander' class frigates) to a factor of 15 (the difference between an early post-war 'A' class submarine and a modern 'Swiftsure' class submarine, or between the Vanguard Battleship and the new Anti-Submarine Warfare Cruiser).

If anything, these figures are understated since they refer only to production costs. Total life-cycle costs would be even more striking, since maintenance and support costs tend to increase faster than production costs because of the unreliability of highly complex and sophisticated equipment. In particular, there is the increase in electronic equipment, which accounts for much of the overall cost: a rise in requirements here involves a disproportionate increase in what are called logistic support costs.

Increases in cost are thus related to increases in the sophistication and complexity of military equipment designed to meet more stringent and elaborate performance targets. Obvious recent examples are the variable geometry aircraft (i.e. the swing-wing) designed to increase flexibility and multiply an aircraft's possible functions; gas-turbine engines for surface warships, designed to increase speed; nuclear propulsion for submarines, to increase their speed and range; and the various types of electronic equipment which are to be found across the whole range of weapons systems and which vastly improve communications, navigation, the identification and detection of enemy targets and the guidance of weapons, and which can reduce the effectiveness of enemy electronic equipment. These are the major technological changes. There are also continual developments in existing technologies – increases in the thrust of a given type of jet engine, improvements in the accuracy of a particular kind of guidance system. Over time, marginal developments of this kind have tended to involve a disproportionate increase in costs.

Technical advance can always be explained in strategic terms. In peacetime, however, the assessment of any threat is necessarily subjective. Changes in technology, by inducing new perceptions and counter-perceptions of the threat, can propel planners into further changes in an autistic fashion. The idea of defence as a race against technology originated during the Second World War, and it has come to have an existence apparently independent of changes in the political and economic environment. Representatives of the arms sector tend to view technology as an end in itself, and regret cancelled projects simply because they represent a loss of technical leadership. One well-established writer, for example, explains that because the Ministry of Defence were sceptical of the variable geometry aircraft concept 'on the grounds of both time and price', and were unwilling to finance continued research by Vickers during the 1950s, it 'became imperative that some manned military application be found for all the theories and tests'.[3] This was the origin of Tornado. And yet, as we shall see in Chapter 4, it is questionable whether increased sophistication really represents technical advance and improvement, whether, in fact, it genuinely represents the optimum approach to strategic problems. Several defence critics are concerned about the unreliability, vulnerability and lack of manoeuvrability of modern weapons systems. The fact that such doubts exist, and have even manifested themselves in thwarted attempts to control costs by cancelling major projects, suggests that the problem lies less with hardware and strategic perceptions and more with the underlying military-industrial structure.

This is not to say that the underlying structure of the defence industries is the sole influence on decisions about military procurement, but it does remain the major consideration. Other significant influences have also been suggested. J. R. Kurth, examining the factors influencing a particular procurement decision in the USA, identified four main types of consideration that could be important: (i) strategic or geo-political considerations, (ii) bureaucratic self-interest and inertia within the compartmentalized defence hierarchy; (iii) electoral calculations by politicians (vide the increase in the 'Soviet menace' shortly after Mrs Thatcher assumed the Tory leadership); and (iv) the needs and interests of the defence contractors.[4] There is every indication that the same considerations are the main determinants of procurement in Britain.

The Structure of the Defence Industry

The capacity to develop and produce military hardware may in general terms be defined as the amount of plant, machinery and labour available for development and production over a given time period. It may be measured, in abstract monetary terms, as maximum output in the given period. In specific terms, however, it represents an infrastructure of skills and techniques, and a set of relationships between the services and the sub-contractors needed to manufacture a particular type of military equipment.

With the nationalizing of the shipbuilding and aerospace industries, much of the defence industry is now under public control. It is to be hoped that the newly nationalized companies will not operate according to the principles of private enterprise, which necessarily mean the principle of independent viability, and hence of profit making. In the past, this has meant that individual defence companies were responsible for financing manufacturing capacity and could not afford long periods of idleness. In a centrally planned economy, where private criteria of efficiency are not the sole determinant, alternative uses can be found for industrial capacity in the pauses between military orders. In the USSR, for example, buffer production has been a central feature of the armaments industry since 1955.[5]

But with the increase in the sophistication of military equipment, military technology has become increasingly divorced from civilian technology, and consequently industrial capacity has become increasingly specialized. The Government is meanwhile the main customer for military equipment in such a way as to avoid excess capacity. As Vice-Admiral Clayton explains:

> We have to give the shipbuilders a regular rolling programme of orders. We depend on the specialist warshipbuilders. They have a very carefully balanced selection of trades – drawing office, steel workers and outfit trades – which are required especially for warshipbuilding and not for commercial shipbuilding, and of necessity we have to keep a flow of orders going to them.[6]

Avoiding surplus capacity is not simply a matter of employment, but also involves expansion. The principle of independent viability entails the principle of profit maximization and the constant striving after technical change. In a competitive situation – competition being defined in its broadest sense – firms must innovate, they must introduce new ideas, designs and products, if they are to maintain their markets. This applies as much in the military as the civilian sphere: firms must keep up with wider international developments in military technology if they are to continue to receive orders and if the armed services are, in turn, to justify those orders. In this way, of course, they contribute to the strategic developments they are trying to match.

Yet new technological developments must be paid for, and that means an increase in orders and hence an expansion of capacity.*

* It might be objected that companies could compete through process innovation, through developing cheaper solutions to military problems. The problem is that such developments might undermine the existing market, leading the Government to reduce military spending and abandon the current inventory of sophisticated equipment as well as removing the *raison d'être* of the specialized defence firm.

The striving for technical progress becomes the more extreme the more limited the market. We have seen how the cost of competition has risen in recent years in the automobile and merchant shipbuilding industry. The same is true of defence. The more the Government has reduced the numbers of types of weapons, the greater has been the effort to achieve technical progress and the greater the compensating cost increase.

In the last fifteen years, Hawker Siddeley has produced or developed seven military aircraft, including two that were cancelled; and BAC has developed or produced six military aircraft, including two that were cancelled. In the previous fifteen years, the companies that amalgamated to form Hawker Siddeley and BAC developed or produced twenty-eight and eighteen military aircraft respectively, as well as several other research aircraft. The number of aircraft produced of any given type has likewise declined substantially, yet the increase in the cost of each individual aircraft is such as to compensate for the decline in numbers. The same phenomenon is to be found in shipbuilding. In the period 1965–74, about half as much warship tonnage was launched as during the period 1945–54. Yet real costs have increased by factors ranging from 10 to 15, leading to a substantial overall increase in warship-building capacity.[7]

The explanation of this trend need not be seen simply in crude capitalistic terms. It is also a matter of redeployment following numerical reductions in orders. Designers shifted from one project to another, bringing with them pet ideas that could always find supporters within the services. The alternative to a rigorous elimination of unnecessary ideas and choosing between competing ideas is the compromise of doing everything.

Attempted Solutions

Governments of all arms-producing countries whose defence sectors play a major economic role are well aware of the problems inherent in the existence of surplus capacity in the defence industries. Current attempts to rectify the problem identify three possible solutions: first, to cultivate expanding export markets; secondly, to reorganize the defence industry; and thirdly, to set up international collaboration in the development and production of military equipment. It may be said with reasonable confidence that none of these solutions has worked in the past or is likely to work in the future.

Export orders account for between 20 and 30 per cent of British arms production. Since this proportion is relatively stable and tends to change primarily in response to world demand, exports can rarely provide a substitute for increased government orders. Neither do exports cover the costs of maintaining development capacity. The Ministry of Defence has had considerable difficulty in recent years in recovering the levy on the sale of military equipment developed at government expense. Moreover, only about a third of the Ministry's sales of military equipment in the financial year 1975–6 achieved a 'target price of full economic cost including all overhead expenses and return on capital'.[8] (There are also political and strategic problems attached to a dependence on the export trade; these and a fuller examination of the economic aspects will be considered in Chapter 6.)

The second solution, the reorganization of the defence industry, has had only a limited success. This is because amalgamation has not involved rationalization. Attempts to reduce the number of defence firms and the number of weapons systems have led to increased complexity and inefficiency as a result of cooperation between competing plants, and hence to increased cost and industrial capacity. Such attempts include the 1957 'Suicide' White Paper (which advocated the replacement of manned aircraft by missiles), the failure of the first of the great multi-role aircraft, the TSR-2[9] (itself a fitting prelude to the even more costly Tornado), and the decision to phase out aircraft carriers in 1966 (which was followed by a costly generation of destroyers, frigates and anti-submarine warfare cruisers). With the progressive reduction in the number of types of weapon delivery systems, the process of further rationalization will eventually reach a logical limit, at least on a national basis.

International collaboration in the manufacture of weapons, on the other hand, reproduces the problem of excess capacity on a grander international scale. So long as individual nations continue to protect their military manufacturing capacity, international collaboration must remain little more than a cooperative form of duplication. From the point of view of industry, international projects have the great advantage that they are less liable to be cancelled. From the point of view of the public, international collaboration means even less parliamentary control, and even greater cost escalation. The difficulties of amalgamating the design teams of two separate companies and reconciling a number of service requirements at a national level pale into insignificance when compared with the difficulty of amalgamating design teams from several different countries and of reconciling the requirements of distinctly separate national armed forces. Hence the dramatic cost increases in Jaguar, the Anglo-French helicopters and, above all, Tornado (not to mention Concorde). When the Anglo-French variable geometry aircraft was cancelled, General Gallois, French Director of Military Affairs, told *The Times* that, on a co-operative project, 'you divided the cost by two and then had to multiply it by three to take account of the difficulties of building it in two countries'.[10]

The Economic Effects of Defence Decisions
Apologists for military spending suggest that the problem of resource allocation is magnified. They argue that defence production yields economic benefits in the form of technological spin-off, employment and exports. The conclusions of the present study, however, suggest that this argument is fallacious. Military spending tends to be inversely correlated with industrial investment, and hence with economic growth. There is a good deal of evidence to show that the expenditure of a given amount of resources in the military sector tends to generate less employment and exports than an equivalent deployment of resources in the civil sector of the same industries. And the advantages of spin-off are likely to be offset by the disadvantages of 'militarizing' science and technology.

Nevertheless, in the short term, a direct transfer of resources would not necessarily rectify the problem. This is especially true in a period when the economy is stagnant and transitional problems could be considerable.

When evaluating any given defence programme, the money totals are often a misleading indication of the true economic costs. For instance, it might be cheaper in terms of direct expenditure to run a conscript army, but in conditions of full employment it would be wasteful of resources to tie up a sizeable proportion of the labour force solely on grounds of cutting direct spending. It is therefore necessary also to consider the opportunity and resource costs of the expenditure. David Greenwood defines the former as being a 'reminder that spending entails allocation and choice among competing opportunities'.[11] The concept of opportunity cost therefore raises the question of priorities, but it is necessary in addition to consider the extent to which the resources available can actually be substituted or transferred between competing programmes or uses in the short term. This will depend on a variety of factors, such as whether the workers and equipment which the project would use are currently being employed at all. If they are unemployed, the resource cost may be very small. And apart from how specific the available factors of production may be, alternative options for (in this case) public expenditure must take into account whether, for example, it would be better to raise the school leaving age than continue with the Tornado project.

An assessment of resource cost involves a choice between real substitutes. The factors of production used for the Tornado programme, for example, are quite specific: skilled labour, R & D facilities, specialized machine tools and so forth. They could not provide the classrooms, teachers and educational materials needed to raise the school leaving age in the short run. On the other hand, the same resources might be in short supply elsewhere in the economy, in the machine-tool sector, for instance, and the resulting short-term bottlenecks could have long-term consequences, through loss of markets or technical leadership among other factors. The absorption of specific kinds of resource needs elsewhere in the civilian economy may be called resource cost.

Ron Smith has identified five major factors which determine resource cost:[12]

(1) The degree of utilization in the various sectors.
(2) The multiplier and linkage effects between sectors.
(3) The degree of substitution possible between commodities, both in the pattern of production and the pattern of expenditure.
(4) The costs involved in adjusting from one pattern of input and output competition to another.
(5) The rates of growth in the different sectors.

Taken together, these factors show the availability of resources; the cumulative effect of changes in use; the degree of specificity in production and the area of choice available in the light of social and political priorities; the costs of conversion; and the fact that growth brings an increase in flexibility.

In perspective, therefore, the major distinction is not really between 'opportunity cost' and 'resource cost', but rather between the resources which could be transferred in the short term, and those which could be transferred in the long term. We now examine the different kinds of resources.

(a) *Labour*

The Labour movement's concern in tackling the problem of defence expenditure is to ensure that reductions do not involve any rise in unemployment. The fact is, however, that defence has a lower immediate impact on domestic employment than any other form of government expenditure. This is because defence production is more capital-intensive than other kinds of production. Three per cent of the economically active population in Britain, some 625,000 people,[13] is employed by the military sector.

International comparisons show how countries with high shares of military expenditure also tend to have higher than average unemployment rates. Britain and the United States, in particular, show this characteristic, while Japan and Germany have lower rates of unemployment coupled with lower shares of military expenditure. This is partly explained by reasons already outlined, but is also because the pernicious effect of defence expenditure on investment, productivity, growth and the balance of payments has meant that states with high levels of military expenditure have had to deflate the economy more, and so create more unemployment in attempts to get their balances of payments into equilibrium.

The Labour Party's programme of defence spending cuts would create the need for new jobs; a detailed examination of how this would be done is contained in Chapter 5. The basic point is that a greater long-term potential for creating more jobs with secure employment leads more from reducing the military budget than from maintaining a large military budget in a weak economy. There are considerable problems in shifting the allocation of public expenditure from one place to another, and the short-term effects of such a reallocation could only be beneficial if stringent planning procedures were adopted.

(b) *Investment*

It has been seen how investment and military spending are directly competitive, and how a high level of defence spending is therefore likely to produce a low level of industrial investment, a factor which in turn inhibits economic growth.

Military spending is competitive with investment in two senses. First, it is competitive with investment in expenditure terms. In advanced industrialized countries, the share of national income devoted to consumption is fairly stable. Therefore investment tends to compete with public expenditure, notably with military expenditure. This holds true for the British historical experience as well as in cross-country comparisons. Since the war, investment in Britain has tended to rise when military spending has fallen and vice versa.[14] Table 10 shows the share of GNP devoted to military spending and investment in advanced industrial countries. It shows that the highest military spenders, the United States and Britain, are the lowest domestic investors, and that the highest domestic investor, Japan, is the lowest military spender. Military spending is also competitive with investment in resource terms since arms are produced in the same industries as capital goods. Whereas military spending accounted for only about 5 per cent of GNP, domestic military purchases represented, in 1971, about 7 per cent of mechanical engineering output, nearly 30 per cent of electronics and telecommunications output, over half of shipbuilding output, and three quarters of aerospace output.

A key determinant of economic growth is engineering output. In the 1950s, Germany and Japan were able to devote a greater proportion of engineering output to investment and exports than we were. In approximate terms, it has been estimated that in the UK it needs an additional £1·00 of investment to produce an additional £0·30 of output.[15] It is therefore reasonable to assume that if this investment were taken away from the military sector and put into the civil sector, it would create increased output, jobs and exports. And this assumes merely a constant level of productivity. It does not take into account the increases in productivity that might be expected from increased investment. Nor does it take into account the effect on prices. Such inflation is fuelled by the failure of productivity to keep up with wage increases, but a faster rate of productivity would slow inflation and thus would increase exports, output and real wages. Hence military spending, when compared with other factors, represents a *loss* of jobs, exports, income, productivity, growth and lower prices.

TABLE 10. INVESTMENT AND MILITARY EXPENDITURE: OECD COUNTRIES, 1974

Country	Military expenditure* (US $ × 10⁶)	Military expenditure as percentage of GNP* %	Investment as percentage of GDP† %	Average annual growth-rate in GNP 1963–73‡ %
United States	85,900	6·15	18	3·9
United Kingdom	10,000	5·24	20	2·7
France	10,600	3·63	25	5·7
West Germany	13,800	3·58	22	4·7
Netherlands	2,320	3·45	22	5·4
Sweden	1,780	3·10	22	3·4
Norway	671	3·13	32	4·7
Italy	4,630	2·93	23	4·8
Belgium	1,460	2·77	22	4·8
Denmark	728	2·37	22	4·5
Canada	2,790	2·05	23	5·2
Switzerland	856	1·91	27	4·0
New Zealand	237	1·75	26	3·4
Finland	255	1·31	29	4·9
Austria	292	0·91	28	5·2
Luxembourg	18	0·87	26	3·4
Japan	3,670	0·83	34	10·5

* US Arms Control and Disarmament Agency, 1976 (Washington DC: US Government Printing Office, 1976): *World Military Expenditures and Arms Transfers 1965–74.*

† United Nations Department of Economic and Social Affairs Statistical Office, *Statistical Yearbook, 1975* (New York: United Nations, 1976).

‡ US Arms Control and Disarmament Agency, 1974 (Washington DC: US Government Printing Office): *World Military Expenditures and Arms Transfers 1963–73*

(c) *Research and Development*
Research and development (R & D) activity could be described as 'future investment' because the amount of attention paid to this aspect can have a crucial impact on the development of the economy. Military spending absorbs investible resources, and, in particular, some of the most valuable of these resources – the trained engineers and scientists who contribute directly to technical progress and competitiveness. The success of the

German machine-tool industry, for example, is not just a result of greater manufacturing capacity, it is also due to the German ability to innovate – to design the package machine-tool, for example. British R & D expenditure is more biased towards the military than that of almost any other Western industrial nation. The true implications of this cannot be gleaned from the figures given. Technical progress foregone is not measurable. Furthermore, it is not just a matter of diverting resources from military to civil purposes. The prevalence of military work has imposed a certain mentality on engineers and scientists – a preoccupation with sophistication and emphasis on complex and elaborate pieces of equipment instead of the cheap and simple products which people can actually use. Even in the civil field, much British R & D effort has gone into nuclear energy and Concorde. One important reason for the failure of British merchant ship-building since 1945 has been the cost of sophisticated facilities designed for naval use. In addition, this mentality has reduced the supposed benefits from spin-off. The increasing sophistication of military hardware in general has actually lessened the possibility of applying such technology in the civil sphere.[16]

The situation seen on a sector-by-sector basis shows an even more alarming picture of the uneven deployment of resources. Almost half the total government-sponsored R & D is devoted to defence.[17] In cash terms, the 1977/8 allocation for military R & D is £826 million, while the total agricultural R & D allocation for 1976 was £34 million and the Medical Research Council will be receiving only £27·5 million.[18] Sixty per cent of qualified scientists and engineers in the mechanical engineering industry work on arms, which represent less than 7 per cent of output. Similarly, while half of shipbuilding output is purchased by the navy, 90 per cent of R & D expenditure in shipbuilding is devoted to naval work.

In the discussion of occupational conversion later in this study (pp.103-4), we will also consider the effect of the whole ethos prevailing in the arms industries which produces attitudes that are incompatible with the aims of an efficient civil sector. Unless an effort is made to redress the balance between research and development in the civil and military sectors, no amount of new capital will produce the kind of real boost needed to regenerate British industry.

(d) *Exports*
Chapter 6 is devoted to the arms trade, and so takes an overall

look at the implications of this trade; the purpose of this brief section is therefore primarily to show how, contrary to much current propaganda, the export of arms is not the most efficient aspect of our overseas trade. The defence-related industries are the most export-intensive industries, i.e. machinery and transportation. In 1971, the total military and non-military exports of the defence-related industries was 40 per cent of total output.[19] This is probably a representative figure as 1971 does not seem to be an exceptional year. In contrast to this figure, arms exports as a percentage of total arms production (domestic procurement plus exports) is about 25 per cent. In other words, the very same resources – machinery, plant and people – when devoted to defence produce fewer exports than when they are devoted to civilian production. This difference has probably increased in recent years since exports of machinery and transportation have increased faster than output as a whole, while the ratio of military exports to military production has changed little.

Professor Kurt Rothschild has applied these facts to the various theories of export-led economic growth (currently popular in Treasury circles). He argues that high levels of military expenditure reduce export opportunities, which in turn slows down growth in GNP.[20] This hypothesis is difficult to prove conclusively, but it is worthy of consideration even if all it eventually shows is that export performance could hardly be any worse with a reduced arms sector, and is likely to be much better.

(e) *Defence and the Balance of Payments*
Another rarely observed aspect of the impact of military spending and its relation to the economy is the net loss suffered by the balance of payments as a result of stationing troops in Germany and elsewhere, and from importing American military aircraft. Table 11 presents the figures given in the Statement on Defence Estimates of the balance of payments effects of military expenditure.

TABLE 11.

	1976–7 £m	1977–8 £m
Cost of stationing troops abroad	690	696
Other military services*	160	147
Purchases of military equipment	167	258
Total debits	1,017	1,101
Receipts from US forces in Britain	30	42
Other receipts†	40	32
Sales of military equipment	670	834
Private expenditure by US forces in Britain	(80)	95
Total credits	820	1,023

SOURCE: *Statement on Defence Estimates, 1977*, Cmnd 6735; figure in brackets is an own estimate.
* Includes contributions to infrastructure projects, R & D levies, and contributions to international defence organizations.
† Includes such items as R & D levies.

This juxtaposition of credits and debits is to some extent misleading, since the item 'sales of military equipment' is not, strictly speaking, directly dependent on the scale of UK military expenditure. How would a reduction in UK military spending affect both the credit and debit sides? Certainly, under some of the options discussed in this study, the cost of stationing troops abroad would fall. (There would, of course, still be some balance of payments cost to their maintenance in this country.) The cost of purchases of military equipment would also be reduced. The effect that expenditure cuts might have on sales of military equipment would depend – among other things – on whether the recommendations about arms sales made elsewhere in this study were accepted. However, one of the contentions of this study is that, given time, the resources used at present in the production of arms for export could very well be transferred to the production of other engineering goods for export, and that the credits from this item could consequently be replaced.

Conclusion
It may be argued that the economic considerations which arise from a discussion of defence expenditure tend to overlook the general benefits inherent in such forms of expenditure. Such benefits would, it is suggested, relate to the need for domestic

security, global considerations of bargaining strength, protection of overseas commercial interests and so forth, as well as the need to maintain full employment when there is ideological resistance to non-military forms of public expenditure, and other matters relating to international obligations. These questions are considered in detail in the following chapters, but even at this stage the economic problems raised pose some fundamental questions about the extent to which the British economy, and those who create the wealth of our country, can and should be expected to bear the burden of excessive levels of defence expenditure.

Mary Kaldor sums up the nature of the inherent problems which have led to the situation in which the defence sector plays such a dominant role in the economy in these words:

'In some senses, the growth of defence spending can be seen as a symptom of the anarchy of our economic system. Partly because of the inherently subjective nature of strategic thinking and partly because of the unwillingness to admit the need for direct control over the economy, defence has come to be seen as a convenient instrument for economic policy-making – creating jobs, saving companies, and supporting the institutions of technology. At the same time, defence production has acquired a momentum of its own, exercising, in turn, an undue influence on military decision-making.'[21]

Such a momentum is built into the very structure of the decision-making system of the defence hierarchy, and has become so entrenched that any attempt to break away from this stultified manner of thinking will meet considerable resistance. Nevertheless, the fact is that the diversion of resources for the purposes of military production has proved inefficient in stimulating employment and exports, or generating new technology, and may even have helped to accelerate the process of industrial decline. It is also worth emphasizing that military expenditure is a particularly inefficient tool for demand management: because it is capital intensive, it has very little employment effect, and since the lead times for projects are so long, it cannot be used for stabilization policy. There is therefore an urgent need to reassess priorities and to see defence spending in its true context as a parasite on the economy as a whole.

If a true reassessment is to take place it must start with a genuine re-examination of future equipment requirements. It should also look towards an abandonment of the principle of independent viability, and hence of profit maximization for armament manufacture, to ensure that military requirements are not subordinated to the needs of the industry. The public ownership of the aerospace and shipbuilding industry is an important first step in this direction. Finally, there must be a fundamental restructuring throughout the defence industry, as any measures which would otherwise be attempted could only lead either to unemployment of workers and their equipment or to preserving the very structure which has proved so resistant to change in the past. The potential for the orderly conversion of the defence industries exists today. But the problem of industrial decline is of a structural nature, so the failure to take basic decisions about restructuring will surely lead to a situation in which there is even less room for manoeuvre.

Notes and References
1. Second Report from the Expenditure Committee, Session 1974–5, 'The Defence Review Proposals', HC 259.
2. SGP No 8. M. Kaldor, 'Defence Cuts and the Defence Industry'. The calculations are based on estimates made by the Stockholm International for Peace Research Institute (SIPRI) of the real resource cost in 1973 US dollars of individual aircraft. Warship costs are based on sources provided by the Annual Appropriation Accounts; *Janes Fighting Ships Annual.*
3. Derek Wood, *Project Cancelled*, Macdonald's and Janes, London, 1975.
4. J. R. Kurth, 'Why We Buy the Weapons We Do', *Foreign Policy*, No. 11, Summer 1973.
5. Dr Michael Checinski, 'The Cost of Armament Production and the Profitability of Armament Export in Comecon Countries', The Hebrew University of Jerusalem, the Soviet and East-European Research Centre, Research Paper No. 10, Jerusalem, November 1974.
6. Evidence of Vice-Admiral Clayton, Fifth Report from the Committee on Public Accounts, Session 1975–6, HCP 556.
7. See SGP No 8
8. Report of the Controller and Auditor-General, Appropriation Session 1975–6.
9. See SGP No 8. for a history of this project, which shows the high degree of similarity between the TSR-2 and the Tornado experience.
10. Arthur Reed, *Britain's Aircraft Industry, What Went Right? What Went Wrong?*, Dent, London, 1973, p. 114.
11. David Greenwood, 'Budgeting for Defence', *RUSI*, 1972, p. 7.
12. Appendix VIII: Ron Smith, 'The Resource Cost of Military Expenditure'.
13. Ministry of Defence, Written Answers, *Hansard*, 19 October 1976.
14. SGP No 7. Smith describes the method used to produce the conclusion like this: 'This hypothesis was put into an operational form as a regression of the share of investment on the share of

military expenditure in GDP. Our theory would suggest that the slope coefficient should be -1. This was confirmed both by a time series regression for the UK, and a cross-section regression for 14 NATO countries. The results did not seem to be sensitive to the source of data, sample of countries within NATO, or the choice of time period. However, some qualifications had to be made with respect to exports in the immediate post-war period in the UK and with respect to intra-country regressions for some other NATO members. It also appeared that there were a number of areas which deserved further investigation.' A further discussion of this theory is to be found in R. P. Smith, 'Military Expenditure and Capitalism', *Cambridge Journal of Economics*, No. 1, 1977.

15. Based on a capital/output ratio of 3 (possibly a rather favourable figure for the UK).
16. This is the conclusion of the Shipbuilding Inquiry Committee 1956–66 Report ('The Geddes Report'), Cmnd 2937; see p. 129.
17. *Daily Telegraph*, 15 December 1976
18. Sources for totals: Military – Cmnd 6735; Agriculture – *Agriculture Research Council Report 1975–76*; Medical – *Labour Research*, May 1977.
19. *Input–Output Tables of the UK*, Department of Industry, 1975.
20. Kurt Rothschild, 'Military Expenditure, Exports and Growth', *Kyklos*, vol. XXVI, 1973.
21. M. Kaldor, 'Defence, Industrial Capacity and the Economy', paper submitted to the LP–NEC Defence Study Group.

4

THE STRATEGIC AND POLITICAL
IMPLICATIONS OF DEFENCE CUTS

The strategic and political implications of defence cuts should not be seen in a purely national context. The modern world is a dangerous place. The advanced industrial countries devote increasing resources to the development and manufacture of means of destruction; poor countries are buying more conventional armaments; the risks of nuclear proliferation are hideous; new technologies are spreading. How might British defence cuts mitigate these dangers, and how could they contribute positively to international security? This is as important a question as the negative one of how we can cut defence spending without affecting national security.

Obviously the strategic and political implications of defence cuts will depend on how they are carried out – the concrete items affected and the co-ordination of defence and arms control policies with those of other countries. The three main ways considered here in which defence cuts might be carried out are by no means mutually exclusive.

First, our allies might decide to take over or finance some of our current military commitments. This would result in some change in the balance of power within NATO.

Secondly, we might try to adopt more cost-effective methods of carrying out current military functions. This could be possible through less emphasis on sophisticated expensive weapon platforms like Tornado and a more efficient use of new technologies.

Again, this would primarily affect relations with our allies and would depend on how they viewed such a change of tactics.

Thirdly, we might reduce real capabilities in the expectation that this would represent a significant contribution to détente and arms control.

Whatever the practical outcomes of these proposals, the strategic and political consequences of a £1,825 million cut in British military spending must be placed in the context of total military spending in the NATO countries with whom we share a common defence pact. In this perspective we find that a cut of £1,825 million would represent a gross reduction of less than 2 per cent in total spending.[1] So the strategic consequences of this action should not be overstated.

Before discussing ways and means by which defence cuts could be achieved 'painlessly', we need to examine critically the whole basis of prevalent strategic assumptions in the Ministry of Defence and elsewhere. It is not proposed to put forward an alternative strategic scenario, but rather to identify the problems inherent in current assumptions about the nature of the Soviet threat, and therefore to find a more rational basis for decision-making on military questions.

What is the Nature of the Soviet 'Threat'?

During a debate in the House of Commons on the 1977 Defence White Paper, Sir Ian Gilmour, the Conservative front bench spokesman on Defence, asked the Secretary of State whether there was really 'no evidence that the Soviets have aggressive intentions? If he really believes that', he continued, 'will he explain why the Russians are devoting all that effort to bring about that (military) explosion in their offensive capabilities?'[2] At a later stage in the debate, the Conservatives returned to the offensive in the shape of Mr Winston Churchill, who told the House that 'the only valid yardstick' by which to measure the necessary level of defence expenditure was an estimation of 'the perceived level of threat . . . Despite the ritual howls for uni-lateral defence cuts,' continued Mr Churchill, '. . . it is signi-ficant that no Hon. Member, even from the left wing of their party, is suggesting that the Soviet threat has diminished.'[3] The comment in the White Paper which seems to have excited Conservative fears appears to be: 'There is no evidence to suggest that NATO's policy of deterrence is failing and that the Warsaw Pact is contemplating aggression against NATO.'[4] It is our view that this is a perfectly reasonable assessment and that there is plenty of evidence to support it.

In analysing the nature of the strategic problems facing this country, we start from the assumption that if there were a genuine possibility of foreign invasion it would be necessary to devote the greater part of our resources to repelling such an attack. In the current situation, however, we are constantly being told that the source of potential aggression is the USSR. If this is so, then it is necessary to consider the problem in terms of:

(a) Soviet intentions;
(b) Soviet interests; and
(c) Soviet abilities.

(a) *Soviet Intentions*

It is a common military aphorism that one should look only at capabilities and not at intentions, since capabilities can be observed and intentions cannot. Unfortunately, this is not a logical suggestion. If we simply considered capabilities, we might as well prepare ourselves for attack from the USA or from the USA and USSR in combination, or from France and the Federal Republic of Germany in combination. We do not do this because, although the USA has the capability, we do not believe it has the intention.

The first point to make about Soviet intentions is that decision-makers in the USSR no more form a monolithic bloc with a single set of intentions than do decision-makers in Britain or the USA. There are very real and well-documented divisions within the bureaucracies of the Soviet and Warsaw Pact countries, and their attitudes tend to follow a fairly predictable pattern. Put broadly, the most influential groupings would appear to be the Party hierarchy (probably rather more 'dove-like' than 'hawkish'), the military and heavy industry (tending towards 'hawkishness'), the state bureaucracy, light industry and agriculturalists (tending towards 'dovishness'), and the internal security apparatus whose tendencies in this respect are hard to define.[5] Within these groupings there are further sub-divisions and divergences, and shifts in the internal balance of power often cannot be explained except by reference to these smaller sub-groups. Furthermore, Soviet decision-makers are very much influenced by Western actions. Therefore if the West were seen to be reducing its military strength, the 'dovish' factions would be strengthened, and *vice versa* if Western military activity was increasing.

Despite the complexity of the question, a general pattern of indications of Soviet intentions can be identified. The recent Helsinki Conference on Security and Cooperation in Europe is a good starting point to examine this question. The Soviet objective at Helsinki, one which Moscow has long sought to achieve, was to obtain international recognition of the *status quo* in Eastern Europe, an area which marks the confines of Soviet territorial ambitions. In return, they have had to offer very limited concessions on the question of human rights, a gesture which was aimed at the West, as well as recognizing the growing tensions in the over-extended 'empire' of Eastern Europe.

Helsinki demonstrates the depth of Soviet concern to stabilize the situation in the territories already under Soviet hegemony. It reflects their intention to consolidate their existing sphere of dominance rather than any drive to create fresh problems by extending territory. Russian expansion is more concerned with the creation of a limited land-based 'cordon sanitaire' on her Western front than with territorial adventures further afield.

We should not, however, underestimate the deeply felt desire of the USSR to achieve parity with the United States in all fields, and particularly in arms. 'Despite years of Herculean effort and indisputable Soviet success in many fields of strategic development,' writes Robert Kaiser, 'the new American President still claims "superiority". It is not difficult to imagine how the Pentagon would react if it had presided over a similar history.'[6] The struggle for parity has been a central characteristic of the world military situation. The general pattern has been for the USSR to follow US technological developments with a varying time-lag. Writing in the US *Foreign Affairs* journal, Alexander R. Vershbow comments: 'The US has too often been the leader (or culprit) in the introduction of new, more deadly technologies.'[7]

The USSR trailed in the development of atomic weapons; in the 1960s it lagged in building up the number of inter-continental missiles (despite the famous, and mythical, 'missile gap' of the late 1950s); it was several years behind in the development of MIRV (multiple independently targeted re-entry vehicles). And now the United States threatens to bound ahead again with the long-range cruise missile, with its sophisticated computer guidance and 'incredibly high degree of accuracy'.[8] The occasions in the post-war period when Soviet troops have actually taken military action outside their borders were when they thought that their hold on their Eastern European colonial empire was threatened.

In world-wide interventions they have been far more cautious than the Americans. There were no Soviet troops in Korea; none in Vietnam; there is no Soviet analogue to the landing of American marines in Lebanon. They even came late to the supply of arms as a technique of trying to win the allegiance of Third World countries – the first supply of arms outside the Soviet bloc was not until 1955.

There are compelling reasons for regarding the USSR as basically a *status quo* power in Europe. Its pressure has always been for the legitimization of the frontiers existing at the end of the Second World War, and for the acceptance of the dividing-line across Germany. It must be remembered that, during the long abortive negotiations over a peace treaty, it was the Western powers who refused to accept the division of Germany, and in Soviet eyes the rearmament of West Germany could well have looked like the prelude to an armed Western intervention to reunite Germany, an alarming prospect given the devastation suffered by the USSR during the Second World War. Indeed, the European policies of the Soviet Union since 1945 contrast heavily with the policies of Germany up to 1939 (which are often cited as an ominous parallel); Soviet pressure, culminating in the campaign leading up to the Helsinki Conference, has been for acceptance of the way things are, whereas it was clear, from the Ruhr occupation on, that the objectives of Nazi Germany were of a different order entirely.

In many parts of the world outside Europe the USSR, like other Great Powers, has given political, material and military support to various organizations and countries but always in pursuit of her own national interest.

Those who are committed to the assumption that the USSR is an enormous threat are forced to view this passivity and this defence of the *status quo* as an enormous trick. Western Europe will be lulled into complacency and, like a thunderbolt from the clear blue sky, the Eastern hordes will swoop across north Germany. This scenario contains severe military problems: even if the USSR had a standing-start attack capability it is most doubtful whether she would wish, or have the capability, to sustain an attack into hostile territory.

Those who argue that Soviet policy is one big trick often assume that Moscow's 'ideological struggle' contradicts and exposes its détente policy. A more dispassionate approach suggests that the Soviet leaders fear – probably rightly – that easing international tension will increase pressure to relax domestic controls; thus, precisely because they cannot utterly control the population, they are determined to prevent the penetration of foreign ideas and influence into Soviet society. In short, the intensification of the ideological campaign is 'related exclusively to internal needs and situations and has exceedingly little to do with Soviet international behaviour'.[9] It is no evidence of aggressive intentions abroad, and in no way supports the 'con-trick' theory of Soviet policy.

(b) *Soviet Interests*
Indications of Soviet intentions become more readily understood once the question, 'What does the USSR stand to gain by invading Western Europe?' has been answered. The reply does not demand an analysis of Soviet benevolence, but rather a practical consideration of whether such an action could produce beneficial results.

As we have noted, the USSR's main present concern is the preservation of her hegemony over the Eastern European Warsaw Pact states – recent events in Poland and Czechoslovakia serve to emphasize the costs and difficulties of Soviet control, as well as the vast problems inherent in deploying thousands of garrison troops outside her own borders. It seems unrealistic to assume that the USSR would wish to compound these problems by taking on the additional problems of a conquest of Western Europe.

There is also a school of thought which accepts that the USSR does not intend to attack us, but argues that it will tend to use its superior military strength to 'lean on' Western Europe in other ways. It is very difficult to discover what ways this school has in mind. First of all, the simple possession of greater military strength is not much help unless one is prepared to incur the odium of a clear and definite threat to use it. And if the USSR is prepared to use its superior military strength to lean on European countries, why has it not done so in areas where that superior strength is obvious? Why has it not used it in negotiations with Sweden, for example? Why has it not used it with Yugoslavia? Indeed, why does it not lean on Britain in the matter of extending fishing limits to two hundred miles, from which it stands to lose considerably?

The reason is that military power does not provide unqualified and simple advantages to any state, and that it is in Soviet interests to pursue a policy of détente rather than of aggression.

The USSR's primary objectives in the European arena, which

underlie its advocacy of détente, hinge on the maintenance of the *status quo*, not its disturbance.[10] It has long sought Western recognition of the inviolability of existing frontiers and non-recourse to force; acceptance of the division of Germany; and expanded economic cooperation between East and West. It is relying heavily on the importation of advanced Western technology and the impact of foreign trade to stimulate a growth-rate which has been declining since the early 1960s. The Soviet leadership is heavily committed to economic growth and steadily rising living standards.

Until recently, the USSR has relied heavily on quantitative sources of growth – the deployment of increasing quantities of the basic factors of production. However, since resources are not available to power quantitative economic growth at the necessary rate, future increases in Soviet growth rate will depend upon a more efficient integration of basic factors of production – upon qualitative growth. The over-centralized Soviet economic system, dominated by bureaucracies which make no adequate response to consumer demand, acts as a drag upon productivity, and rather than contemplate a radical overhaul of the system, the Soviet leadership seems to prefer increasing imports of sophisticated Western technology to foster qualitative growth.[11] Dependence upon Western technology – as the Soviet leaders, no doubt reluctantly, accept – draws the Soviet economy into the capitalist-dominated world economic system. Above all – and this may well be a point in its favour for many Soviet policy-makers – it means that the country's economic interests are best served by, and are indeed becoming reliant upon, improved political relations with the West.

The hallmark of the Soviet leadership, and the main cause of its neurotic reaction to any sign of internal dissent, are caution and conservatism. Both within and without the borders of the USSR, there is a clear *status quo* that the Soviet leaders wish to preserve. Many of the apparent ambiguities in their détente policy stem from occasional contradictions between the demands of their internal and external policies.

Détente, in other words, symbolizes the growing interdependence between East and West. Ironically, it is also the very existence of outward hostility between the two super-powers which holds together their respective spheres of influence. Victory or defeat, total destruction or total reconciliation, would destroy their domination and hegemony.

(c) *Soviet Abilities*

The USSR possesses massive and well-equipped armed forces. The fact that the country devotes about 12 per cent of the Gross National Product[12] to defence, and that she is developing sophisticated weapons delivery systems like the Backfire and the Foxbat, are the main reasons why there is concern about Soviet military intentions. The military balance is discussed in more detail below, but it will be useful here to indicate certain influential factors which could affect our perception of the balance.

First, it should be emphasized that whereas the West identifies its sole potential aggressor as the Warsaw Pact countries, the USSR sees itself as facing two threats: from China as well as the West. The USSR is consequently obliged to divide its deployment of military resources accordingly. The Chinese, with a population of 850 million and a military force of some 3·5 million incorporating a nuclear capability, are seen as a serious opponent. Military planners in Moscow are constantly exercised by the fear of a two-front war. As a respected American author on Soviet diplomatic history has written: 'The rapid buildup of the Soviet stockpile and delivery system in the 1960s must be understood as reflecting the need not only to reduce America's superiority but to establish a crushing Soviet superiority over China's budding nuclear power.'[13] Similarly, a Rand Corporation expert on Soviet military policy has stated that it is 'a real question whether the Soviet leadership still feels that the primary military threat to Soviet interests is posed by [the West] or . . . [by] Communist China'.[14]

Apart from the Chinese dimension, there is also the lesser factor of the deployment of Soviet forces in Eastern Europe. The fact that Soviet troops are virtually, if not actually, garrison troops, means that they cannot readily be moved to another front. Indeed, any movement of troops could have severe internal repercussions. There is also some debate on the degree of homogeneity within the Warsaw Pact. While the NATO countries could almost certainly be expected to act in concert, several observers (notably the CIA in the 1960s) have questioned the extent to which the troops of some Eastern European countries could be relied upon to fight outside their own borders. As the former US Defence Secretary, Robert McNamara, remarked: '. . . we are no longer convinced that the East European forces, which constitute more than half of the Warsaw Pact's combat ready strength in Central Europe, would be fully effective in an unprovoked attack on NATO.'[15]

The Military Balance

The military balance is open to a great variety of interpretations; consequently, perceptions of this balance are highly subjective. Official sources (concerned to raise defence budgets) tend to emphasize Warsaw Pact strength. Other sources (including those in semi-official circles) present a different picture. It does not take great imagination to suppose that the Soviet Ministry of Defence's presentation of the military balance is different again. The picture can change according to the way in which the geographical area is defined, or qualitative factors are taken into account, and so on.

But a more important reason for not giving undue weight to detailed presentations of the military balance is the fact that they too often imply that security is a technical matter (and, indeed, a matter of how well you can do your sums). Yet this is fallacious, for security is created by economic strength and political will as well as military power. The balance can be conceived of as relatively stable, and impervious to minor shifts in one direction or another, since it is impossible to calculate it in any but the crudest terms. While it may be true that détente rests on a military balance which precludes the use of force in East–West relations, it is wrong to see the military relationship as the only (or even the dominant) one.

Methods of assessing the military balance are themselves a matter of concern. On both sides, hawks and doves are involved in the process of assessing and drawing conclusions from the state of the military balance; the data themselves thus become matter for political dispute.

It is relatively easy for a hawk to justify the case for an unlimited arms race. All that is necessary is to argue that the nation should be ready for the 'worst case'. With two sides doing a 'worst-case' analysis, there is, of course, no limit to an arms race. The position of the doves on either side is weaker. They can point to the insanity of the process, an argument which never sounds as powerful as it should, and to the waste of valuable resources. But the momentum behind military policies on both sides is formidable.

It is therefore unfortunate that most sources of information are in the hands of the hawks. The people who advise the civil power on the size of the military threat are the military themselves; this inevitably leads to bias, since the military, like all other bureaucracies, are obviously looking for justifications to increase their resources. There is nothing particularly malign about the tendency of the military to select the facts which fit their cause. All such groups tend to do this. In most cases, however, there are good sources of counter-information. But in this case – the assessment of the threat from the potential enemy – reliable independent sources of information are almost non-existent.

It is nevertheless essential to try to gain a more objective perspective of the balance of military power. Such an attempt necessarily involves a consideration of the rather repetitive numbers game, and even though this is an unsatisfactory way of calculating the comparative effectiveness of opposing military forces, it is on numbers that much of the public justification for defence policy depends. In the sections which follow we look at the arguments as they relate to each branch of the armed forces.

(a) *Naval forces*
The main concern of those who talk about Soviet naval expansion is not its absolute size. The numbers of ships have not actually changed in recent years, but rather has the Soviet navy changed from being a coastal defence force to an ocean-going fleet. It is suggested that the USSR is aiming for an offensive capability to enable it to intervene and interdict Western shipping. Yet the evidence for this viewpoint is scant. The Soviet naval infantry is extremely small, numbering 14,500, and has been declining. The USSR has no sea-based air power to give cover for inter-vention. The new VTOL aircraft carriers are of little use in any role but anti-submarine warfare. Finally, the USSR has a very limited sea-based support system, and her overseas naval facilities, particularly following political shifts in Egypt and Somalia, are quite inadequate. Some authorities have explained the changed deployment of the Soviet fleet as a response to the US nuclear attack carriers which appeared in the mid 1950s and Polaris submarines which appeared in the early 1960s.[16] Moreover, the Soviet press emphasizes the way in which naval 'visibility' pro-vides a political underpinning to détente.

It is this last factor which is probably the primary cause of concern over the Soviet Navy for NATO's leaders. According to Michael Klare:

By gaining 'new visibility in areas of the world which have traditionally been within the Western sphere of influence' (to use Admiral Rectanus' words), Moscow has called into question the invulnerability of Western fleets there and thus their utility as instruments of coercion and influence. Thus, it is not Western *shipping* that is threatened by Soviet naval deployments, but Washington's strategy for continued Western hegemony in remote Third World areas.[17] (Italics in original.)

(b) *Ground forces*

The 1976 Defence White Paper states categorically that 'in conventional ground and air forces, the imbalance in Central Europe has moved further in the Warsaw Pact's favour'.[18] It is significant that this kind of grandiose claim is not repeated in the White Paper for 1977.

Above all else, the problem troubling Ministry of Defence strategists seems to be the sheer size of the 1·8 million-strong Soviet army. Yet the total army strength of European NATO nations, together with US troops stationed in Europe, is about 2 million, and the USSR must also be uncomfortably aware that the Chinese army numbers some 3 million. Thus the size of the Soviet army does not seem over-large given the size of its potential adversaries. Much has been made of Warsaw Pact numerical superiority in the European theatre; the presentation has often been misleading, occasionally descending to the level of counting up divisions without mentioning that the average NATO division strength in the theatre is considerably greater than the Warsaw Pact average strength.

The International Institute for Strategic Studies has estimated that NATO has 635,000 combat and direct support troops available in Northern and Central Europe, while the Warsaw Pact has an estimated 910,000. A recently published survey of forces shows a different picture, estimating effective NATO strength at 725,000 and Warsaw Pact strength at 780,000.[19] This disparity is probably accounted for by the fact that the Warsaw Pact uses military people to perform functions which are carried out in NATO by civilians. The IISS figures do not reflect this position (although they do qualify their figures in the accompanying text). However, this superiority is not of itself of a magnitude likely to be essential for a conquest of Western Europe; the NATO figure excludes all French forces, and also US 'dual-based' brigades which could be flown in relatively quickly. In Southern Europe, the numerical advantage is NATO's, by 540,000 to 395,000.[20] The Warsaw Pact does, however, have an advantage in terms of reinforcement, as a land link between its forces ensures a more speedy availability of heavy equipment. Nevertheless, the picture is by no means as alarming as the more extreme 'worst-case' analysts would have us believe, and limited reductions or increases on either side would affect the analysis only minimally, if at all.

Besides the concern about manpower, there is alarm over Soviet tank superiority. The factual basis of this assertion is undeniable. However, a purely numerical comparison is misleading. The former US Assistant Secretary of Defence, Alan C. Enthoven, shows why this is so:

> It is not clear that this numerical superiority in Pact tanks is a decisive advantage. It reflects Soviet tradition, which stresses tanks heavily. NATO armies have deliberately chosen to place less emphasis on tanks than do the Soviets. We could increase the emphasis on tanks if we thought the total effectiveness of our forces would be increased thereby . . . Studies show that the NATO tanks and anti-tank weapons have a high kill potential against the Pact tank force . . . In addition, one must consider the additional large tanks kill potential of our tactical aircraft.[21]

The decreased emphasis on tanks by NATO is, in our opinion, a proper decision; and it does make the concern about inferiority in tank numbers seem a little misdirected. If NATO's anti-tank weapons are effective, then the tank balance matters less, quantitatively or qualitatively. Finally, it should be emphasized that NATO sees itself as having different strategic objectives from the Warsaw Pact; its forces should not therefore be a copy of Warsaw Pact forces. A military force should be structured according to its own objectives, not according to some other force's assumed objectives.[22]

(c) *Air force*
There is little conclusive evidence to reveal the true balance between the rival air forces. Although the Warsaw Pact appears to have maintained a consistent numerical superiority in tactical aircraft, much of this is accounted for by its far greater numbers of interceptor aircraft for air defence. In 1973, the Pentagon estimated that, in an actual European battle, given likely reinforcement, the tactical air forces of each side would be 'about even at 5,000 to 6,000 each'.[23]

(d) *The military budget*
There is an increasing tendency for military planners to cite disparities in the military budget as a means of comparing

military strength. Such comparisons have a poor track record for accuracy, particularly with regard to Soviet figures from Western sources (and no doubt *vice versa*, if we had access to this information). The point is, however, that the level of spending tells us very little about what is being purchased. The whole process of making budget comparisons is obscure at every level, and is so contentious that firm conclusions are difficult to reach. The sources of information range from the CIA data included in the *Military Balance* to alternative figures from the Stockholm International Peace Research Institute (SIPRI). The problem is illustrated by a comparison of Soviet military expenditure between 1970–75, the CIA estimating that it has *increased* by 4 to 5 per cent, and another American source estimating a *9 per cent increase*. SIPRI, on the other hand, estimates a *fall* of 3 per cent.[24]

The differences between these various estimates can be explained by different methods of calculation. The CIA estimate is an attempt to calculate how much it would cost the US government to 'buy' the Soviet military establishment, in dollars, at American prices. The calculation immediately falls into awesome methodological problems, since conscription makes Soviet military manpower much cheaper than American manpower, whereas general technology is cheaper to the USA than to the USSR. If the calculation is reversed and the rouble cost to the Soviet government of 'buying' the American military establishment is estimated, the US appears the bigger spender. As a US congressman, Les Aspin, points out, since the American arms inventory includes items much more sophisticated than the Soviets have (in the field of computers and advanced electronics), a calculation such as this is bound to *underestimate* the cost to the USSR of 'replicating the physical dimensions and operational capabilities' of the US military. As he rightly concludes, 'the answer to the question, "who is spending more on defence?" depends on the price system used.'[25] The SIPRI figures are based on the official Soviet budget, taking account of hidden costs included in the Soviet science budget and of the difficulties of choosing an appropriate exchange rate.

Yet another method, used by the American source quoted above, would include an unaccounted-for residual in the output of the machine-building and metal-working industries.

(e) *The Nuclear Balance*
The USA is estimated to have approximately 8,530 strategic nuclear warheads, the USSR approximately 3,250.[26] Various

other measures of the balance qualify this appearance of an American lead. For example, in megatonnage the USSR is believed to have a considerable lead, though this is partly offset by the greater average accuracy of the US arsenal. It is extremely difficult to arrive at an adequate assessment of the relative capabilities of the two forces, and this is in any case really a rather pointless exercise. As Henry Kissinger once asked, 'What in the name of God is strategic superiority? What do you do with it?'[27] The only kind of strategic superiority which would have meaning is if one side were to develop a first-strike capability – the ability to destroy all or most of the other side's strategic nuclear weapons. Neither side possesses this ability; nor is it likely to for some time, if ever; nor is it in any way desirable for either to seek to attain it. Short of that, the staggering degree of overkill in each side's nuclear arsenal makes talk of strategic superiority meaningless.

Britain possesses 192 strategic nuclear warheads on Polaris missiles, a trivial figure in comparison to the super-powers' arsenals and evidence enough that the British contribution to the strategic nuclear balance is irrelevant.

Although the figures are not known with any precision, it is generally estimated that NATO's European theatre nuclear forces consist of about 7,000 tactical nuclear weapons for delivery by aircraft, missile and artillery, or in the form of atomic demolition munitions. The Warsaw Pact is believed to have about 3,500 tactical nuclear weapons.[28] Recently both sides have strengthened their theatre nuclear forces: NATO through the increased numbers of American F-111 bombers based in Britain, and the Warsaw Pact through the development of the SS-X-20 medium-range missile.

The study group views this build-up with alarm. There is already a considerable degree of overkill in each side's forces: it seems impossible that NATO could have 7,000 potential targets, or anywhere near that number, for its theatre nuclear forces; and while the Warsaw Pact has fewer weapons in this category, they are believed to be more powerful though considerably less accurate. Any war involving the use of large numbers of these tactical weapons would have, so far as Europe is concerned, strategic effects killing millions of people and rendering vast areas uninhabitable. At the Mutual Force Reduction talks in Vienna, NATO proposed withdrawing a large number of tactical nuclear weapons in exchange for the Warsaw Pact withdrawing a number of tanks. This would seem to be a recognition that NATO has tactical nuclear superiority (for what it is worth), and that it has more tactical nuclear weapons than it can possibly need.

The nuclear arms race presents an incredible risk of warfare in the modern age. To gain some impression of the size of the problem, we should remember that the atomic bomb dropped on Hiroshima unleashed an equivalent of between 13,000 and 20,000 tons of TNT. It has been estimated that the USA has warheads aimed at the USSR with an equivalent TNT explosive power of some 4,200,000,000 tons, and that the Soviet arsenal is even larger.[29]

After examining the problem of making comparisons about the military balance, Dan Smith comments:

> Most of the conventional comparisons of military force between NATO and the Warsaw Pact are very dubious, serving to confuse rather than clarify. No clear picture can be provided and firmly held. So much depends upon estimation and interpretation of disputed data, that only the most limited conclusions are really possible.[30]

We have considered some of the problems of examining the military balance. We do not conclude that the NATO countries are militarily either stronger or weaker than those of the Warsaw Pact. Any reconsideration of expenditure priorities must incorporate a reconsideration of both the purpose of our military expenditure and the strategic context. If calculations of essential expenditure are to be based on the magnitude of the Soviet threat, a rather more objective method of making a true assessment is needed. There seems to be nothing in this area to dissuade the Labour Party from its policy of reducing military expenditure.

The Arms Race, Détente and Arms Control
This study's proposals for reducing defence expenditure are put forward during a period of intensive international reassessments of political and strategic questions. The contradictory pressures of an escalating arms race, as against a growing awareness of the need for arms control, are tempered by the emerging political environment in a period of East–West détente. While it is possible to identify some areas of achievement on human rights and trade matters in East–West relations, it is not possible to point to any significant success in the field of military détente.

Seen in global terms, it is clear that not even a reduction in

British military expenditure of around £1,800 million would radically alter the overall picture, though such a move should be viewed in the context of these developments.

Détente between East and West is a far more complex and ambiguous relationship than that which prevailed during the Cold War. The argument advanced by each side that a balance of military power provides the basis for détente presupposes, in the West, that the USSR is fundamentally aggressive; and in the USSR that the West is fundamentally aggressive. At the same time, each side accepts that common interests exist in avoiding confrontation and extending economic relations. These ambiguities are deeply rooted, not only in international relations but also in domestic institutions. Disturbing though they are, they are preferable to the certainties of the Cold War. Moreover, there are powerful reasons why the two super-powers should wish to maintain this ambiguous relationship, and in this their interests are, to a considerable extent, complementary.

The arms race is not merely continuing; it is becoming both more extensive and more intensive. Three worrying features may be discerned. First, since 1973 there have been substantial real increases in military spending in both NATO and Warsaw Pact countries, in particular in the USA, France, Britain, the two Germanies, Poland and the USSR. Secondly, there has been a proliferation of military technologies of all kinds. The trade in conventional major weapons with the Third World has increased nearly five-fold over the past ten years. Six countries have now exploded nuclear devices, while a seventh, Israel, is widely believed to possess untested nuclear weapons. Since 1965, the number of countries capable of producing nuclear weapons at short notice has increased from six to twenty-two.[31] Thirdly, there have been major technological changes in non-nuclear weapons. The development of precision-guided munitions has received much attention, but while their defensive potential has been lauded, their development could provoke the development of new 'area suppression' weapons, such as napalm, binary gas weapons and environmental modification weapons. One more hopeful sign in this respect is the new ENMOD treaty on the control of hostile uses of environmental modification techniques, which represents a limited but necessary first step towards reducing the dangers of environmental warfare.

Against this background, talks proceed in Vienna about mutual balanced force reductions, the super-powers discuss strategic arms limitation (SALT) and the nuclear powers engage in attempts to prevent nuclear proliferation. Tangentially, the reduction of military tension was on the agenda of the Helsinki Security and Cooperation Conference, and is on that of the follow-up Belgrade Review. The striking fact which emerges from all these high-level deliberations is the lack of progress towards a reduction of military forces; nor, with the possible exception of the SALT talks, is any real progress likely. This suggests that while international negotiations may provide the *forum* in which arms control agreements are reached, they will not be the *agency* for achieving effective arms limitation, because they do not tackle the domestic sources of the problem. Defence cuts, by weakening domestic pressures, can make a major contribution to arms control. As we have said, East–West relations are complex and ambiguous, but it is most unlikely that even swingeing defence cuts would make these relations unstable. But defence cuts might well contribute to a policy of emphasizing the cooperative rather than the hostile elements in the relationship.

British unilateral action will not in itself (obviously) alter dramatically the course of arms control talks. It will, however, be a contribution to something more productive than long-drawn-out international gatherings, and it is an opportunity for Britain to play a role in bringing about a less tense international climate.

Options for Reducing Defence Expenditure
As we have demonstrated, there are, at the very least, wide areas of contention in prevalent strategic thinking. Questioning some of the deeply held assumptions of military planners shows that, if there are to be reductions in military spending, there is more room for manoeuvre than might be expected. We now look at the strategic and political implications of the various options for cuts suggested in Chapter 3. Once again, the options discussed are not prescriptive. There is an infinite variety of combinations of cuts that could be made, and what is involved here is an illustrative exercise, assessing the nature and consequences of the kind of changes which a significant reshaping of our military effort would require.[32]

In Chapter 2, we showed how a remodelling of the armed forces might include a mixture of the following alternatives:

1. Phasing out Polaris.
2. Paying off *Ark Royal, Hermes, Fearless* and *Intrepid,* can-

celling the second and third ASW Cruisers and Sea Harrier, running down patrol submarine activity and 'stretching' building programmes for fleet submarines, destroyers and frigates.

3. Reducing BAOR to about 30,000 men, with appropriate equipment reductions and a proportional cut in home forces to reflect a reduced requirement for rotating units to and from BAOR.

4. Cancelling both versions of the multi-role combat aircraft, Tornado, with reductions in tankers and air transport, and increased expenditure on Harrier and Jaguar.

These options would have to be considered together with proportionate reductions in support programmes and the withdrawal of all ground forces stationed outside Europe.

1. *Polaris*

The option discussed is a progressive phasing-out of all provision for the Polaris force.

As we have seen, the contribution of Polaris to the strategic nuclear balance is trivial; there is, in fact, some doubt, from the Pentagon at least,[33] about the role of Polaris in NATO's strategic forces. The political arguments in favour of keeping Polaris have generally been more comprehensible than the military ones, though they owe much to lingering delusions of grandeur. Some have argued that there are dangers in abandoning Polaris and leaving France the sole strategic European nuclear power. Others suggest that retaining Polaris ensures the (already certain) commitment of the USA to NATO. If this latter argument implies that Britain scrapping Polaris would result in the USA leaving NATO, then it is absurd; and if it does not imply that, it is hard to see what it does mean.

There is no overwhelming financial advantage in phasing out Polaris, and our options therefore do not rule out maintaining the fleet for the rest of its natural 'life' (ten to thirteen years).[34] However, retaining Polaris must not be allowed in any way to undermine the clear commitment of the Labour Party and Government against developing a further generation of strategic nuclear weapons. In any case, the replacement of Polaris would impose a very severe burden of expenditure.

The study group has found, however, no compelling arguments against abandoning Polaris, and the opportunity might be taken to use the scrapping of Polaris to stimulate wider and faster moves towards arms reductions and disarmament. The attempt might fail, but if Polaris is to be abandoned because of the need to reduce military spending, and if there are no *adverse* consequences, then the attempt should be made to wring all possible advantage from the decision.[35]

2. *Naval General Purpose Forces*

Britain's navy, large by comparison with the navies of our European allies, is structured around big ships with deep-water, long-cruise ability. It provides 70 per cent of NATO's East Atlantic forces. The effect of the reductions proposed here would be to transform the structure of the Royal Navy. This would continue to fulfil its main roles, but without large surface ships, such as the new anti-submarine warfare (ASW) cruisers, and would abandon some roles, such as amphibious warfare, upon which less emphasis is already being placed.

There is a general consensus that the main naval threat to NATO in the areas in which the Royal Navy operates comes from the Soviet submarine forces, though there is also some concern about Soviet surface ships. However, it is not at all clear that the best way to counter this threat is the current strategy of using ASW task forces based around ASW cruisers. Not only are larger ships larger and more inviting targets, but their expense limits the total number of them available. A navy based on smaller ships would seem to offer an equally effective and cheaper way of carrying out the main tasks for which, if these proposals were adopted, the Royal Navy would be exclusively structured. Although the navy would lack both sea-borne aviation and amphibious warfare ability, its range of operations for the missions which should receive priority would not be changed.

3. *European Theatre Ground Forces*

The reduction of BAOR from 55,000 to 30,000 men is a major cut, but looks less significant in the overall context of NATO European ground forces. As a percentage of combat and direct support troops, it corresponds to a cut of 3·9 per cent in North and Central Europe, a 2·1 per cent cut in the whole of Europe, and a 3·4 per cent cut in the area covered by the Mutual Force Reduction talks. If French ground forces were also counted, the percentages would be lower. Directly proportionate equipment reductions might mean, for example, reducing the BAOR tank force from 650 to 350, corresponding to reductions of 4·3, 2·7 and 4·7 per cent in NATO tank forces in the same regions, again excluding French forces.[37]

If such reductions were undertaken, their effect on BAOR's combat strength might be lessened through further reorganization.[38] They could also profitably be accompanied by tactical alterations designed to take advantage of the new types of anti-tank guided weapons (ATGW) which seem likely to reduce the tactical advantages of tanks. The accuracy of ATGW, combined with modern sensor devices, could, it has been argued, provide an individual company with 'a greater anti-tank potential than a whole mechanized battalion before 1970'.[39] While it must be recognized that the effectiveness of ATGW is a subject for debate, it may be that their deployment, and appropriate tactical changes, would make possible a reduction of manpower without loss of defensive effectiveness. BAOR would then be a relatively small force, capable of strong mobile defence and based on small units which possessed the advantages of relative speed of movement and ease of concealment.[40]

Such changes in BAOR would, of course, have political repercussions (see below). Many will also argue that it would lower the nuclear threshold – hence, that in a conflict, nuclear weapons would have to be used earlier. This objection is based on the assumption that a limited war in Europe is possible. Yet this assumption seems questionable so long as tactical nuclear weapons are deployed widely in Europe. Those who are concerned about the nuclear threshold should focus on the size of the nuclear arsenal rather than on the consequences of marginal changes in conventional forces.

Finally, no discussion of British ground forces can ignore their major active role – that in Northern Ireland. It has not been within the study group's terms of reference to discuss this issue in any detail. Nevertheless it is clear that if the Northern Ireland role is to continue, there should be adequate provision for air transport so that, should a reduced BAOR be drawn on for some contingency in Northern Ireland, then the speedy return of troops to Germany remains practicable. On the other hand, reductions in this programme could go hand in hand with a major reassessment of the utilization of troops in Northern Ireland.

4. *Air Force General Purpose Forces*
The central feature of this option is the cancellation of the multi-role combat aircraft Tornado. The proposed reduction in air transport creates some difficulties, as air transport has already been reduced. Decisions about it should be made in the light of

the transport requirements of ground forces and the possibilities of commandeering civil aircraft.

Cancelling Tornado would effectively remove from the RAF one of its most important missions: long-range bombing. Missions which involve deep penetration of enemy air space have been made more difficult by the development of precision-guided surface-to-air missiles (SAMs), though terrain-following radar does make low-flying missions more feasible. Electronic counter-measures can also reduce the effectiveness of SAM, but it can be restored by the use of electronic counter-counter-measures. It may be argued that the expensive deep-penetration mission – whether to strike at airfields, troop concentrations, installations or cities – should not be a priority for a country like Britain which needs to cut back its military spending.

Also lost would be the Tornado air defence variant, which, together with SAM, is intended to provide air defence for the United Kingdom. The proposal in the event of Tornado cancellation would be to use Phantom as an air defence interceptor throughout the 1980s, with improvements to their Sparrow air-to-air missiles.

The consequent air force would be one whose priority missions would be battlefield interdiction and the attainment, so far as possible, of local air superiority over the battlefield. For the first of these the RAF has two fine aircraft, the Harrier and the Jaguar; the latter is also capable of tactical reconnaissance and air combat. Expenditure upon either or both could be increased if thought necessary. Limiting the RAF's roles in this way might in the future eliminate the need for multi-role aircraft, which are not only extremely expensive, but in general tend to perform less well than aircraft designed for specific tasks.[41] It may become possible to concentrate upon cheaper and more specialized aircraft.[42]

5. *Additional Cuts*

Reductions in support programmes would not involve political or military consequences apart from those already discussed. Reducing or removing overseas bases would have political consequences, but it is not clear what positive role these bases have. During the 1974 Cyprus crisis, British troops stationed on the island were confined to barracks. It is hard to see the point of stationing troops where they cannot be used.

Burden-sharing

A possible way of reducing British defence expenditure would be to share the defence burden. The principles of collective burden-sharing are well established; they are the same principles which underlie progressive taxation. The *proportion* of resources devoted to a common purpose should be greater for the rich than for the poor. This principle is, of course, fully accepted, not only by the Labour Party but also by the Conservative Party in its approach to the taxation of incomes. The only question at issue is the degree of progressivenesss.

The same basic principles should apply in burden-sharing between countries which are devoting resources to a common purpose. Income per head in the United Kingdom is, at estimated purchasing-power parities, now some 25 per cent below income per head in Western Germany and France.[43] The gap is widening, and there is every indication that it will continue to widen. It widened further in 1976, when national output (at constant prices) is estimated to have risen 5 per cent in Germany, 4·5 per cent in France, but only 1 per cent in the United Kingdom. It is virtually certain that it will widen further in 1977, when national output is forecast to rise 5 per cent in Germany, 4 per cent in France and 1·5 per cent in the United Kingdom. Thus, year by year, it becomes more and more difficult for Britain to devote the same proportion of its resources as France and Germany for collective defence purposes, let alone a larger share. It is clearly time that we admitted to being a relatively poor country and had our share of the collective burden consequently adjusted. This argument is simple and straightforward, and it is a big argument. It is a manifest absurdity that, when we are so much poorer than France or Germany, we should be devoting a larger proportion of our resources than they are to military expenditure.[44]

Why then has the United Kingdom been so weak in putting forward in NATO this essentially open-and-shut case? There are a number of reasons. First, politicians dislike admitting that Britain is economically weak, and persist as long as they can with the fiction that we are an economically powerful country suffering temporary adversity. There is still, in their minds, the old correlation between military power and importance and standing in the world.

The second main reason is that most such negotiations in NATO obviously involve a strong military influence. And the British military could hardly be expected to put much weight behind the presentation of a case which would lead to a diminution of their power and influence.

Political Consequences

Much of the objection to reducing defence expenditure is derived from unthinking political dogma, but it is certain that a reduction of the order envisaged in this paper would have serious political repercussions throughout NATO as well as at home.

Apart from domestic considerations, which are very important and form the bulk of this paper's content, the reaction of our NATO allies is the major factor which has so far dominated discussion of this subject. Two main arguments are advanced: first, that defence cuts would weaken the confidence and cohesion of the NATO countries; secondly, that a cut in the British NATO contribution would alter the balance of power within NATO, and this would have undesirable international repercussions.

One particular problem would arise if the Federal Republic of Germany, which has Europe's strongest economy, should decide to follow the logic of the burden-sharing argument and increase her level of military spending. Besides the fact that such a move would alarm the USSR out of all proportion to its intent or real strategic significance, it could also have internal repercussions in that it might provide a justification for increased authoritarianism in German society. As internationalists, the Labour Party shares this concern and recognizes its relevance for ourselves. However, we do not believe that reductions in British military spending will have a decisive effect one way or the other.

The concern about possible shifts in the balance of power within NATO is a valid one, but as it is presented in some quarters it makes it appear that we should be aiming to keep pace with our friends rather than our enemies. This is an element in the fallacy that we can compensate for our economic weakness by military strength. It is, in any case, worth questioning whether the pursuit of status is a particularly worthwhile objective, and whether it justifies the expenditure of a high proportion of our resources. One might as strongly argue that status will not result from hankering after past and discredited glories, and is probably more a function of economic than of military strength.

Conclusion

Let us suppose that the United Kingdom informed its NATO allies that, over the next five years, it was intending to bring down its share of national product devoted to defence to match their average figure. Would the strategic and political consequences be so damaging as to outweigh the economic benefits? The evidence clearly points to the answer: no. This is not a plea for other European countries to make good the deficit caused by a British reduction. On the contrary, it must be stressed that the Labour Party would not see this as a desirable outcome since it views defence cuts as part of a process towards the lessening of military tension. While the current military balance in world affairs is dominated by the actions of the super-powers, it seems unfortunate for the European nations to be dragged into an arms race, the outcome (and indeed direction) of which is entirely beyond their control. Should the super-powers wish to pursue these dangerous policies, they must be firmly told that Europe will not be helping to finance their extravagance.

The desirable political and strategic approach to defence cuts is therefore that they should contribute to multilateral disarmament at best, or at least make way for a more rational assessment of the military threat and ensure the maintenance of the necessary strategic posture at a more realistic cost. In European terms, this would involve the deployment of new military technology as well as specialization among the allies to rationalize the defence effort. There is no suggestion of disarming Britain, and we certainly hope that the debate on these proposals will recognize that we are proposing to defend ourselves in a more rational way, at a more appropriate cost. On the basis of our work, we believe this to be eminently possible as well as desirable.

Notes and References
1. Based on 1976 expenditure figures given in the *Military Balance, 1976–77*, which shows a total of £159,975 million for the military expenditure of all NATO countries.
2. Debate on the Defence Estimates 1977, Cmnd 6735; *Hansard*, 22 March 1977, col. 1113.
3. *Hansard*, 28 March 1977, cols. 58–9.
4. Statement on the Defence Estimates 1977, Cmnd 6735, February 1977, p. 2.
5. This is rather a complex question. For further discussion, see, e.g., Vernon Aspaturian, *Process and Power in Soviet Foreign Policy*, Little Brown, New York, 1971; and M. Schwartz, *The Foreign Policy of the USSR: Domestic Factors*, Dickenson Publishing, New York, 1975, Ch. 6.
6. Robert Kaiser in the *Guardian*, 14 March 1977.
7. Alexander R. Vershbow, 'The Cruise Missile: The End of Arms Control', *Foreign Affairs*, No. 55 (i), October 1976, p. 146.
8. ibid., p. 135.
9. R. G. Horn, 'Détente Myths and Soviet Foreign Policy', in Potichnyj, P. J., and Shapiro, J. P. (eds.), *From Cold War to Détente*, p. 104. See also Schwartz, *The Foreign Policy of the*

USSR, pp. 146, 148–9, 154–6; and Levgold, R., 'The Soviet Union and Western Europe', in W. E. Griffith (ed.), *The Soviet Empire, Expansion and Détente*, Lemington Books, New York, 1976, pp. 230–31.

10. For fuller discussion of Soviet motives in pursuing détente policies, see Levgold, R., 'The Soviet Union and Western Europe', in Griffith (ed.), *The Soviet Empire*, pp. 234–6, 243–4; and G. A. Flynn, 'The Content of European Détente', *Orbis*, vol. 20, No. 2, Summer 1976.

11. See J. S. Berliner and F. D. Halzman, 'The Soviet Economy: Domestic and International Issues', in Griffith (ed.), *The Soviet Empire*.

12. *The Military Balance 1976/77*, IISS, London, 1976.

13. Adam B. Ulam, *Expansion and Co-Existence*, Praeger Publishers, New York, 1974, p. 771.

14. Thomas W. Wolfe, 'Military Power and Soviet Policy', in Griffith (ed.), *The Soviet Empire*, p. 150. See also Ulan, *Expansion and Co-Existence*, p. 678.

15. Robert S. McNamara, *The Essence of Security*, Hodder & Stoughton, London, 1968.

16. Michael McGuire, 'Soviet Naval Capabilities and Intentions', paper presented to the RUSL Conference on the Soviet Union in Europe and the Far East, Milford-on-Sea, 1970.

17. Michael Klare, 'Super-power Rivalry at Sea', in *Foreign Policy*, Winter 1975–6.

18. Statement on the Defence Estimates 1976, Cmnd 6432, ch. 1, para. 20.

19. Henry Stanhope in *The Times*, 31 May 1977, gives the NATO figure; and the Warsaw Pact figure is quoted from Professor William Kaufmann, a Pentagon consultant analyst from MIT.

20. All figures from *The Military Balance 1976/77*, p. 99.

21. Alan C. Enthoven, *Review of a System Analysis Evaluation of NATO v Warsaw Pact Conventional forces*, 90th Congress Report of the Special Sub-Committee on the National Defense Posture of the Committee on Armed Services, US House of Representatives.

22. See SGP No 5. Dan Smith, 'The Political and Strategic Implications of Reduced Defence Programmes'.

23. 'Study Insists NATO Can Defend Itself', *Washington Post*, 7 June 1973.

24. See *The Military Balance*, p. 110; and *SIPRI Yearbook 1976*, p. 151.

25. Les Aspin, 'How to Look at the Soviet–American Balance', *Foreign Policy*, No. 22, Spring 1976.

26. *The Military Balance*, p. 106 (figures refer to mid 1976).

27. Quoted in 'Planning for the Day the Yanks Go Home', *Guardian*, 17 March 1977.

28. *The Military Balance*, p. 103.

29. *Guardian*, 17 March 1977.

30. Smith, SGP No 5.

31. A. Wohlstetter, *Foreign Policy*, No. 25, Winter 1976–7.

32. The basis of this discussion is to be found in SGP No 2. David Greenwood, 'Defence Programme Options 1980–81'. It should be

noted that these suggestions were first made in 1975 and a few have been overtaken by events; as an illustrative exercise, however, it remains valid.

33. See Annual Defense Department Report FY 1977 (US, DoD) Sec. III/A, p. 99; and Sec. III/D/1, p. 106.
34. The case for the retention of Polaris is argued in SGP No 3. Paul Cockle, 'Note on the Budget Approach and a Programme Option Which Retains the UK Nuclear Strategic Forces'.
35. See SGP No 5. for a fuller discussion of the implications of scrapping Polaris.
36. See SGP No 5. for a fuller discussion of naval roles and alternatives to present planning.
37. Calculations based on *The Military Balance 1976/77*, pp. 97–105.
38. The scope of reorganization is discussed in SGP No 5. Appendix XI.
39. Steven Canby: *The Alliance and Europe, Part IV: Military Doctrine and Technology*, Adelphi Paper No. 109 (London, IISS) 1974/5, p. 24.
40. For a fuller discussion of these points, see SGP No 5.
41. See William D. White, *US Tactical Air Power* (Washington DC, Brookings Institution), 1974, for a discussion of the problems of multi-role aircraft.
42. See SGP No 5. for a more detailed discussion.
43. *A System of International Comparisons of Gross Product and Purchasing Power*, UN, World Bank, and University of Pennsylvania. The 1970 comparisons are brought up to date by the movements of real GNP, 1970 to 1976.
44. Note that if the comparisons were made at current exchange rates, instead of at purchasing-power parities, the gap between income per head in this country and in France and Germany would be very much wider.

5

CREATING NEW JOBS

Introduction

If a worker is faced with the choice of producing armaments or joining the dole queue, quite naturally he will opt for producing armaments. He would also take this option if he were offered an alternative job which he regarded as being of lower status, or one which required him to move to another part of the country. That he believes this to be the only choice is understandable. There has been no systematic attempt by the Government to identify an 'alternative future', and there has been no methodical investigation of the employment consequences of a lower defence effort. As a result, defence cuts are always associated with redundancy, and working people know exactly what this means. Redundancy highlights the powerlessness of workers, and demonstrates how influences wholly outside their control can completely disrupt their lives, sense of security and expectations. When, for example, the Government decided to cancel the Blue Water guided missile project in 1962, half of those made redundant were forced to accept new jobs which paid less. Six per cent of them were dismissed without the prospect of another job. No magic market mechanism is going to give the redundant armaments worker a new job as of right.

At the same time, it must be remembered that defence expenditure is not designed to maintain employment. If it were, there would be no problem, for it would be cheaper to continue the payment of salaries to workers in the industry and stop other expenditure on arms. It seems strange that those who advance the erroneous proposition that the defence budget is necessary to maintain employment tend to be the same as those who are calling for public expenditure cuts so as to release resources for industrial expansion. Yet more resources and fewer employees are tied up in defence than in any other public sectors. The freeing of resources for other purposes should be seen as an asset rather than as a potential unemployment statistic and liability.

Our argument is that it will be possible over not too long a period to release some of the resources at present tied up in the arms programme for more socially and economically useful production, and without causing unemployment or more than a small minority of workers to change their place of work. Not only is this transformation possible, but it is necessary if we are to restructure and regenerate the British economy. If we do not divert capital formation into the non-military sector, and if we do not redeploy defence workers into productive work, the anticipated upturn in the economy will be retarded, just as it was in 1951 and 1973. In particular, the shortage of skilled workers outside the defence sector is liable to hinder growth itself. Moreover, the ability to convert the economy to civilian work may itself be a precondition for a Government to consider peace or disarmament proposals on their own merits.[1]

In the present chapter we demonstrate the possibilities for conversion in the defence industries and show what has actually been achieved in comparable circumstances. Alternatives to the wasteful policy of maintaining a high level of expenditure do exist; the problem of their implementation is one of planning and organization. There is need for planning at a national level, since the choice of appropriate products must be co-ordinated with government priorities in such diverse fields as health, agriculture, transport, energy and manufacturing generally. And there is a need for reorganization at a local level, since the current structure of military institutions, particularly the defence companies, is geared up to military-oriented innovation and tends to preclude domestic civilian investment.

Two terms are used here which should be understood separately. *Conversion* refers to the process by which part of our military industrial capacity would move into a different field of manufacture – a once and for all change. *Diversification* implies a widening of the base of activity – alternating military and non-military work for unconverted capacity.

The Effect of the Proposed Cuts
The money tied up in the defence effort represents resources – industrial capacity, human skills and manual labour. Just under a million people are directly employed in the defence sector. A third of a million servicemen and women are backed by the same number of civil servants. In Chapter 2, we saw that some 200,000 are employed on armaments work for the Ministry of Defence, and about 75,000 on export work; that a further 350,000 are indirectly involved, to some degree, on armaments work.[2]

What then would be the likely consequences of (a) the cut of approximately £1,800 million from the defence budget which would be required by equalizing defence/GNP proportions by the early 1980s, or (b) the reduction of approximately £1,000 million for which precise programme options have been identified in Chapter 2 (pp. 41–45)? A detailed analysis for the smaller reduction suggests that, if effected over a five-year period, the options developed in Chapter 2 would entail an actual release of manpower as shown in Table 12.

TABLE 12.

	Servicemen	MoD civilians	Contractors' employees	Total
Option 1	55,000	37,500	37,500	130,000
Option 2	32,500	27,500	55,000	115,000
Option 3	52,500	37,500	45,000	135,000

SOURCE: Study Group Paper No. 10

Allowing for the multiplier effect, one might double these numbers so that over a five-year period around 50,000 jobs would be lost each year in the defence sector. This is not a trivial adjustment problem, though the number itself is very small indeed when compared with the six million who voluntarily quit their jobs each year to start a new job or retire, or with the two and a half million who simply change jobs. Frank Blackaby[3] has estimated that there are about 180,000 jobs changed each year in the defence sector alone, so the overall number of jobs reduced in the defence sector would not be overwhelming, though the impact would, of course, depend on whereabouts in the country those jobs were located.

Contrary to popular belief, defence contracts are not placed predominantly in regions of high unemployment. In fact, whichever option for cutbacks is taken, in the absence of an alternative employment strategy the region which would suffer most would be South-East England in terms of both displaced servicemen and civilians. The South-East has the lowest average unemployment rate, and yet, with 30 per cent of the population, some 40 per cent of defence contracts are placed in this region. By contrast, Wales, with a relatively high rate of unemployment and 5 per cent of the British population, receives less than 3 per cent of defence contracts.[4] This illustrates how the aggregate problem of conversion is not one that would fall disproportionately on those regions least able to cope.

This is merely to put the matter in perspective: the Labour Party does not for a moment deny the extent of the problem within localities. We fully recognize that defence work may, within a region, be concentrated very heavily in a small number of firms and localities; and that a whole town may at present be entirely dependent on the defence industry. These facts clearly underline the need for effective and detailed planning. Much work must be done to ensure a smooth transition. But no one can seriously suggest that land, equipment, labour and skills should be committed to military purposes solely on the grounds that transitional measures would be troublesome.

Learning from Experience

In an exercise of this kind we must be guided by experience. Britain's own previous experience of coping with cutbacks – the successes as well as the failures – is relevant, as is the experience of other countries. In 1945 the Labour Government redeployed nine million people within twelve months. No one suggested that the war should continue to keep people in employment. Similarly, the experience of the railways, the docks, the cotton industry and the mining industry in the post-war period indicates that large reductions in employment in given sectors can be achieved without large-scale unemployment.

In the immediate post-war period, when the main problem was less people than materials, particularly enormous stockpiles of sheet metal, aircraft companies manufactured a wide range of production in such fields as construction and civil engineering, household and office equipment. Shorts, for example, became the largest producer of milk cans in Europe. De Havilland's adapted their vibration test equipment to remove bubbles from milk chocolate. More important than individual examples, however, was the spread of skills and techniques to other industries. Particularly significant from this point of view were high-strength aviation plastics, kinetic heating, heat-resistant materials, light alloys, radar, titanium, glass fabrics, hydraulics and servo-mechanisms. British Rail, notably its establishment at Derby, have made use of design talent from the military aircraft sector, and this has been of particular importance in the development of new rail technology, particularly the high-speed train. English Electric, before it was merged with General Electric, used aero-dynamicists in the development of steam turbines. The civil engineering sector of Hawker Siddeley, best known for the manufacture of diesel engines, has benefited from the transfer of engineers from the military side of the group. (Further examples of conversion in the Naval Dockyards are to be found in SGP No. 13.) But while the post-war period demonstrated the potential for transferring production away from military supplies, it also demonstrated the problems which arise when there is an early reversion to military production. For this reason, among others, some of the examples given above proved not particularly successful. Their significance lies in the tangible experience of the potential for converting the arms industries.

Even more information on industrial conversion has been published in the United States, where the scale of the problem is much greater.[5] One significant initiative has been in the field of 'Community Planning'. To help communities overcome the consequences of military cutbacks, the American Department of Defense established an Office of Economic Adjustment (OEA). In each locality subject to a defence cutback, the OEA has been instrumental in establishing the participation of all concerned in planning for conversion. Each community has naturally taken a different stance, but they have all evolved a development strategy. This strategy has consisted of the following six elements:

1. Identification of community assets: plant, land, equipment, people.
2. Market survey, demand forecasting, consumption patterns.
3. Identification of new products.
4. Consideration of constraints, e.g. transportation.
5. Overcoming of constraints, e.g. road-building.
6. Conversion of defence installations.

Local communities have been encouraged by the Defense

Department (through the OEA) to use their knowledge to plan for a smooth transition in an environment which, it should be remembered, is particularly unsympathetic to centralized planning, and with little government aid. As military bases have closed they have been converted, and in some cases where bases have been contracted rather than totally closed, sharing arrangements have been negotiated. For example, Topeka air base in Kansas is now open for both civil and military air transport. Of some forty-six recently converted defence installations, the new uses include industry (in 40 cases), education (33 cases), aviation (27 cases), recreation (25 cases), housing (21 cases), commerce (13 cases), and agriculture (9 cases). A cutback in military activity which was at first viewed as a threat became, thanks to community planning, a challenge and an opportunity. The chief advantage found by the OEA was that a community generally chose prosperity. Several small employers tended to replace one large employer, so giving the area a greater sense of stability. As the OEA said in its 1975 Report: 'The imagined disasters that a base closing portends can actually become catalysts for community improvements never before thought possible.'[6]

These American findings are, for Britain, particularly relevant in the Preston area of Lancashire. Preston very much depends upon work at the BAC factory (where MRCA will be assembled) for its local prosperity. Defence cuts without conversion, or a de-emphasis on military aircraft, could hit Preston hard. However, near Preston, the Central Lancashire Development Corporation is in the process of a new town development which includes industrial development. This will provide the opportunity needed to diversify the basis of Preston's local economy and inject greater stability. Equally, reductions in defence work could release the skilled workforce necessary for industrial development, which could in turn contribute to both local prosperity and revitalizing British industry.

It may be argued that the American experience is not in general especially relevant to Britain since conversion was taking place there in the context of a more dynamic economy, better able to absorb surplus capacity. Yet the fact is that on a long-term basis economic growth rates in the United States and Britain are very similar, apart from the level of unemployment being much higher in America.

One final example of previous experience gives an indication. During the four years 1953–7, the Conservatives cut the defence budget by one third. They did this without any great disruption. In 1953, unemployment stood at 1·5 per cent; in 1957 at 1·4 per cent. In terms of 1976 prices, the Tories cut £1,650 million off the defence budget over four years. Thus, not only can it be done, it has been done.

The Need for Planning
Converting military industrial capacity cannot be left to chance. The measures needed to ensure the smooth transition of resources from military to non-military production are basically the same as those required to solve the economic crisis.

There is no question of leaving the task to market forces or making free handouts to industry. The approach put forward in this study is centred on the need to produce socially useful and necessary products. We know, from bitter experience of the performance of British industry over thirty years, that these products will not emerge without strong planning and control of the process. We must also be aware that the experience and habits of management in military industry will in many ways be unsuited to non-military production; the conditions of military industry are very different from those of civil industries, reflecting different objectives. This aspect is looked at further at a later stage in this chapter. Obviously industrial conversion is not a simple matter of transferring production from the military to the civil sphere since there are often profound problems of over-capacity in certain areas of civil production. On the contrary, the transfer of production should be directed to areas of industry where there are (a) gaps in existing capacity, (b) export demand and (c) the possibility of providing import substitution. Not all alternatives to military production are necessarily desirable, and those likely to increase unemployment are especially to be avoided.

The right approach must therefore be a series of obligatory planning agreements, accompanied by energetic public investment and participation in profitable manufacture. We must, however, be careful to distinguish two types of 'profitability'. The first type is the one with which we are all familiar: returns should be greater than investment. The second type could be described as 'social profitability': the gain to society of producing, for example, well-made medical equipment to improve the NHS, or cheap and effective methods of energy generation. This would introduce a new criterion of 'profitability' into the economy. Such manufacturers, where there would be significant public investment and control if not full public ownership, should not necessarily be expected to perform 'profitably' by showing surpluses in their accounts since society would be profiting from their products. It is also important to ensure that the way in which the defence industries are to be run down should not determine priorities in public spending. In other words, the existence of new industrial capacity should not of itself influence priorities for the choice of products to be purchased by the public sector. As the Government is likely to be the most important customer for many of these products, it will be in a position to play a leading role in planning, both by defining its requirement and by ensuring the stability of the market for the new products.

A party and a Government already committed to planning the general direction of the economy, and to controlling important sectors, should find nothing especially daunting in the task of planning to convert the defence sector. The opportunities provided by conversion can, with effective and imaginative planning, take our aspirations for society a further stage towards achievement.

Planning will be necessary at both local and national level. The Government's main role would be to co-ordinate local plans. It would need to supply financial assistance to the localities concerned, and to industries establishing themselves. Aid to those industries may also need to include selective import controls for a period to allow new products to establish themselves on the market. Finally, the Government will need to keep an eye on the whole process – a kind of 'watchdog' role.

The details of planning must be expected to emerge at local level. As noted below, the best ideas for alternative products have emerged from the work-forces involved, including both development and production staffs. It is they who best know what other uses exist for their plant, skills and ingenuity. It is they who can best form the creative basis of conversion planning. Without their input to the planning process by way of a system of planning agreements, we cannot expect the opportunities which present themselves to be fully utilized.

It is not only the workforce of affected plants that must be involved at a local level. The wider local communities also have a stake in the conversion process. Through trade councils, community organizations and local authorities, they must also have a real say in the planning. Knowledge of the facilities available in the area, and of the constraints there may be upon the introduction of certain industries, is an equally important ingredient of proper planning. Many of the most imaginative and practical suggestions for conversion have come from the shop floor. In one case, at the BAC factories in Preston, the joint shop stewards have even offered to forgo 1 per cent of a proposed wage claim if the firm would devote the amount saved (£300,000 per annum) to research and development into converting production to civil products.[7]

Decentralization is entirely consistent with a general planning approach. Indeed, widening participation in the planning stage should inject a needed urgency, creativeness and concern for the interests of those affected in the process, as well as avoiding the dangers of bureaucratization. We must now, however, examine two elements in such a plan which are of particular importance. One concerns planning for the conversion of the product, the other planning for the conversion of occupations.

Production Conversion

The objective is to plan for conversion so that workers no longer make armaments which are no longer wanted by society. It would be absurd if such workers should now commence to make some other equally useless product. Any new product which is to be produced by former armaments workers must be one that is needed, and the first stage in a product conversion plan must be the 'product search'. Long before it announces a particular defence cutback, the Government should require the contractor involved to establish a joint management–union committee to evaluate the possibilities for alternative products. Rather than dismissing the firm – and the workers – with a cancellation fee, it should subsidize the market research. The joint committee will need to consider all the distribution outlets, and to negotiate licences and patents. As we shall see, workers in the armaments industries are already developing new technologies in anticipation of possible cuts. Their ideas are ones which would be economically and socially useful, improve Britain's competitive position and reverse the trend towards boring and dehumanizing work. Yet only proper planning can successfully bring about the conversion of 'threats' to 'opportunities'.

Two alternative routes for product conversion are distinguishable. One is the commercial route: to improve the efficiency of current product lines. For example, one of our tasks is to revitalize Britain's existing industries, from motor-cycles to textiles, from electronics to machine tools. The capital resources released from defence could be invested to modernize those industries which are heavily under attack from foreign competitors. We have seen that the armament sector is also the capital goods sector, and that a reconversion of armaments production to the production of capital goods could therefore make an important contribution towards progress in manufacturing. Our existing problems are not the result of a lack of demand for the products but of a lack of efficiency in the industries concerned.

It should nevertheless be recognized that, in a deteriorating international economic climate, the cost of improving the efficiency of certain commercial product lines, such as merchant shipbuilding, may prove prohibitive. Furthermore, any gains made in Britain's competitiveness will be at the expense of workers elsewhere. It might therefore be more fruitful to concentrate on the second alternative for product conversion: the search for new products which satisfy social needs not currently being filled by either ourselves or imports. (See pp. 104-112 for a list of examples.) In the first instance, some of these would be commercially viable in that they would be purchased by private customers, for example, power-packs for cars to reduce fuel consumption and polluting emissions, heat pumps for private homes, and so forth. For others, the Government would be the main customer. Such social expenditure would cost the Government no more, and probably much less, than current expenditure on the dole or on defence. Nor is there any reason to suppose it would reduce foreign exchange earnings: as we have seen, the armament sector is rather inefficient at exporting compared with civilian sectors in the same industry, while there would be some saving of military costs. There would also need to be changes in government policy in other areas, since many of the suggestions made may run counter to current government priorities. For example, suggestions for transport place more emphasis on railways and canals than on roads, while the energy projects aim to provide alternatives to oil, coal and nuclear energy.

In summary, it may be said that proper planning can eliminate the problems which concern the demand for the new products. In the words of a United Nations report: 'There are so many competing claims for usefully employing the resources released by disarmament that the real problem is to establish a scale of priorities.'[8]

Occupational Conversion

We have argued that the market would not pose insurmountable problems to a conversion of production. But what of the workers themselves? There is a need for the Manpower Services Commission to undertake detailed 'manpower audits' at each locality where conversion is planned. The information needed will cover the types of skills possessed by the workers, their turnover levels, the travel-to-work catchment areas and so on. Some retraining or upgrading of skill levels might be desirable, but it is probable that this would be smaller than is generally assumed. On the other hand, planning is necessary, not only to avoid unemployment, but also to avoid bottlenecks in the form of shortages of skilled workers. It might be possible to de-skill some of the work, but it would be far better to upgrade the level of skills. Let us consider some of the occupational groups which would be involved.

The production occupations would cause little problem. In the 1960s in California, where there is a high concentration of military aerospace production, a detailed examination was carried out by the State Department of Employment on 127 production occupations.[9] Of these, twenty-eight were of the 'basic' craft-types, for example, 'carpenter', 'plumber' and so on. A further ninety-three could be matched to one or more non-military production occupations. Only six occupational groups were found to require retraining. Similar conclusions were reached by US Federal Government Survey.[10] In Britain, research has produced comparable results in relation to, for example, the Clyde submarine base[11] and the Vickers shipyard at Barrow-in-Furness.[12]

Civil servants – the administrators – would probably constitute the hardest conversion challenge because of their lack of outside experience. But the task would not be impossible, and with suitable training they could develop management skills suitable for the new industries. Among the ranks of existing civil servants are the Defence Sales Organization export salesmen. The release of these skills would be a boon to our manufacturing industry, which is very short of competent export salesmen.

Conversion also demands a change of attitudes. Experience has taught military scientists that the main target is not cost-effectiveness. We certainly would not want to see the cost of, for example, medical equipment spiralling upwards through unnecessary extra sophistication for marginal or illusory benefits.

Many administrators would be sure to find posts in the public programmes that would be expanded after arms cuts, and the management sectors of converted industries could be expected to absorb many. As with occupational conversion in general, this should not be left to chance, and where different management techniques are appropriate, retraining should be made available, possibly for periods of around six months.

Numerous skills are available in the lower ranks of the armed forces, in civil and mechanical engineering, communications, electronics and so on. Some of these skills are directly applicable to civilian employment, while others would provide a sound basis on which to build a short period of retraining. Given the likelihood that arms cuts would create more jobs than are provided at present, it should be possible to utilize these skills in converted industry to everybody's advantage. Many other servicemen, however, will not have these skills, and will need to start in retraining almost from scratch. Adequate provision for this should be guaranteed, and could most valuably include opportunities for these servicemen to enter apprenticeships.

A reduction in recruitment levels would be sufficient to overcome the problem caused by a rundown in servicemen. Yet the matter does not end there. It would also be necessary to consider the working-class youth in the depressed areas who traditionally join the services. Once again, the solution to the problem is the same as the solution to the present crisis: a vigorous regional policy as part of an overall strategy of industrial and social regeneration.

Case-studies in Alternative Employment Possibilities

If we are to overcome the natural hostility of defence workers to plans for conversion we must be more than theoretical. The alternatives must be concrete and practical. There is no specific blueprint, but the following pages show that practical planning for conversion is possible. It is up to the Government to provide the facilities and the general policy which can translate the possibilities outlined into practical reality for the workers in the industries concerned.[13]

(a) Alternatives in the Aerospace Industry

The major military aerospace project for Britain at present – and for Italy and Germany – is the development of the multi-role combat aircraft (MRCA), the Tornado. Although only 10 per cent of British aerospace workers are at present engaged on Tornado, a major project like this generates other associated projects.

Tornado is an unusually expensive aircraft. The latest government estimate puts the unit production cost (in 1977/8 Estimates

Prices) at £6.3 million for the IDS interdiction/strike or 'common' variant), of which 220 copies will be produced, and £7·72 million for the ADV (Air Defence Variant), of which 165 copies are expected to be produced.[14] For the study group it has been calculated, using a widely accepted convention, that the total unit cost, including development and maintenance, will be £16·7 million for the IDS and £20·3 million for the ADV – a total cost for the whole project of £7,000 million in the same prices. The high cost is caused by the multiplicity of roles which the aircraft is expected to perform; it will thus, by definition, have many wasted characteristics. It is being built as an all-weather, long-range aircraft with a large weapon load, accurate weapon delivery, manoeuvrability, rapid acceleration and climb, and subsonic, short take-off and landing capabilities. The Tornado is a classic example of tremendous sophistication providing over-specialization on the one hand and under-utilization on the other.

While it is questionable whether it would be feasible to scrap the Tornado project, it is worth looking at the nature of the resources devoted to its production, mainly to demonstrate the skills and potential in the aerospace industry as well as to caution against embarking on similar projects in the future. Some 17,000 workers are at present directly and indirectly involved in Tornado. This is expected to rise to a maximum of 36,000.[15] Half of these workers are employed by BAC in Lancashire and Rolls-Royce in Bristol. The others are employed by such firms as GEC Marconi-Elliott, Lucas Aerospace, Dunlop, Decca, Ferranti and Plessey. This study's task is therefore to consider what alternative work could be found for these people in the event of the project's cancellation. But we might equally consider their plight once Tornado production has passed its peak, for it is questionable whether Britain could afford a comparable successor. Even without defence cuts, there is an urgent need to consider diversification possibilities simply to give greater stability and security to an industry which has serious structural problems. Two measures may be expected to assist in this process: one is the nationalization of the aircraft industry, the other is a measure of industrial democracy. Together these will help to facilitate effective planning for conversion.

(b) *The Airframe Industry*
Tired of the fluctuations in work levels, and the hiring and firing that go with it, the BAC Shop Stewards Combine Committee has advanced a claim for 'Job Protection Agreements'. In this they

are seeking alternative work during the troughs in military aero-space production, like those which occur during the gaps between the peaks in Jaguar and Tornado production. However, there are many flaws in the concept of 'Job Protection Agreements', and these have been highlighted in a paper written by Peter Ward, a BAC Preston Shop Steward, which is now before the Combine Committee and entitled 'Alternatives to Arms Production'.[16]

The problem of merely filling in with alteratives during a lull in military production is this: in a competitive market, once one has put a foot in the door with an alternative product, a continuous effort is needed to keep it there. It is no good making machine tools for a few months and stopping when more military work comes along. There is, first, a need to establish the level of military aircraft development and what production capacity will be compatible with our economic capacity. Secondly, the remaining aircraft capacity should be permanently converted. Thirdly, the first part should be diversified. This will be a major planning risk, but the alternative is the continuation of present threats of redundancy, waste of resources and the surplus capacity which provides the impetus for the weapons procurement policy.

Many options are available for diversification. As McDonagh and Zimmerman comment: 'The primary resource of the [airframe] industry is its ability to design, develop and manufacture new and advanced products.'[17] Diversification has been suggested in such new areas (where there is no harsh market competition) as barrage schemes, nuclear material disposal schemes, solar-panel generators, rolling stock and prefabricated parts for the construction industry.

(c) *Aero-engines*
The main aero-engine manufacturer, Rolls-Royce, does, of course, already diversify its operations between military and civil aircraft production. In particular, Rolls-Royce is internationally competitive in the civil aviation field. Further diversification options are possible. The company has found a market for adaptations of engines for civil non-aerospace use. Processing plants, which have long lead times, could be built using BAC infrastructure and Rolls-Royce engines. These would make an ideal diversification option. Rolls-Royce workers – and managers – would be well advised to consider partial conversion in addition. This would mean that military production would continue, but that there could be civil alternatives. The American competitor,

Pratt & Whitney, has already converted some of its capacity to manufacturing machine tools.

(d) *The Equipment Suppliers – Lucas Aerospace*

The most celebrated proposals for alternative production are those which have been drawn up by the Lucas Aerospace workers. The Shop Stewards Combine Committee, faced with a permanent threat of redundancies, decided to put forward their own corporate plan which would, if implemented, provide their members with work on socially useful products.

> The object of the Corporate Plan [say its authors] is twofold. Firstly, to protect our members' right to work by proposing a range of alternative products on which they could become engaged in the event of further cutbacks in the aerospace industry. Secondly, to ensure that among the alternative products proposed are a number which would be socially useful to the community at large.[18]

Lucas Aerospace employs some 13,000 highly skilled design and manual staffs in thirteen sites in Britain. It is the largest manufacturer of electrical generating systems in Europe, and also produces gas-turbine engine starting and control systems, power units, instruments, medical equipment and so on. However, a large part of its business is connected with military aircraft projects, including the Tornado. The Combine Committee thought that there would be no point in organizing 'sit-ins' to maintain the production of something that nobody wanted. They decided therefore to seek ideas for alternative products from trade union headquarters and from academics. The result was disappointing, and the shop stewards turned instead to their own workforce, distributing questionnaires to all the workers. As a result, some 150 designs for alternative products were submitted to the Combine Committee. All the products aimed at using rather than displacing people; all involved the adaptation of existing design techniques and experience. The Corporate Plan is, consequently, not an aspiration or a moral assertion. It is a series of concrete proposals which have been widely acclaimed in the specialist press.[19] Prototypes for a whole series of socially useful products are already in existence, and six large volumes of technical data have been collected.

The alternative products include power packs and microprocessors which would assist the development of low-energy housing, telecheiric machines for use in dangerous environments, and submersibles for exploring the ocean bed. Alternative products which would make transport more efficient, safer and less of an environmental hazard include battery cars which could be recharged by a diesel or petrol hybrid power pack (thus overcoming the present problem with such cars), braking systems which would adapt the eddy-current dynamometer technology to existing coach retarders, and the hybrid road/rail vehicle featured on the television programme, *Tomorrow's World*. The prototype of this is being considered for use by the Highlands and Islands Development Board; with its gradient ability of 1 : 6, the hybrid vehicle would be very cheap to run, particularly when compared with the cost of constructing conventional low-gradient railways. The plan also proposes a whole range of alternative energy-producing equipment. In the field of medical equipment, there are three products listed: first are kidney machines, which are already manufactured by both Lucas and Vickers. A second product which the workers want to manufacture is the 'hobcart'. This was originally designed and built by apprentices at Lucas's Wolverhampton factory for the benefit of children suffering from spina bifida. Finally, the aerospace workers would like to develop aids for blind people. Aerospace technology could easily be adapted and applied to assist those who have no sight, using exactly the same principle as the 'blind landing system' in the Tornado.[20]

Lucas stewards envisage a phased transition from military to civilian work. Some of the products could be manufactured immediately, while others are longer term. The object would be initially to arrest the contraction of the industry, and then to reverse it gradually as diversification increases. Lucas would be an ideal company for a tripartite planning agreement. Thus, the local initiatives could be co-ordinated and assisted, maximizing the benefits to society as a whole.

The management reply to the Combine Committee was unfortunately predictable. As the *Engineer* reported:

Management based its reply on a reaffirmation of its established business strategy. In so doing it paid no regard to the damage to personnel morale inflicted on highly qualified senior engineers, technologists and shopfloor engineering workers.[21]

The initiative now lies with the Government.

(e) *The Shipbuilding Industry*

In considering shipbuilding, we are conscious of the way in which entire localities have come to be dependent on particular naval orders. For example, about 40 per cent of the male employees in Barrow-in-Furness[22] are employed by Vickers, and about one fifth of Vickers's shipbuilding and engineering capacity in Barrow is taken up with the design, development and production of a large aircraft-carrying warship known as the 'Anti-Submarine Warfare Cruiser'. The cruiser is likely to cost around £100 million over a period of five years.

Vickers is one of the oldest armaments companies in Britain. After each war it has faced the problem of diversification. The company's experience indicates the technical possibility of conversion. The Barrow shipyard has manufactured such diverse items as cement kilns, sugar-beet crushers and irrigation systems. It also indicates the need for central government control and planning, since Vickers has used profits made in the armament sector to diversify into such fields as office equipment and lithographic plates, often through the acquisition of foreign subsidiaries.

In terms of the equipment and skills which would be released at Barrow by a cutback in defence expenditure, and considering such factors as the marine experience of the workers and the poor road and rail access, a study of the available resources points in the direction of alternative production being of a sea-based nature. Shipbuilding skills and facilities are commensurate with any kind of large-scale and relatively labour-intensive construction and assembly activities which involve heavy metal fabrication and materials handling, and complex logistic problems of supply, storage and scheduling. A study commissioned by the US Government identified fifty-five conventional product ranges into which shipbuilding could be converted, and demonstrated that, over a three-year period, a third of total US naval building capacity could be converted by capturing 10 per cent of the annual *growth* of the market for these products.[23]

The prospects for merchant shipbuilding are poor because of the huge over-capacity in world shipbuilding which is forecast. But there do seem to be opportunities in the new technologies now being developed to make use of the sea and the sea-bed. A high proportion of designers and electricians employed at a naval shipyard could prove a positive asset in such fields as marine agriculture and mining, oil drilling, marine-based energy sources and new forms of marine transport. Already, Vickers Offshore Engineering, based in Barrow, is in the forefront of development and submersibles. Other products which have been suggested include wave-power generators, fish-farming tanks, ocean-going tub-barge container ships and the Morecambe Bay barrage system (for tidal power).

The point has been made earlier that the problem of conversion to socially useful production is not technical, but one of political will. The Vickers Company aggravated the situation by hiving off its non-shipbuilding activities into a separate company to minimize the extent of nationalization in advance of the Aircraft and Shipbuilding Act. This development makes the diversification and conversion of the military shipbuilding activities difficult. In addition, Vickers has already implied that it intends to spend its compensation money from nationalization on its private-enterprise activities abroad. Thus, to cope with the problems of under-investment and conversion, there is a need for active government involvement in this particular area. The idea of leaving industry alone is diametrically opposed to the idea of a planned conversion for socially useful production.

(f) *Alternative Employment in Other Industries*

Industrial sectors which would greatly benefit from a release of resources from military use are mechanical engineering, machine tools, electronics and telecommunications. In Sweden, for example, computers which were initially developed for fast decision-making in complex military situations have been adapted for hospital management. But of more importance than the release of equipment is the release of human skills. A real shortage of skilled workers is hindering the development of vital sections of British industry, including the orderly development of the North Sea oilfields. Some 60 per cent of qualified scientists and engineers in mechanical engineering are at present engaged on military work, and yet this work accounts for less than 30 per cent of the total domestic output. The skills tied up in the military sector are not being used efficiently.

(g) *Alternative Employment at the Clyde Submarine Base*

The Labour Party and the Government have given a Manifesto commitment not to develop a second generation of nuclear warheads.[24] Thus, even without a new cutback in defence spending, there is a need to consider the plight of the Clyde Polaris Submarine Base in Faslane. The base, which generates about 8·5 per cent of the economic activity in West Dunbartonshire, was the subject of an investigation in 1975 by the Scottish Campaign for Nuclear Disarmament, for whom an economic profile of the base and an analysis of the employment (and other) consequences of rundown were prepared by the University of Aberdeen's Defence Studies Unit.

The base provides work for 3,100 servicemen, most of whom are 'posted' to the area, as well as 2,800 civilian support workers and 2,000 indirect workers in the surrounding service industries. The composition of the workforce is as follows (the total of 100 per cent allowing for rounding):

Managerial, scientific, technical	21%
Clerical	19%
Skilled craftsmen	14%
Semi-skilled and unskilled	45%

Some 80 per cent of the skilled craftsmen are mechanical and electrical fitters.

The survey team took this job mix and tried to see what alternative industries would use such a range of skills, and considered a number of options.

The first option was that of gradually closing the base completely, with no alternative employment being created. By the time the base closed, it was estimated, all the servicemen and some 1,000 of the 'posted' civilians would have moved on as a matter of course. Another 500 might be expected to migrate, leaving 3,000 workers to receive unemployment benefit. However, it was felt that even the payment of very high levels of unemployment benefit would be unacceptable since the object of the exercise was to release manpower for useful work rather than the dole queue. A second option involved the migration of workers to the developing Scottish east coast. (This structural change is already occurring; the total population of the Strathclyde region fell by 33,000 between 1971 and 1973.) A third option envisaged the cancellation of Polaris, but the continuation of conventional naval activity at the base. A fourth option involved the creation of new jobs in the same location as the former military base.

This fourth option might be of one major project, or smaller diverse projects, which would use existing resources. The survey team looked at the present job mix and decided that petrochemical work would not 'fit'. Oil platform construction work was considered to be an uneconomic proposition on the west coast of Scotland, since even Marathon found itself with empty order books at the end of 1976. However, two other alternatives seemed viable. One was electrical instrument engineering, and the other the development of submerged oil production systems like those being developed at present in the USA by Exxon. If there is too great a loss of skilled workers in the transitional period, the level of skills of the 45 per cent semi-skilled and unskilled workers in the transitional period could be upgraded by special MSC training courses, or the skilled work itself could be broken down into semi-skilled component and assembly work. This has already occurred on Tees-side, where a shortage of skilled welders has led to the development of friction bolting.

The American experience of 'Community Planning' described earlier in this chapter is clearly relevant in this sort of base conversion.

Conclusion

In this chapter we have advocated the permanent conversion of a large part of the military sector. We have illustrated that it is possible, and that it has been done before, both in the United Kingdom and elsewhere. There are, of course, problems arising from conversion which must be taken very seriously, but these are distinctly soluble. This chapter has shown how the obstacles might be overcome and, indeed, turned into golden opportunities. All that is required is the will to act.

There is something paradoxical about the way in which we devote valuable and highly talented human resources to the manufacture of complex and sophisticated instruments of destruction while so many basic needs remain unrealized. For example, even in an advanced industrial country like Britain, old people suffer from hypothermia and children with chronic diseases are not adequately treated. And yet there are plenty of workers currently unemployed or employed in the defence sector who have the skills, ability and motivation to provide the means of solving these problems.

It is up to the Government to show determination and to start the necessary detailed planning immediately. They should set the pace in the public sector through the National Enterprise Board. Nationalization of the shipbuilding and aerospace industries presents opportunities which should not be missed. It is crucial to protect the right of these industries to diversify their operations, and equally important for this opportunity to be taken up. In the private sector, the pace must be forced through by the vigorous use of planning agreements. In both sectors, the planning process must harness the imagination of the defence workers themselves, through the implementation of our policies for the joint control and public accountability of industry.[25]

Notes and References
1. See Seymour Melman, *The Defence Economy*, Columbia University Press, 1970.
2. *Hansard*, 19 October 1976.
3. SGP No 10. Frank Blackaby, 'Note on the Employment Consequences of a £1,000 million Cut (at 1974 prices) in Military Expenditure over Five Years'.
4. Secretary of State for Defence: Parliamentary Written Answers, *Hansard*, 22 May 1975 and 9 June 1975.
5. See SGP No 15 Dan Smith, 'Community Planning and Base Conversion', for a further discussion.
6. *Economic Recovery*, Office of Economic Adjustment, Washington DC, 1975, p. 4.
7. The claim is in addition to the 5 per cent annual pay claim; as reported in the *Lancashire Evening Post*, January 1977.
8. 'Economic and Social Consequences of Disarmament', United Nations Department of Economic and Social Affairs, 1962, para. 169.
9. 'The Potential Transfer of Industrial Skills from Defense to Non-Defense Industries', Californian Department of Employment, 1968.
10. Camborn, J. R. and Newton, D, 'Skills Transfers', in *Monthly Labor Review*, US Department of Labor, June 1969.
11. 'Replacing Employment at the Nuclear Bases', Scottish CND, February 1975.
12. SGP No 13. Mary Kaldor, 'Alternative Employment for Naval Shipbuilding Workers: A Case-Study of the Resources Devoted to the Production of the ASW Cruiser'.
13. A list of alternative products for manufacture in the industries no longer making military goods is given in Appendix II.
14. *Hansard*, 9 November 1976, col. 199.
15. See SGP No 14. Dan Smith, 'Tornado – Cancellation, Conversion and Diversification in the Aerospace Industry', for breakdown.
16. Peter Ward, 'Alternatives to Arms Production', paper presented to Preston North Constituency Labour Party Two-Day School, 2 April 1976.
17. McDonagh, J. J. and Zimmerman, S. M., 'Mobilization for Peace: a Program for Civilian Diversification of the Airframe Industry', unpublished thesis, Columbia University, 1961, p. 181.
18. 'Corporate Plan – a Contingency Strategy as a Positive Alternative to Recession and Redundancies', published by the Lucas Aerospace Combine Shop Stewards Committee, 1976.
19. See, for example, the *New Scientist*, 3 July 1975 and 16 September 1976; *Engineer*, 5 February 1976 and 13 May 1976; *Industrial Management*, July 1976.
20. See SGP No. 16

21. *Engineer*, 13 May 1976.
22. The basic research for this section is drawn from Mary Kaldor, SGP No. 13, and evidence from the MP for Barrow, Mr Albert Booth.
23. 'Final Report on Industrial Conversion Potential in the Ship-building Industry', Mid-West Research Institute Contract No. ACDA/E, 1966.
24. *Labour Party Manifesto*, October 1974, p. 29.
25. See *Labour's Programme 1976*, pp. 33–6.

THE ARMS TRADE

About a quarter of British arms production goes to the export trade.[1] In so doing, it makes this country one of the 'Big Four' arms exporters, probably slightly behind France but considerably behind the two super-powers. Any serious programme for cutting defence expenditure must therefore have implications for the nature and extent of our continuing trade in arms. The two problems are interrelated, and any strategy directed towards the overall reduction of military production must take into account the export dimension. Indeed, the very presence of demand for military hardware abroad, and the existence of industrial capacity to meet this demand, constitutes one of the most frequently cited arguments for maintaining a high level of defence expenditure, which is in turn a concomitant of maintaining a high level of military capacity. If no other reasons existed, the political and moral case against the arms trade ought to be sufficient cause for reducing our arms manufacturing capacity.

The Political Argument

While few would question the wisdom of limiting and controlling the international arms race, there are considerable disagreements about how this could be achieved and to what extent it is desirable. Dr Iklé, Director of the United States Arms Control and Disarmament Agency, has summarized the case for his agency's work succinctly:

The goals of arms control, simply stated, are to prevent war and to reduce the destruction of war if it should occur. Limiting the availability of weapons can lessen the chances of war by promoting a stable balance, and mutual reductions in arms can make wars less destructive. Moreover, in the long run, it is to be hoped, arms control will permit us to lower the economic burden of defence.[2]

There is a consensus of opinion which advocates the ideal solution of a multilateral arms control agreement, possibly operating through the UN. As in the great debate over nuclear disarmament, a division exists between those who believe in a unilateral start and those who are disinclined to accept anything less than a multilateral agreement.

There is without doubt some scope for unilateral action by Britain, to set a principled example as well as to respond to our pressing economic need to reduce the level of arms production. The advent of the Carter Administration in the United States has certainly signalled the start of a major initiative towards arms control. Shortly after President Carter's inauguration, Cyrus Vance, the US Secretary of State, said:

We are reviewing our own arms sales policy because we must know what are its objectives and tailor them accordingly. Secondly, our arms sales should be determined by our foreign policy objectives and not by economic ones. Thirdly, once we have determined our own unilateral policy, we may then move on to the question of international agreements to make the policy effective.[3]

The British Labour Government can hardly do less than respond in kind to the Secretary of State's proposals, even if the US will be playing the leading role in this sphere.

The critics of unilateral action argue that British influence in bringing about an international agreement would be more effective as a major supplier nation than as one which had already played its cards by taking independent steps. It would also, they feel, mean that the supply of arms could then fall into the hands of less responsible exporters. And, they argue, British arms supplies to Yugoslavia, for example, help deter a possible Soviet attack, as, similarly, they deter a Guatemalan attack on Belize.

Whatever the truth of these arguments, they tend to ignore the basic issues of the arms trade. First this trade is, by its very nature, dependent on a high level of military hostilities, without which the demand for armaments would surely fall. Suppliers of arms to countries involved in conflicts, such as Israel and some Arab states, can hardly claim to be working towards a peaceful settlement in those regions. There is no single government involved in the Middle East arms race which is not publicly committed to a peaceful solution, yet it is hard to deny that the activities of the arms salesmen, who provide the means by which hostilities can be continued, negate the often-proclaimed objectives of foreign policy. This is not to suggest that the supply of arms is *of itself* the cause of unrest, but there can be little doubt that the levels of military supply can have an effect on the intensity, duration and even initiation of hostilities. The suppliers cannot blandly stand aside and disclaim all responsibility for the ultimate use of their weapons. This is particularly the case where arms sales are provided as an alternative to the direct military involvement of the supplier nation. Countries engaged in this kind of behaviour are simply hypocritical in their earnest declarations of non-involvement in wars fuelled by their actions.

Yet it is not only in areas of international tension that weapons are in demand. They are also needed by the leaders of repressive régimes as a substitute for democratic consent to their rule. It is therefore no coincidence that Iran, Saudi Arabia and Oman should be among the biggest purchasers of British arms. Indeed, it is clear that the very existence of certain repressive régimes is largely dependent upon the force of arms. A Labour Government – or any government – should not find itself in a position where it provides the means of repression. Moreover, the inherent instability of such régimes will often lead to their demise, and it is an open question as to how their new rulers will view trading relations with those who supplied weapons to their former oppressors. Even if the arms exported to countries with repressive régimes are not actually used, they serve as a deterrent to the opposition and contribute to the prestige and status of the rulers.

Most governments claim that they allow arms sales only to 'friendly' nations, and that this in itself provides a contribution to the objectives of their foreign policy. In practice, this concept is highly questionable, since there is little guarantee that the friends of today will not become the enemies of tomorrow. There are innumerable examples of arms sales to countries which have, after a change of régime, turned their weapons against the supplier nation. Even during the First World War, British guns supplied to Turkey were responsible for the deaths of British soldiers in the Dardanelles; and, today, the British provision of submarines to fascist Chile[4] is a constant reminder of the transitory nature of certain alliances. This experience is not unique to Britain. The Americans in South-East Asia have become virtually the main military supporter for some of the new Communist régimes, and the USSR has found its weapons backing the increasingly pro-Western régimes of the Egyptians, Sudanese and Syrians.

It is also necessary to consider the effect of arms sales to Third World countries in more general terms. Arms sales create a state of military dependence. They draw Third World countries into the global confrontation, and play an important role in imposing or preserving the international division of labour in peripheral economies. It was in general as a consequence of armed force that the objectives of production were shifted from local self-sufficiency to the world market in colonial countries. And today, the political and repressive role of armed forces is often important in carrying out a 'development' strategy which involves heavy emphasis on industrial growth and foreign investment at the expense of meeting basic needs.

Typical examples of the correlation between high military spending, foreign dependence, industrial growth and extreme inequality (which might be measured crudely by reference to the rate of infant mortality) are Brazil, South Korea, the Philippines and the Middle Eastern countries. This political aspect of arms sales is just as important in its impact on 'development' as the direct absorption of resources, particularly scarce foreign exchange, which might otherwise have contributed to the elimination of poverty. We would not, however, suggest that the Third World should be subject to special arms sales control before similar action had been taken in relation to the industrialized nations. The whole question is very complex, and there is a need for a wider study of the effect of arms sales on Third World economies.

Another feature of the arms trade which has recently come into sharp focus is the high level of attendant corruption. The Lockheed and Northrop scandals, to name but two, have had tremendous political repercussions in the USA, Japan, Holland and Germany. The relationship between corruption and the arms trade is hardly coincidental, for few comparable commercial transactions are surrounded by such secrecy, involve such vast sums or are so dependent on the decisions of a single purchaser. This is not to say that the problem of corruption is necessarily endemic, but there needs to be some serious thinking about the way in which arms procurement decisions are taken in some countries.

Finally, it would be quite inadequate to pursue a discussion about the arms trade without considering the moral aspect. Ultimately, the arms trade is nothing more or less than the export of instruments of death, and for this reason it is really not good enough for the arms traders to argue their case in purely economic terms. The economic and moral arguments are inseparable. A court of law does not accept the plea of a drug pusher who says 'If I did not sell them, then others would', and neither should we.

A temporary loss of some military exports is a small price to pay if it helps to avoid the much greater losses that could be caused by fuelling regional disputes which could develop into a Third World War.

The Economic Argument

It is estimated that 70,000 to 80,000 jobs[5] in the United Kingdom are directly involved in the export of defence equipment, and that many other jobs are indirectly affected. The total value of arms exports in 1975/6 was £530 million,[6] which is substantial indeed when set against the budget for domestic procurement in that year, estimated at £1,853 million.

Chapter 3 of this study showed how there are many misconceptions about the economic benefits of arms exports. In the first place, current levels of military expenditure produce an overall deficit on the foreign exchange side of our military account. Secondly, the same industries which produce both civil and military goods have proved themselves to be more efficient at exporting in the civil sector, despite the consumption of valuable resources by the military. Thirdly, the deployment of high levels of investment and R & D resources by the military sector is partly responsible for hindering the development of British industry in general. It can also be seen that the firms which export most are least likely to be affected by defence cuts. The main items of military equipment included in our procurement programme are the least successful in the world market. The basis of our arms trade consists of such items as fast patrol boats, light aircraft, including helicopters, trainers and armoured cars and so forth. The problem is therefore not so much one of concern about the possible loss of economic opportunities from reducing the arms trade but of concern with the present loss of export opportunities in the civil sphere.

Against this point of view, it is argued that Britain has a clear comparative advantage in the production of arms for the world market and that we would be foolish to relinquish this position. There are said to be three main benefits of the arms trade. First, that arms exports lower the unit costs of production for our own needs. Secondly, that arms exports help to maintain a high level of industrial capacity and R & D capability which would be either idle or non-existent without the arms trade. And thirdly, most sales of military goods tend to aid the sales of civil products like commercial vehicles and earth-moving equipment, particularly where the military play a large role in the governments of recipient countries.

While it is certainly true that arms exports help to maintain a high level of industrial capacity, this in itself poses some serious problems. As Chapter 3 demonstrated, it is the existence of a capacity to produce armaments that distorts a rational approach to procurement decisions. The real reason for the tremendous effort directed towards military exports is the acute and continual problem of surplus capacity in the British armaments industry. It is significant that the formation of the Defence Sales Organization (the Export Promotion Department of the Ministry of Defence) in January 1966 coincided with the last serious attempt to restrain defence expenditure. If this is the real reason for promoting arms exports, does the contention that such sales subsidize the defence budget by reducing unit costs stand up to analysis? The answer must be a qualified 'no', since no evidence has ever been produced to back this claim.

The Ministry of Defence not only refuses to place a value on its own procurement contracts, but declines to answer questions on either the value or content of individual sales to foreign governments. So long as they withhold such information, the argument that longer runs significantly reduce unit costs must remain an assumption rather than a judgement based on evidence. Such evidence as does exist in the public domain tends to confirm this impression. A study by the Stockholm International Peace Research Institute (SIPRI), for example, has shown that in general the cost of producing a typical military aircraft is only marginally reduced in the early stages of a production run, and that any benefits at a later stage (after, say, 500 units have been produced) are probably exaggerated. This is particularly so when it is remembered that most export models require some form of modification which increases unit costs.[8] Another study by the US Congressional Budget Office supports the conclusion that R & D recoveries and lower production costs are mainly associated with the purchase of new aircraft and missiles which account for a relatively small proportion of arms sales both in the US and Britain.[9] Furthermore, such savings have to be offset against the additional costs which lead from the strain on resources from meeting foreign demands at an early stage of production. In Britain, the pressure of export demand was one reason Vickers had to sub-contract a Type 42 destroyer, HMS *Cardiff*, to Swan Hunter for completion at considerable additional cost to the Ministry of Defence.[10]

Setting aside the arguments relating to the situation as it exists at present, there is some doubt about the future levels of demand for British arms abroad. On the one hand, there is a tendency for purchasers of British arms in Iran and Egypt, for example, to buy not the products themselves but manufacturing plant, design and technical know-how, thus reducing the actual demand for British-made hardware. On the other hand, there is evidence that internal pressure by US arms manufacturers on their government will lead to an insistence by the United States on a larger American share of the NATO weapons market. Should arms sales fall off as a result of these factors, there will be a strong incentive to increase domestic spending to keep these industries in activity. A further irony of the situation is that some recipients of arms-producing capability then use their acquired know-how to compete with, or embarrass politically, the original exporting nation in third markets – as has happened, for example, in the attempted export of Israeli-made military aircraft (the Kfir) to Ecuador.

This view is challenged by those who see a running down of arms purchase as being most unlikely, particularly by smaller nations unable to maintain their own industries; and as far as NATO is concerned they see moves towards a collective Euro-group procurement policy as a guarantee for maintaining the British share of the market. Yet, far from being beneficial to our domestic requirements, there is evidence that the export market's demands are in fact of no help whatsoever, and may indeed adversely influence procurement decisions. Is the £85 million being spent on the Maritime Harrier, for example, justified solely on grounds of domestic requirements? 'For a long time no government spokesman claimed it to be anything more than a "useful additional capability" – it has now been promoted to the status of "essential" '; yet this aircraft is considered to have tremendous export potential. The fact is that we shall ourselves have to use anything that we wish to sell to others, and so foreign confidence in the Maritime Harrier, for example, can be gained only if it is ordered by our own Ministry of Defence.

This argument can be taken a stage further when we consider that it is because of a high level of export orders that it is possible in the first place to consider domestic purchases of equipment which would not otherwise be suitable, especially in the light of the serious decisions at present facing us over priorities in public spending. Thus, while it may be argued that we may save money by spreading costs, the saving is illusory since it ignores the high levels of initial investment necessary to create the kind of economies of scale that are worth while. The proposition can even be turned on its head, and we shall see that purchasers of equipment already in production may well be receiving subsidized goods, the historic costs of R & D and production having already been met by the supplier nation.

The Labour Party acknowledges the possibility of some diminution in the overall level of exports in the transitional period. Set against this, however, is the potential for import substitution in other spheres which should lessen the impact on our overall balance of payments. There is also a good prospect of freeing existing industrial capacity devoted to military production for expansion into new export markets. Seen in this light, releasing British industry from the defence burden can only be a benefit; the arms trade may provide some incidental bonus to our balance of payments, but it is a most inefficient means of achieving such an objective.

How Can the Arms Trade be Controlled?

Current British Government policy for achieving arms control emphasizes the need for a multilateral agreement involving arms importers as well as exporters, possibly on a regional basis.[11] This seems to be a sensible approach as far as it goes, but we believe that the most effective way of introducing criteria for the limitation of arms sales would be to examine the following suggested ways of controlling the arms trade on a unilateral basis.

There should be two major criteria for an embargo on arms sales to overseas buyers. First, we could follow the German and Japanese practice of not supplying arms to states either engaged in international disputes or likely to become involved in them. Secondly, and more problematic (in terms of definition), we should consider limiting arms sales to all countries which use torture and repression against their subjects. In other words, governments with a continuous record of human-rights violation should not be given the means to perpetuate their activities.

Another step which could be taken as part of the process towards control of the global arms trade would be to create a Register of Arms Sales, possibly as part of an international register, under UN auspices. Such a register would include full details of the sales as well as the names of agents and the level of their fees. A register will not in itself guarantee a reduction in the scale of the arms trade, but it should have beneficial effects both by generally breaking down the unnecessary secrecy surrounding these transactions, and by making a useful contribution to public debate on this question.

At present information on arms sales in the United Kingdom is very difficult to come by. Were it not for the activities of the Stockholm International Peace Research Institute (SIPRI) in Sweden and the Campaign Against Arms Trade in Britain, we should probably not even have access to the limited information that is available. The Director of the United States Arms Control and Disarmament Agency has stated the case for this kind of disclosure very clearly: 'My Agency,' he says, 'has made a special effort to inform the public about the flow of armaments and trends in military expenditure throughout the world . . . We feel the world public has a right to know for its vital interest is at stake.'[12]

Britain has already taken some unilateral steps towards arms control with the introduction of a licensing system. This has meant that an embargo has existed on the sale of arms to South Africa since 1974, and we are no longer accepting new orders from Chile. Even stricter controls have been in operation for much longer prohibiting sales to Warsaw Pact countries and most of their allies. Therefore proposals for extending this kind of action would not mark any major change of policy. Nevertheless, it would mean a tightening of existing regulations, since there is now, particularly in the case of South Africa, considerable evidence of British companies (notably Marconi, ICI and Raçal) circumventing the arms control legislation.[13]

None of this is to suggest that the supply of arms is in all cases wrong, or that we should withdraw entirely from the arms trade. On the contrary, we should reserve the right to provide supplies (even as aid) to countries with whom we maintain friendly relations (even accepting the fact that governments change and so do relations), especially when they are threatened by hostile powers. The central point of our proposals is to emphasize that no arms sales policy can be disinterested in its political implications, and that it is therefore essential to establish workable criteria for such sales.

There will, of course, be problems of defining what is and what is not military equipment, but it would surely not be beyond the capability of the Ministry of Defence to ensure the regulation of all sales requiring a licence under the 1970 Customs and Excise Exports of Goods (Control) Order. It would also be necessary to include those sales not requiring a licence, but whose use is likely to be military: radar, sonar detection equipment, microwave components, ionospheric and meteroric scatter radio relay communication equipment, to give a few examples.

Conclusion

The programme for a reduction in arms sales will have to be combined with a programme for industrial conversion. Implicit in the strategy for the latter is the need to cut down on arms exports. The fundamental objective of such planning should therefore be both to release economic resources and to contribute towards a lessening of tension in what is, at present, a militarily overstocked world.

Notes and References
1. For a breakdown of Britism arms exports, see Campaign Against Arms Trade, Factsheets available from CAAT, 5 Caledonian Road, London N1 9DX.
2. Dr Iklé, speech at the Conference of the Committee on Disarmament, 29 July 1976.
3. Interview with Henry Brandon, *Sunday Times*, 30 January 1977.
4. The original order was placed with the 'friendly' Christian Democratic government of President Frei. Between the time of ordering and final delivery, there have been two significant changes of government.
5. *Hansard*, 19 October 1976, cols. 421–2.
6. *Hansard*, 24 January 1977, col. 502.
7. *Statement on the Defence Estimates, 1976*, Cmnd 6432, p. 48. (This figure is given at 1976/7 Estimates Prices.)
8. See *The Arms Trade and the Third World*, SIPRI, Stockholm, 1971, p. 400. Further discussion of this question is to be found in ibid., Ch. 14.
9. 'Foreign Military and US Weapons Costs', Staff Working Paper, 5 May, 1976, Congressional Budget Office, Washington, DC.
10. See SGP No 8.

11. See Speech by the Rt Hon. Lord Goronwy Roberts in the First Committee of the United Nations XXXIst General Assembly, 2 November 1976.
12. Dr Iklé, as no. 2 above.
13. See *Black South Africa Explodes*, Counter-Information Services, London, 1976, p. 49.

PART TWO: THE STUDY GROUP PAPERS

A. POLITICAL AND STRATEGIC ISSUES

(The papers in this section were mainly written during 1975 and 1976. Were the authors to write on the same subjects today, they might develop similar arguments but could be expected to amend their presentations to take account of recent developments. Accordingly the papers should be read with this qualification in mind.)

1 OBSERVATIONS ON THE PROPOSAL TO ALIGN UK DEFENCE
EXPENDITURE WITH THE AVERAGE PERCENTAGE SPENT ON
DEFENCE BY THE FRG, ITALY AND FRANCE

by Paul Cockle

I Technical Observations

(a) The 1974 Situation

If we accept the figures in the 1975 Defence White Paper we
find the following apparent relationship between defence
shares of GNP:

UK   5.8%
FRG   4.1%   (4.9% with Berlin Aid)
France   3.8%
Italy   3.0%

These figures would seem to support the view that Britain
is paying a much larger relative share of its GNP to
collective defence than its major European allies.   The United
States share was 6.6 per cent.   The inequity seems greater
when one notes that West Germany's GNP is twice as large and
Italy's just a fraction smaller.   These figures seem to
suggest that the burden within NATO falls short of either a
proportional or progressive 'ability-to-pay' principle.

(b) The Catch

The above figures contain a well disguised trap.   The defence
expenditure figures follow the 'common' NATO definition and
naturally one supposes they are directly comparable.   This is
not the case.   The NATO definition offers no more than a list
of items and activities that should enter into the calculation.
It only lays down guidelines for the scope of defence expenditures.   There are no guidelines on costing these items.
    Manpower costings present a major problem for those wanting
to make international comparisons.   Except for the United
States, Canada, the UK and Luxembourg all other NATO countries
have some form of conscription.   In 1974 around 49 per cent of
the FRG armed forces were conscripts whilst the French percentage was 54 per cent.   The conscript undoubtedly provides
'cheap' labour and subsidizes the taxpayer by providing his
labour at wage rates far below those that would induce him to
serve freely.   Bearing in mind that for France and West
Germany between 40-50 per cent of their budget covers manpower
costs and that even these costs are grossly understated in
British terms, it becomes a moot point whether they are
spending proportionally more.   The Expenditure Committee
(Defence and External Affairs) quoted a West German Study that
claimed a volunteer force would add a further 1.5 per cent to
the 4.7 per cent share of GNP devoted to defence in 1973.
This would exceed the UK defence share.   The MOD made their
own estimates on the roughest basis and concluded that only
0.3 per cent for West Germany and 0.6 per cent for France of
their respective GNPs would be added to their 1974 budgets.
However the methodology looks more than a little vulnerable.

(c) Conclusion

Politically it is dangerous to express a principle in figures,
for the entire case is then seen to rest on the validity of
the statistics.   In this instance the integrity of the
statistical comparison is rather too easily challenged.   It
is unsafe to assume that alignment with European spending
shares will bring a significant decrease in defence expenditure
if conscript costs are given their due.   It is safe to assert,
however, that political opponents at home and abroad will make
this very point.   If the conscript cost adjustment is
completely ignored then one's case for alignment begins to
look a shallow pretence to justify cutting alliance commitments.   Under these circumstances allied resistance will
probably be fierce and may not be limited to just the defence
sphere but could foul all other areas of interaction.
     In addition to this specific point on conscription costs a
general observation should be made.   To arbitrarily establish
a limit for defence expenditure, without regard to the
consequences that may arise from implementing that ceiling, is
not rational for a government partitioning its resources.
Rational choice involves weighing the benefits of various
ends against their respective cost, all being subject to an
overall constraint imposed by the countries disposable
resources.   Whilst there seems to be a keen appreciation
that resources saved on defence have desirable alternative
uses, there is very little awareness of the possible
consequences.   The rest of this paper is devoted to
observations of the political and strategic consequences of a
major defence cut.   This is intended to provide discussion
points rather than impose a particular structure of the issue.

II Political and Strategic Observations

Throughout much of this century British strategy has had to
face two sets of requirements.   One dealt with the colonial
and Commonwealth interests, the other with the balance of
power on the continent.   For a time Britain was successful
in deploying its forces, primarily naval, to safeguard its
colonial interests whilst adeptly playing balance of power
politics on the Continent.   As time went on Britain was
drawn ever deeper into European politics with the result that
she had to develop a considerable army to meet European
contingencies.   Academics may compete in pinpointing the
date when it became obvious that Britain could no longer
sustain both requirements but the period after 1945 was undoubtedly a period of realization.   Since 1945 there has
been a long retreat from the colonial commitments with a
corresponding affirmation of the strategic importance of
Europe.   The last defence review tidied up some of the ends

left from that retreat and more sharply defined the nature of the European commitment.

Thus in the past when one has spoken of major defence cuts they have tended to centre on 'East of Suez' commitments but now any large cuts are going to vitally affect the European commitment. Now only the minutest amount can be saved by axing non-European commitments, for example Belize, Falkland Islands, Oman, Hong Kong, Cyprus. So anyone planning major cuts in UK defence expenditure must address the problem of European commitments.

Four Areas of UK Commitments to NATO:

(a) The Central Region. Britain has an obligation under the Brussels Treaty to provide the 'equivalent' of four army divisions and a tactical air force on the mainland of Europe. This treaty expires in 1998. It contains a clause which 'invites' NATO to review the extent of the commitment if Britain has 'external financing' problems. This particular clause has weathered many economic vicissitudes and may not be easy to invoke. Furthermore, the reduction of forces in the Central Region is being discussed at the MBFR negotiations between East and West. The MBFR forum may be where the child of détente will be born. The negotiations are as much a test of mutual understanding as of will. If unilateral reductions occur in either western force levels or in their quality then the Warsaw Pact will see little point in sitting down with NATO to negotiate 'mutual' force reductions. Britain would be acting most irresponsibly if she torpedoed the chances of a successful conclusion to the MBFR negotiations by reducing her central region forces.

One must also take into account the military balance in the region. Quantitatively, the Warsaw Pact forces outnumber the NATO forces on the ground in terms of manpower, tanks, field guns and tactical aircraft. Observers feel that this quantitative advantage is mitigated by certain qualitative advantages on the NATO side but perhaps not sufficiently when one recalls that the main reservoirs of reinforcements are more readily available to the Warsaw Pact than NATO. An uncompensated reduction of forces in this region would tilt the balance further against NATO. This might not produce any visible consequences but it will reduce the subjective probability that the Soviet Union attaches to western willingness to resist diplomatic pressure. They will reason that each country individually may feel a little less willing to take a stand on important issues. Events in Norway illustrate the process quite well.

If Britain reduces its forces in this region it might produce a multiplier effect. The US Congress has produced an 'ensemble' of Senators who can harmonize with startling virtuosity around the theme of 'European backsliders'. Whether such a choral filibuster could force the Administration

133

to make troop reductions is an open question.   The Administration would surely resist it but that no longer guarantees it will not happen.

Smaller European countries with an equal distaste for having to spend money on armed forces might cave in under popular pressure to emulate one of the cornerstones of NATO. No doubt there will be recriminations all round.

If the gap left by Britain were filled by Germany it would raise her status in US eyes as an ally and increase her influence within Europe.   Neither of these prospects would be to Britain's advantage.   Britain's close relationship with the US gives her a special role as a diplomatic conduit between the United States and Europe.   This affords Britain a degree of influence which could be eroded by closer ties between West Germany and the US.

Finally, British troops in Europe on their current scale confer upon Britain a certain level of influence on European affairs.   An economically weak Britain in or out of the Community, might well need such a lever for some time.

(b) The Eastern Atlantic and Channel Area.   Britain after the United States is the main naval power within NATO.   The prime role of the naval forces in this area is to ensure that all sea-borne supply and reinforcements from the United States can get through to Europe.   They would also play a part in detecting and tracking Soviet strategic submarines.   Their role is therefore to enhance the credibility of the ground forces by ensuring that the massive US reinforcements can actually get to Europe.   This role is probably one of the more controversial roles for our forces.   The ground figures are configured to fight a relatively short war - absorbing a major conventional onslaught to gain time for political action before the next phase of tactical nuclear weapons is invoked which could lead to an exchange of strategic nuclear weapons. Some argue that by the time US sea-borne forces have arrived in any force NATO will have been already forced past the tactical nuclear stage.   So what is the Navy doing protecting a capability that will not be used?   I raise this argument because it is often heard but do not pretend to have enough expertise to evaluate it.

Of course there are subsidiary roles and uses of naval forces.   Some argue that it is an ideal way of bringing pressure to bear on a country without the danger of a major conflagration.   Now that North Sea oil is throwing up its production platforms one will hear of scenarios whereby the Soviet Navy could exert pressure on Britain by 'inspecting' rigs for listening devices.   This would be too risky a proposition for the Soviet Union if it had to confront a sizeable modern navy.

Finally, there is another argument advanced to persuade Britain to maintain its naval capability.   It can be called the European argument.   It is based on the fear that the US

134

commitment will become increasingly weak and that Europe
will have to form a latter day EDC.   In this context Britain's
Navy would provide a principal component.

Under these circumstances Britain would be able to exert
considerable influence over the nature of the EDC.

(c) The Security of the Home Approaches.   This approaches
the role of territorial defence but protects the means of re-
supplying and reinforcing the forward based forces, sometimes
with specialist forces that were developed during the days
when Britain had colonial commitments but that now are
designed for NATO.   Most of these forces are closely tied to
central region forces and some observations made there apply
here.

(d) The NATO Nuclear Deterrent.   Britain's Polaris fleet is
available to NATO's strategic nuclear forces.   Its maintenance
costs are only a fraction of the defence budget and it has a
life well into the 1980's.   The deterrent credibility of the
force lies in its integration with US strategic forces.
Independently it might deter a first strike on a British city
by the Soviet Union although it is not likely to deter any
other form of aggression.   The real deterrence capability
rests with US forces and their configuration.   The credibility
of that deterrent depends on Moscow's perception of the
political commitment to Europe by the United States.   It is
argued that Britain's possession of strategic nuclear weapons
enhances the credibility of that US commitment.

It is also true that the Polaris fleet confers upon
Britain a major voice in NATO affairs, particularly in
strategic policy.   It also gives Britain influence in any
nuclear disarmament conference and the ability to assure its
security.   It would be a more influential critic of any
super-power agreement that might establish a nuclear hegemony
- a fear that sometimes rises in Europe.

Finally, there have been proposals from time to time that
eventually Europe will have to provide for its own defences
completely.   This argument is based on the assumption of a
gradual but inevitable erosion of the US guarantee.   The
argument then goes that Britain's nuclear fleet in association
with the French strategic forces would provide a building
block of a European deterrent.   If the need came about,
Britain would have a powerful voice in the structure of this
resurrected EDC.

III Conclusion

These observations on political and strategic matters have
been listed rather than co-ordinated to form a single theme,
as I felt it would serve discussion better.   All the same I

would like to draw together some ideas.   Defence expenditure
is not as one White Paper once claimed 'irreducible'.   It can
always be cut.   There are, though, consequences of such cuts.
   If they occurred within the framework of East-West
negotiations it would generate pressures for another round and
providing NATO was not bargaining from a weak position then
they would be an ideal way of reducing defence expenditure.   I
would find this course preferable although it still leaves
open the question of sharing out the cuts between allies.   In
the absence of such a framework some judgment is necessary on
the political costs to Britain of a unilateral cut and the
political and strategic costs to the West as a whole.   If
the consequences of a unilateral cut was the undermining of
MBFR talks, general discussion within NATO tending to erode
western political will and a visible military weakness of
NATO, then it would be an irresponsible act.   If the cuts
could be negotiated so avoiding all these consequences by
getting other Europeans to fill gaps left by Britain it would
cost Britain a loss of influence in European affairs plus the
political price that Europe would exact.   However, if the
plea is for Europeans to shoulder Britain's inherited burden,
on the grounds that it is excessive, it will not have any
credence unless account is taken of the cost of conscription.

2 DEFENCE PROGRAMME OPTIONS TO 1980-1
(and Postscript)

by David Greenwood[*]

---

*David Greenwood was not a member, but acted as a consultant to the Study Group.

Introduction

Before the General Election of February 1974 the Labour Party
pledged itself 'to reduce the proportion of the Nation's
resources devoted to defence so that the burden we bear will
be brought into line with that carried by our main European
allies'.   This, it was stated would mean achieving 'annual
savings over a period ... of several hundred million pounds.'
   Accordingly the Labour Government that took office in March
1974 initiated 'the most extensive and thorough review of our
system of defence ever undertaken by a British Government in
peacetime'.   The exercise was nearing completion at the time
of the second 1974 Election campaign, in which the manifesto
commitment was reaffirmed.   It was completed before the end
of the year, an outline of proposals being given in a
ministerial statement in December.   After three more months,
to allow for consultation with allies and both parliamentary
and public reflection on the proposals, a new defence programme
for 1975-84 was formally presented in March 1975 as part of
the 'Statement on the Defence Estimates 1975' (Cmnd 5976).

Outcome of the Review

The new programme prescribed far-reaching changes compared
with previous (i.e. Conservative) plans for the later 1970s
and beyond.   These included abandonment or rundown of
deployments outside Europe;   substantial reduction in specialist reinforcement forces maintained primarily for operations
in Northern and Southern Europe;   and some diminution, or
dilution, of contributions to NATO forces in Central Europe
and the Eastern Atlantic.   Such redefined commitments were
the basis for a revision of force structures and levels,
making possible manpower rundown and the cancellation or
stretching of certain procurement plans.
   For all that, the budgetary projections associated with the
programme foreshadowed continuing 'increases' in real defence
spending until 1977-8, and a stable level of expenditure
thereafter with outlays of c. £3,800 millions a year (at 1974
prices). (1)
   Provided economic growth is resumed the revised budget
targets will, however, involve allocation to defence of a
diminishing share of Gross National Product (GNP) and thus -
on this definition - a reduction of the 'burden' of defence.
But even if the annual growth of GNP averages 3 per cent to
1983-4 the proportion is unlikely to fall below 4 or 5 per
cent or thereabouts.   This would imply a priority for defence
higher than that assigned by France, Italy or West Germany at
present and higher than these countries are unlikely to be
ascribing in the 1980s (on all realistic expectations of growth
performance and defence expenditure trends).

'Savings' of several hundred million pounds over a period
will be achieved too, in the sense that for the forthcoming
decade intended expenditures now amount to substantially less
than the Conservatives envisaged — in all some £4,700 millions
less (at 1974 Survey prices again).   But no reduction is
anticipated measured against current provision.   In fact the
Defence Review yielded a programme for 1975—84 which, in
resource terms, is on a par with that of the present and
recent past.   The funds are allotted differently and to fewer
places.   But no major diminution of the national defence
effort is involved.

Be that as it may the Government has argued that the new
programme represents fulfilment of its pledges.   There are
to be 'savings' — in relation to the inherited programme.
Resources will be released for other purposes: indeed the
manpower rundown has started and industrial capacity is
already becoming free as modified procurement plans are put
into effect.   And some alleviation of the 'burden' is in
prospect, even though the check to rising real expenditure
will not bring the United Kingdom's defence/GNP proportion
down to the European norm.

There are a number of reasons why the Government may not
have wished to go further in reshaping the defence effort.
In the first place, no Government can (or should) approach
a reassessment of public expenditure priorities with fixed
ideas about either the cost level or the scale and pattern of
provision that is appropriate in a given area.   To do so is
to make nonsense of the essential business of resource
allocation.   The purpose of an exercise like the 1974 Defence
Review is to judge where the balance should be struck between
relevant 'needs' and the real resource costs involved in
meeting them — in the economist's sense of opportunity costs.
There is, clearly, no such balance which is self-evidently
'right'.   The least that can be said about the Defence
Review is that it yielded a defence programme whose merits
cannot be dismissed out of hand.   The programme is one which
concentrates the national effort on core security interests in
Europe (including the United Kingdom itself) and European
waters.   At the same time it allows for a decline in defence's
share of national resources.   Moreover it reflects a coherent
approach to choices and a fairly consistent line on priorities
within the defence domain. (2)

Secondly, there is sufficient ambiguity about the notion and
definition of 'burden' to admit the argument that assuming a
more equitable 'burden' vis-à-vis European allies need not
entail exact equalization of defence/GNP proportions.   It is
beyond dispute that the money spent on defence represents a
larger share of GNP in Britain than in France, West Germany or
Italy (or indeed, on 1974 figures, any other European member of
NATO except Portugal).   Alternative measures of 'burden' tell
a different story, however.   The proportion of men of military
age serving with the armed forces, for example, reflects civil
output forgone because of the defence effort.   For the United

Kingdom in 1974 this proportion was 3.4 per cent, compared with 4.8 per cent in Belgium, 4.9 per cent in France, 4.0 per cent in West Germany, 3.9 per cent in Italy and 4.1 per cent in the Netherlands. In this respect therefore, European allies already bear more onerous 'burdens' than Britain. (3) In any event, as a matter of principle, the equating of 'burdens' per se is not a satisfactory basis for reshaping a defence programme: in public budgeting 'keeping down with the Europeans' is no more rational a basis for decision than 'keeping up with the Joneses' is for household expenditure choices. (4)

Thirdly, it is one thing to seek a release of resources for other purposes and quite another to ensure their assimilation elsewhere in the economy. Adjustment problems may be formidable at the best of times, in practice if not in principle. They are one reason why shifts in the pattern of public spending are typically brought about by prescribing either less than proportionate expansion in a context of generally rising expenditures or, more problematically, differential rates of contraction. Quite apart from the question of striking a judicious balance between security benefits and resource costs, therefore, the Government may have judged it inappropriate to attempt more rapid contraction. As it happens anxiety has already been expressed, in specific industries and localities, about the expected effects of planned changes, the more insistently because of the general uncertainty of the short-term employment outlook. Had more dramatic defence cuts been sought, there would undoubtedly have been an even greater chorus of protest. (Certainly any attempt to cut defence spending by £1000 millions in a year or so, as was implied by a 1973 Labour Party Conference resolution, would have been inconsistent with responsible economic management. If that was seriously advocated it was absurd.) (5)

The case for Exploring Alternatives

It would be surprising, of course, if there were not good reasons for reshaping the defence effort to the extent, in the manner and at the pace that the new programme envisages. But this does not mean that the debate about how much (or little) defence is 'enough' stands adjourned. Rather the contrary: such discussion is, as it should be, a continuing process.

At the present juncture, for instance, there is no lack of support for the sentiment that, with no evident, direct and immediate threats to security but a variety of urgent claims for remedying social ills, a lower level of military expenditure would represent a better balance between defence needs and the resource costs involved. Moreover, as economic distress continues, the United Kingdom's commitment to allot a higher proportion of GNP to defence than many states with

140

stronger economies looks less and less tenable – ambiguities about the notion of 'burden' notwithstanding. Debate about 'how much for defence?' accordingly remains open and active.

Fundamentally any such debate is about priorities and, therefore, about values. It is not matters of fact that are at issue, but of what importance to attach to what facts. Nevertheless uncertainty and straightforward lack of knowledge confuse and distort discussion about values and choices, in relation to resource allocation for defence as elsewhere.

This is certainly the case in the aftermath of the 1974 Defence Review in that the basis for judgment about <u>lower cost defence programmes for 1975–84</u> has not been properly laid. In particular, there has been no systematic attempt to identify specific 'alternative futures' developed from a definition of the nation's core security interests more stringent than that which underlies the Government's programme. Nor has there been methodical investigation of those employment and other economic consequences of an even lower defence effort of which account would have to be taken in any policy reckoning. As a result the 'official' programme is represented as the <u>only</u> one which makes any strategic sense, which may or may not be true. And advocates of substantially less costly provision have pressed their arguments without precise information on the scale and character of important consequences of the course they propose.

The purpose of this paper is to provide a sounder foundation for analysis and judgment in the light of these points. It develops selected <u>Defence Programme Options to 1980–1</u> (that is, alternatives to the programme yielded by the 1974 Review) whose common characteristics is that they entail estimated expenditure in 1980–1 of not more than £3000 millions (at 1974 prices). This is the level to which British defence-spending would have to fall to put the United Kingdom's defence/GNP proportion on a par with those of European allies, on plausible estimates of relative growth performance and future allied defence provision. The scale of reduction involved would also yield 'savings' of several hundred millions a year measured against current outlays.

Each 'package' has been composed on the basis of a more stringent definition of national security interests than that adopted in framing the present programme and a relaxation of certain key assumption simplicit in it. In this respect the options are not 'radical' in the strictest sense; it is recognized that the <u>status quo</u> must be the point of departure for the generation of policy alternatives. Even so, the scale of expenditure reduction under examination implies far-reaching change. Each of the 'packages' sketches a defence posture fundamentally different from that now assumed and one to which the United Kingdom could not move without the most profound political and strategic repercussions.

Such repercussions would, of course, have to feature prominently in any comprehensive evaluation of the options. For present purposes, however, such considerations are set

aside. The limited aim of this exercise is to elucidate the
kind of change in the defence effort that might be required to
bring the level of expenditure down to £3000 millions at 1974
prices over five years. This is a necessary preliminary to
thorough exploration of the employment and other economic
consequences of such a reduction, and the formulation of
policies for transition, to enable these to be weighed alongside the attendant politico-strategic costs and potential
benefits from alternative resource use. But, it must be
stressed, 'costed options' provide a foundation for analysis
and judgment not a substitute.

Is the elaboration of specific programme options really
necessary? Would it not be sufficient to assess the
consequences of reduction on this scale in a general,
aggregated fashion? Alternatively could one not settle for
calculating the implications of, say, a further 25-30 per
cent manpower rundown and/or the 'savings' that might accrue
from cancellation of certain major weapons projects? The
view taken here is that the options approach is to be
preferred. General or macroeconomic analysis can provide a
useful perspective regarding the scale of adjustment
involved. But it cannot take account of the obvious fact
that it is the form of any reduction which determines the
implications for industries and regions and - most important -
for specific plants and particular localities. And it is
precisely these implications that may most require illumination.
The objection to making equipment cancellations or manpower
cuts the basis of assessment is obvious. The procedure is
inherently unreal, unless one is prepared to countenance
troops without equipment or weapons system without personnel
to man them. (In any case, if sacred cows should be viewed
with suspicion, so should sacrificial cows, or corporals.)(6)

Generating programme options has problems and limitations
of its own, however. Detailed information on budgetary
projections is not available in the open literature. Nor are
authoritative cost data. Furthermore, there is obviously an
infinite number of possibilities that might be considered. The
effort is none the less worth making, provided due allowance is
made for the inevitable crudeness of many calculations.

The Present Programme to 1980-1

The starting point must be the existing defence programme to
1980-1. As originally envisaged in the Defence Review the
annual cost of this programme, expressed at 1974 Survey prices,
was £3,700 million in 1975-6 and a constant £3,800 millions a
year thereafter. In the March 1975 Budget the target Estimate
for 1976-7 was reduced by £110 millions (on the same price
basis). Thus the present projections are:

142

£m at 1974	1975-6	1976-7	1977-8	1978-9	1979-80	1980-1
Survey Prices	3700	3690	3800	3800	3800	3800

This profile of planned outlays is undergoing reassessment, along with all other public spending programmes, in the ministerial deliberations on the 1975 Public Expenditure Survey. Downward revision is likely. However, not until the next Public Expenditure White Paper appears (around the turn of the year) will the outcome of this scrutiny be known. There is, therefore, no alternative but to take these figures as the benchmark for the present exercise.

For what purposes are these funds to be provided? The answer lies in the functional analysis of the Ministry of Defence's Long Term Costings, which, in turn, lie behind combination locks in official safes. But at least the functional costing for 1975-6 was published, in the usual way, in the Statement on the Defence Estimates 1975 , Cmnd 5976. A summary of the breakdown is given in Table 1. The total Defence Budget figure here is not £3,700 millions, however, but £4,548 millions. This is because the price basis for the table is not 1974 Survey prices but 1975-6 Estimates prices. The difference reflects pay and prices increases between the planning exercise and the presentation of the Estimates.

Since the analysis in Table 1 is the only available point of reference for the consideration of options it is necessary to adopt its 1975-6 Estimates prices basis for the main argument of the remainder of this paper. The 'benchmark' programme budget totals in these terms are:

£m at 1975-6	1975-6	1976-7	1977-8	1978-9	1979-80	1980-1
Estimates prices	4548	4525	4675	4675	4675	4675

(The 1975-6 figure is the Actual Estimate, cf. Table 1: the remaining figures are conversions, given to the nearest £25 millions.)

The reference figure for the options to be reviewed – £3,000 millions at 1974 Survey prices – corresponds to £3,687 millions at 1975-6 Estimates prices or £3,675 millions to the nearest £25 millions. The latter is the most useful working figure. It happens also to be exactly £1,000 millions less than the 1980-1 'benchmark' value. Hence the programme options represent alternative routes to a defence effort in 1980-1 costing, at 1975-6 Estimates prices , at least £1,000 millions less than is at present planned and c. £875 millions (20 per cent) less than the current defence budget. (7)

Table 1 THE BENCHMARK PROGRAMME
DEFENCE EXPENDITURE FUNCTIONAL COSTING 1975–76 and 1980–81
(Defence Estimates 1975–76: Hypothetical Costing 1980–81)
£ millions at 1975–76 Estimates Prices

Major Programmes (and sub-programmes)	Def. Est. 1975–76 £ millions	Hypothetical Costing 1980–81 £ millions	% 1980–81
A Nuclear Strategic Forces	58	75	1.6
B Navy GP Combat Forces			
(1) Carriers/Amphib Forces	37	25	
(2) Submarines	115	125	
(3) Cruisers/Destroyers/Frigates	307	400	
(4) MCMVs & other ships	104	70	
(5) Aircraft	39	70	
(6) Overseas Naval Bases	30	10	
	632	700	15.0
C European Theatre Ground Forces			
(1) BAOR	584	650	
(2) Berlin	14	15	
(3) Home Forces	228	185	
	826	850	18.2
D Other Army Combat Forces	88	25	0.5
E Air Force GP Combat Forces			
(1) Air Defence	78	100	
(2) Offensive Support	64	)425	
(3) Strike/Attack/Reconn.	269	)	
(4) Maritime, Transport & Other	177	125	
(5) Operational Stns.	119	80	
(6) Gen.Support/HQs/Charter	167	120	
	874	850	18.2
Total 'Mission Programmes'	2478	2500	53.5
F Reserve Forces	79	75	1.6
G R&D not included elsewhere	283	200	4.3
H Training	417	375	8.0
I Prod. Rep. & Associated Fac	340	350	7.5
K 'Other Support' Functions	875	1100	23.5
J,L Stocks/Misc. Rec (net)	76	75	1.6
Total 'Support Programmes'	2070	2175	46.5
TOTAL DEFENCE BUDGET	4548	4675	100.0
(£ millions at 1974 Survey prices)			
TOTAL DEFENCE BUDGET	3700	3800	

Source: Cmnd 5976, Annex B, p.93 (1975–76).
       Author's Estimates (1980–81).

Returning to the functional costing (Table 1), it is now necessary to form a view of the probable distribution of intended expenditures among the several mission programmes and support programmes into which the defence budget is there divided.   Without access to the Ministry of Defence's figures this is necessarily a matter of intelligent inference at best, coarse guesswork at worst.   These are the 'techniques' — and the only ones — that have been used to produce a hypothetical costing for 1980-1 (also in Table 1).

Among the factors borne in mind in generating this distribution are the following:

Mission Programmes

(a) The benchmark programme must incorporate increased provision for the Polaris force, partly to 'maintain its effectiveness' as intimated in the Defence Review, partly to accommodate rising support costs. (8)

(b) The Royal Navy's share of the defence budget is planned to increase because of re-equipment.   The pattern of new construction envisaged means that the share of the cruiser, destroyer and frigate programme elements will rise. (9)

(c) No change is contemplated in the strength of Rhine Army but procurement provision is probably higher for 1980-1 than now (because of plans to introduce LANCE, the FH70 and SP70 guns, MILAN and work on a future tank etc.).   On the other hand there is to be a rundown of Home Forces. (10)

(d) By 1980-1 all the planned withdrawals from outside Europe should have been completed. (11)

(e) Within the air force mission programme the later 1970s will see,
    (i)   the phased rundown of transport forces,
    (ii)   station closures in the United Kingdom and outside Europe,
    (iii) some reduction in the 'general support' area,
    (iv)   high and rising new equipment outlays, associated with the continued deliveries of some aircraft types already in production (e.g. Jaguar and the Lynx helicopter), the remaining development and initial production spending on the 'Common Multi-Role Combat Aircraft (MRCA) and development work on the air defence variant of this plane. (12)

Support Programmes

(f) A reduced R & D effort is planned and the expenses of various forms of training should fall, because of lower numbers and continuing rationalization. (13)

(g) 'Other Support' costs will probably rise through the later 1970s, and assume a higher proportion of the total defence budget, in line with the trend over the last decade. (This item accounted for 13.9 per cent of the budget in 1966-7, 19.2 per cent in 1975-6.) Included here, of course, are the programme elements for Service pensions, family and personnel services, and the Whitehall organization. (14)

General Points

(h) Within each sub-programme involving capital outlays over 1975-84 attention will have been paid to the need to phase spending so as to avoid 'bunching'. (15)

(i) In settling the overall programme during the Defence Review preservation of a measure of balance among expenditures associated with the three Services was sought.

(j) The ratio of expenditure on mission programmes to that on support programmes is likely to remain unchanged (at around 54: 46 per cent).

It must be emphasized, however, that such points as these only provide bearings — and some insurance against gross error — in what remains essentially speculation about the likely structure of a future defence budget. The 1980-1 figures in Table 1 are a hypothetical costing which may or may not correspond closely to the official version.
Even then, an heroic cock-shy at the planned output budget for a single future year does not amount to 'a view of the probable distribution of intended expenditures' of the sort needed as a frame of reference for reviewing programme options for the intervening period. It must further be assumed, therefore, that the Long Term Costings to 1980-1 provide for a steady progression from the present pattern of provision (as shown in the 1975-6 functional analysis) to that postulated for 1980-1; and that the directions of change under particular headings correspond to the revised parameters for the size, shape, equipment and deployment of the Armed Forces that emerged from the Defence Review.
This is the best that can be done to gain an impression of 'present plans'. Governments lift the veil on the detailed composition of their expenditure programmes for only one year ahead, when going to Parliament for 'supply' makes it inescapable. Worthwhile discussion on alternatives is made immeasurably more difficult by this lack of adequate information on exactly where present policy is leading. (16)

One might now put the simple question: what lower cost
defence programmes represent feasible alternatives to present
plans? But 'feasibility' is an elastic notion.   The
possibilities would be endless, embracing the much-quoted
Danish 'proposal' of purchase of a tape-recorder to deliver a
'surrender' message and the numerous hare-brained schemes for
radical reconstruction that occasionally enliven the pages of
defence journals (usually involving scrapping the Navy, Army
or Air Force, reviving the militia and yeomanry, or bringing
back conscription).

More precise terms of reference have been adopted for this
exercise.   They make it necessary to pose a series of
questions:
  - What is the conceptual basis of the present (or benchmark)
  programme?
  - Upon what interpretation of 'essential defence needs' is
  it founded?
  - What judgments or assumptions about 'proper provision' for
  those needs underlie it?
  - If one re-examines the conceptual basis, adopts a
  marginally more stringent definition of 'needs' and relaxes
  some critical assumptions about 'proper provision', what
  are the implications?
  - What programme options suggest themselves at this juncture?

Arising as it did from an 'extensive and thorough' review,
the rationale of the present programme has been thoroughly
explored - in Government exposés des motifs , Parliamentary
Debates and Committee Reports, and also in the specialist and
popular press.

In this process commentators have also pinpointed controversial interpretations of 'needs' and questionable judgments
or assumptions.   There is therefore little difficulty in
dealing with the initial questions. (17)

The present programme is founded on the notion that the
United Kingdom should comfort herself as a substantial medium
power in a European setting.   It has become necessary to forego
some costly capabilities (e.g. for credible nuclear deterrence
or effective amphibious and airborne operations) and to accept
contraction of the geographical spread of commitments (not
only into , but also within , the Atlantic Alliance area).
But NATO is a 'first and overriding charge on the resources
available for defence' and the national contribution, on the
Continent and in European waters, should remain undiminished -
quantitatively and qualitatively - so far as possible.   Further,
an attempt should be made to preserve military stature even
when assuming a more modest defence posture.   Among other
things this means continuing to muster 'balanced forces',
which also represent insurance against the unforeseen.

The interpretation of 'defence needs' follows from this.
The 'essential' requirements are that the United Kingdom
should:

(a) maintain an Army force level of <u>c</u>. 55,000 and a
tactical air force in the Central Region of Allied
Command Europe (ACE) plus some capacity for reinforcing
these and other allied formations throughout the ACE
area,

(b) continue to provide the 'main weight of the maritime
forces immediately available to the Alliance' in the
Eastern Atlantic and Channel,

(c) ensure the security of the United Kingdom itself and
its immediate approaches,

(d) retain a ballistic-missile submarine force as 'a
unique European contribution to NATO's strategic nuclear
capability', (18)

and there remains a felt need to

(e) maintain forces in certain dependent territories.(19)

Among the key judgments and assumptions about 'proper
provision' for these needs four may be singled out for
emphasis. The first is the <u>acceptance of existing force</u>
<u>levels in the ACE area</u> as immutable. The Brussels Treaty
is invoked here; so too is the damage that a unilateral
rundown of strengths would cause to NATO's negotiating
stance at the mutual force reduction talks in Vienna.
Secondly, there is <u>the principle of balanced forces</u> .
This enters in two forms, in relation to inter-Service and
intra-Service 'balances'. On the face of it it is surprising
that, despite the changes in defence policy and posture of the
last ten years, there has been no question of future provision
based mainly on (for example) ground and air forces — with a
disproportionate reduction in naval strength. Nor has there
been examination of radically different force structures viz.
a Navy with a heavier weighting of submarines and smaller
surface ships, an Army designed for maximum early effectiveness in
defensive operations in north-west Europe, or an Air Force geared
primarily to close support rather than the more demanding —
and costly — tactical air missions like deep strike and
interdiction. 'Reduced Circumstances' have not affected
'style', nor is it intended that they should do so in the
future. (20) Thirdly, there is the judgment that, for a few
more years at least, a unique European contribution to NATO's
strategic retaliatory capability will confer benefits that
outweigh the costs involved in prolonging the active life of
the four-boat Polaris force. In short, there is a belief in
<u>the continuing utility of a quasi-independent deterrent</u> even
as it approaches obsolescence. Fourth and last is the
assumption that the security interests of those remaining
dependencies which are under some political pressure require
the stationing of contingents of United Kingdom forces, at a
cost to the British defence budget. There is a <u>residual</u>
<u>imperial policing role</u> (albeit a very minor one).

Rationales for Options

As has been argued earlier, discussion of lower cost defence programmes cast in terms of 'generalized objectives or figures chosen more or less at random' is unhelpful. So too is advocacy which ignores the need to take the status quo as the starting-point for policy alternatives. A specific, but 'incrementalist', approach to defence programme options is needed. Against the background outlined this means seeking an alternative 'conceptual basis', asking whether some needs are not more essential than others, and considering different judgments or assumptions about 'proper provision'.

The alternative to the underlying, fundamental rationale of the present programme is obvious: it is assumption of a position, within the Atlantic Alliance and the European community, congruent with present perceptions of the United Kingdom's political weight and economic strength. This would imply, and would itself confirm, a transformation of the national self-image from one of 'reduced circumstances' to one built around a lower place in the pecking order of military power and influence. Naturally, NATO would remain the first charge on resources for defence, but the scale and scope of the British contribution to force levels would require reconsideration.

Re-examination of key priorities among 'defence needs' in this context might embrace the following arguments:

(a) A continuing commitment of ground and air forces to Allied Command Europe is crucial. But neither existing force structures nor present force levels need be regarded as sacrosanct.

(b) The Eastern Atlantic and Channel areas - including the North Sea and the Norwegian Sea - are equally vital. But the share of the Alliance effort that the United Kingdom provides might be reduced; and the nature of that provision would bear scrutiny.

(c) Ensuring the security of the 'home base' is a third co-equal priority. With the assertion of limited sovereignty over offshore areas the scope of this responsibility will grow.

(d) Neither strategic nuclear forces nor garrisons in remaining dependencies can be classed as 'essential' elements in a national order of battle constructed on the par inter pares basis envisaged.

Within such a policy framework each of the main judgments or assumptions about 'proper provision' identified earlier would be put in question.

Most obviously, the residual imperial policing role could no longer claim resources. As a practical matter a case could be made for continuing provision for dependencies fully matched by host territory contributions of the kind already made by Hong Kong. But no charge to the United Kingdom's

149

budget would be justifiable.   It would also be impossible to
sustain the thesis of continuing validity of a quasi-
independent deterrent, at least in the form of a ballistic
missile submarine force.   If importance were attached to some
unique European retaliatory capability, the onus would be on
all the European members of the Alliance to devise a 'European
solution'.   And developments in cruise missile and remotely
piloted vehicle (RPV) technology suggest that it is to these
sources that medium powers would be most inclined to look for
future long-range nuclear strike potential.

Yet neither stationed forces outside Europe nor the Polaris
force really count for much in the United Kingdom's defence
effort, either now or as at present envisaged.   (See Table 1,
p. 144)   It is the judgments and assumptions underlying
choices about navy general purpose combat forces, European
theatre ground forces and air force general purpose combat
forces that exert greatest leverage on the size of the bill
for defence.   What if these were modified or relaxed: if
existing force levels in the ACE area were not regarded as
immutable, and if there were willingness to think again about
the principle of 'balanced forces'?

In the first instance, upon what criteria could such
modifications or relaxations be based?   Are there rational
grounds for alternative judgments and assumptions?

European Force Levels

The sizes of Rhine Army and RAF Germany are formally related
to the Brussels Treaty as 'modified' and 'completed' by the
Paris Agreements of October 1954.   The obligation hinges on
a concept of 'equivalent fighting capacity', to be assessed
by the Supreme Allied Commander, Europe (SACEUR), and an
ambiguous undertaking to respect the wishes of the majority of
the members of Western European Union (WEU).   It is not an
obligation that has impeded adjustment to force levels in the
past, such as the adoption of a three division structure for 1
(BR) Corps or the redeployment of 6 Brigade and one RAF
squadron (in 1968).   Nor for that matter is it inhibiting
current restructuring: this involves a format of five 'new
style' divisions - even though the Paris Agreements specify
that the United Kingdom's ground troops on the European mainland 'shall not exceed in total strength or number of
formations ... four divisions.' (21)   In short, these
force levels are negotiable.   It is difficult to imagine
collapse of the whole NATO edifice brought about by national
adjustment effected in consultations with allies.   Much would
depend upon the basis and context of the adjustment proposed.

In this connection certain features of current debate about
European security arrangements are pertinent: they relate to
perceptions of 'the threat', speculation about NATO dispositions
and doctrine vis-à-vis operations in the Allied Forces
Central Europe area (AFCENT), expectations concerning force

150

reductions on the Central Front and the economic - especially
manpower - constraints that impinge on all Alliance members.
To summarize:

(a) Military commentators discount the possibility of a
Warsaw Pact offensive on the Central Front.   But there is
anxiety, because of the 'danger that military force may be
used as an instrument of political pressure' bearing on
Western countries and, perhaps, bringing them 'very near to
losing their ... self-confidence'; in which case, the
argument runs, 'the Warsaw Pact would naturally be correspondingly encouraged to test how far it could go in completing
the undermining process'. (22)   Even those who would object to
the 'naturally' here might agree that early euphoria about
détente was indeed misplaced, and that Western Europe would be
ill-advised to dismantle its military apparatus - given the
known proclivities of powers when they can hope for deference
born of weakness.   At the same time definition of a politico-
military threat has two significant implications.   For one
thing it intensifies the security dilemma, since vulnerability
may be a function of frustrated social aspirations as well as
inadequate military preparations.   For another it puts a
premium on appearances, including displays of cohesion, and on
having available ways of signalling resolve.

(b) Reassessment of the character of 'the threat' is, however,
only one among many factors that have prompted speculation
about NATO's doctrine and dispositions in recent years.   Much,
if not most, of it has been prompted by doubts about the
Alliance's purely military competence.   Maldeployment has
been castigated by the WEU Assembly. (23)   More fundamentally,
persuasive arguments have been advanced to suggest that by
modifying their interpretation of flexible response, adopting
more realistic concepts of operations, and radically restructuring their forces, the NATO powers could significantly enhance
their effectiveness <u>without</u> employing additional resources.
(24)   And it has certainly been shown that 'models' for
European defence can be elaborated which would be viable with
reduced formations and equipment. (25)   Recurring themes in
much of this discussion are the inefficiencies associated with
inappropriate stress on staying power (the short <u>v.</u> long war
controversy) and failure to exploit technological opportunities
favouring the defence e.g. in anti-tank weapons.   (In part at
least these are inefficiencies which persist, it has been
argued, because military institutions incorporate 'elements
committed by tradition and instinct to preserving their
expertise in familiar, experience-proven areas'. (27))   The
debate as a whole indicates quite clearly that there are
strong pressures for change in the ordering of Alliance
arrangements and that, with revised force structures and
doctrine, 'undiminished security' might be obtained with
lower force levels.

(c) 'Undiminished security' is the central theme of NATO's stance at the Vienna negotiations on mutual force reductions (MFRs). Behind the insistence on mutual <u>and balanced</u> reductions and on formulae incorporating <u>common ceilings</u> is the conviction that alternative approaches, based on numerical and percentage reductions, would shift the military balance further in favour of the Soviet Union and Warsaw Pact and thus offend against this principle. Yet the Soviet Union holds that the present (conceded) imbalance constitutes an equilibrium, to alter which would be destabilizing. Not surprisingly, therefore, the negotiations have made no headway. But the log-jam is likely to be broken soon. Prior to the negotiations proper, a 'parallelism' was acknowledged between the force reductions talks (a NATO proposal) and the Conference on Security and Co-operation in Europe (CSCE) for which the Soviet Union had pressed for many years. Governments have emphasized repeatedly that the CSCE milestone passed in Helsinki this summer does not alter the position of the negotiating parties at Vienna In fact, however, the mechanisms of 'linkage politics' and the expectations of Parliaments and publics in the West will make it difficult, if not impossible, to prolong the talks without some sub-stantive agreement, however partial and limited. And in order to promote such an outcome, the merits of a NATO initiative should not be discounted. A sharp test of Soviet good faith in the negotiations would clear the air, if nothing else.

(d) In any event manpower imperatives may impel such an initiative whether military authorities like it or not. Recruiting to volunteer regular forces is likely to become more difficult in a climate of détente, real or imagined. Managing conscript systems is proving troublesome for all the continental West Europeans, because of demographic factors (too many reaching military age than can be equipped or paid for), equity and morale factors (the unacceptability of gross discrepancies between volunteer and conscript pay) and the operational problems of assimilating short-term enlistees in increasingly sophisticated organizations. (27)

All in all, therefore, there is no lack of intellectual foundation for the proposition that force levels on the Central Front should <u>not</u> be regarded as immutable. Rather the contrary: if the terms were right, lower force levels could represent a more realistic response to the politico-military threat; sustain no less effective a defensive posture; at best advance and at worst usefully clarify the arms control prospects for Central Europe; and not least, make military manpower problems more manageable. ('If the terms were right' begs most of the critical questions, of course. But the important point in the present context is that existing force levels are not the only ones that might make sense.)

Balanced forces: Air and Naval

What, next, of 'balanced forces'?   Because the forces in
AFCENT include RAF Germany it is convenient to raise this
question in the air force setting first.   For the grounds
for alternative judgment and assumption about the spectrum
of mission capabilities which the Royal Airforce should
realistically strive to cover are, in fact, bound up with
the European theatre forces debate just reviewed.

1. Tactical Air Power

There is no doubt that there are such grounds.   All the
leading NATO countries have assigned substantial resources to
tactical air power and plan to continue doing so in future.
Yet advances in air defence, demands for aircraft of higher
(and more versatile) performance with more complex avionics,
the economic unfeasibility of fielding large numbers of
expensive aircraft and the need to divert many to supporting
missions (air defence suppression, electronic counter
measures etc.) — all these have worked against the effectiveness of such forces.   Accuracy limitations, low sortie rates,
low aircraft numbers and high aircraft attrition combine to
yield unattractive 'cost-effectiveness' estimates.   Cheaper,
simpler aircraft 'improve' the calculations — if expected
losses can be kept in bounds.   But present attrition estimates
are high enough already, given the expected air defence density
in Europe: it is plain to see that compounding survival rates
of $c.95$ per cent per sortie means that an air force ceases to
exist after a short period.   Mission priority dilemmas are
intensified in such a situation.   Is striving for total air
superiority worth the candle?   What is the value of deep
penetration into enemy air space against well-defended and
sheltered aircraft (or other targets)?   And if resources are
heavily committed to deep interdiction — as NATO has done by
complicating the design of its aircraft to give them appropriate
capabilities (high speed, large payload and long range,
elaborate penetration avionics) — may not resources be diverted
from more central and urgent tasks like giving fullest support
to ground forces striving to hold the attacker? (28)
  This devil's brew of problems lies behind the concern that
has been expressed in the United States that it is becoming
too expensive to buy in quantity aircraft with the performance
characteristics needed to carry out the more demanding
missions.   There are doubts as to whether the latest high-
performance fighter-bombers — for example, the F15 Eagle —
offer improvements over existing F4 Phantom IIs to justify
their higher unit cost.   Indeed it is thought that it would
be feasible 'to limit future modernization of the long-range
multi-purpose aircraft to improved models of the F4' and that
'capabilities would be little affected in the near future by
halting procurement of the F15'. (29)   Reinforcing such

thinking is the fear that stressing the capacity for airfield attack and deep interdiction at the expense of generous provision for close support of ground troops on the battle-field and local air superiority above it could prove a decisive error in a 'short war'.

'Alternative judgments' about what aircraft for what missions arise naturally from this analysis. Close air support and local air superiority are the priority missions. As for appropriate aircraft, one commentator has expressed the following views:

> For these two ... missions, the Western practice of designing multi-purpose aircraft is no longer suitable. The requirement now is for two specialized, cheap aircraft — the lightweight air superiority 'dog-fighter' and the sturdy fighter-bomber for close air support. For air-superiority fighters the choice is between numbers and sophistication. Sophisticated aircraft ... may be the best fighter in a one-to-one encounter and for an unconstrained budget, but they are less manoeuvrable than some of their lightweight competitors, too expensive and thus foreclose the numbers and sortie rate necessary for local air superiority ... The characteristics for a good close air support aircraft are simplicity, survivability and lethality from the air force point of view, and responsiveness from the army's ... During crises tactical air power must be capable of very high sortie rates — which is easier for simple, rugged aircraft operating from forward bases. (30)

Underscoring the points, he adds, with a significant general conclusion,

> The penetration/counter-penetration battlefield requirement is unsuitable for sophisticated aircraft like the F.15, MRCA and associated command and control aircraft such as the Airborne Warning And Control Systems (AWACS). Simpler aircraft besides being more effective in support of ground forces, also imply major savings from changes in basing, maintenance and pilot training ... These factors indicate that, even if NATO's present number of aircraft were retained, larger numbers of more effective aircraft could be concentrated upon the penetration battle while procure-ment costs and manning could be readily cut in half. (31)

In sum, the assumption that the only air forces worth having are 'balanced' — with a 'high-low' mix of capabilities — is actually becoming out-of-date. Therefore, to consider the implications for the United Kingdom's defence effort of rep-lacing that assumption is not only permissible but obligatory.

2. The Balanced Fleet

A similar assumption features in naval planning, with
particular reference to the surface fleet.   Deciding how many
ships to include in the new programme and what types must have
been among the more difficult choices that arose in the 1974
Defence Review.   In the event the planners opted for a modern,
 balanced fleet - including some costly new 'ships of high
quality'.
     Practically speaking, however, genuine balance is unattain-
able, given the constraints.   The larger surface units for the
later 1970s and early 1980s will consist of:
     1 attack carrier, Ark Royal   ('until the late 1970s')
     1 anti-submarine warfare (ASW) carrier, Hermes (from 1976)
        (able to act as a Commando carrier, if necessary)
     1 assault ship (plus 1 in refit or care and maintenance)
     1 ASW cruiser, Invincible   (after 1980)
say, 10 guided missile destroyers, Bristol and the Sheffield
        class (plus remaining County class ships).

And 'one of each' is the very antithesis of balance, quite
apart from the operational planning problem it generates.
     Nevertheless there would be a compelling case for a naval
order of battle headed like this if it were demonstrably true
that maritime strategy required the periodic deployment of
'ships of high quality'; or if seaborne strike aircraft, ASW
helicopters and amphibious warfare units could be shown to
have unique capabilities for vital missions.   In fact,
however, it is difficult to sustain either of these propositions.
     What is the naval mission in the Eastern Atlantic (EASTLANT)
area of NATO's Atlantic Command and in the Channel?   The
 Statement on the Defence Estimates 1975 asserts that,
'deterrence on the mainland of Europe would not remain credible
without a parallel strategy in the Eastern Atlantic and Channel
areas'. (32)
     But there is no elucidation of the 'parallel strategy'.
This is not surprising, for it is difficult to develop a
concept of naval operations when a short war v. long war
controversy is going on over the posture for the land battle
and when, if there were a land battle, fighting reinforcement
convoys across the Atlantic (or to Denmark and north Norway)
might not be the top priority for limited naval resources.
There is no alternative in fact, but to define the mission
in terms of 'presence'.   Accordingly the Statement continues
as follows:

     seaborne supply and reinforcement routes ... pass through
     these areas.   If the balance of maritime power were
     allowed to shift so far in favour of the Warsaw Pact that
     it had an evident ability in a period of tension to
     isolate Europe by sea, the effect on Allied confidence and
     political cohesion would be profound. (33)

155

In other words the important thing is to be there , so
that in a crisis NATO's European members are not inclined to
submissiveness arising from a sense of insecurity. This is
not an unreasonable rationale for even a substantial naval
presence in the EASTLANT area. It recognizes that, nowadays,
NATO faces a politico-military threat to 'confidence and
cohesion' rather than the prospect of brutal attack with naked
force. It does not, however, provide firm guidelines for
deciding what ships can best furnish the presence, or for
deciding how much is enough. In particular it is not self-
evident that a smaller ship Navy and/or one with a greater
weighting of submarines would not afford the required re-
assurance.

Do 'ships of high quality' nevertheless provide unique
capabilities for vital missions? There are no clear-cut
answers to this. Uniqueness is a matter of judgment: so,
too, is what is 'vital'. Suffice it to say that,

(a) the increased vulnerability of large surface ships makes
it dubious to use them as platforms for strike aircraft
except when there is no, literally no, alternative way of
attacking high priority targets and when the platform itself
can be given assured protection. The expenses of effective
protection alone put the cost-effectiveness calculations
beyond what is acceptable for all but targets of the very,
highest priority - none of which are likely to be accessible
from the EASTLANT area.

(b) The merits and demerits of alternative ASW systems are
debated at great length and the claims for carrier-borne
fixed- and rotary-wing ASW aircraft cannot be dismissed.
But in general the most effective systems in future seem
likely to be land-based aircraft and hunter-killer submarines.
(There may be something in the idea that, in naval pressure
for 'quality' ASW ships , it was more a matter of needing ASW
for quality than needing quality for ASW.)

(c) The United Kingdom has already elected to forego the option
of a fully-fledged amphibious warfare force, presumably because
the mission is judged marginal. Retention of a residual
capability for such operations may, again, owe as much to
reluctance to pay off 'ships of high quality' as to operational
imperatives for their retention.

Indeed it is perhaps only for the task of providing tactical
command and control for large-scale operations that the major
surface warahip now has a substantial comparative advantage vis
-à-vis other forms of maritime force. Certainly its claims
to an assured place in the Navy of a medium power are disputable.
To contemplate a fleet composed wholly of smaller ships and
submarines is neither outrageous nor irresponsible.

To sum up this extended argument on rationales for options:
it has been shown that a foundation for developing specific
lower-cost defence programme options can be laid by assuming
marginal modification to the 'conceptual basis' of the present

programme, taking a more stringent definition of 'essential defence needs', and revising certain judgments or assumptions about proper provision - especially those relating to the immutability of force levels in Europe and the indispensability of 'balanced forces'. The question now is: what are the budgetary implications? Specifically, with the programme to 1980-1 as a benchmark, what 'savings' would be possible under the various functional headings if provision were altered in accordance with the ideas that have been reviewed?

(Before tackling this question it is appropriate to reiterate that the benchmark - the profile of intended defence expenditures to 1980-1 - is, perforce, a hypothetical one; we do not have the official figures. It may also be timely to enter a reminder about the purpose of the exercise. <u>Its aim is to elucidate</u> the kinds of change in the defence effort that would be required to bring the level of expenditure down to £3,000 millions (at 1974 prices) over a five year period, <u>as a basis for analysis and judgment</u>. There is no claim to prescription.)

Programme option modules

The benchmark programme totals to 1980-1, expressed at 1975-6 Estimates prices are:

	1975-76	1976-77	1977-78	1978-79	1979-80	1980-81
£m at 1975-76 Estimates prices	4548	4525	4675	4675	4675	4675

The estimated functional breakdown for 1980-1 has been given in Table 1 (right-hand columns). It is assumed that the full Long Term Costings for the period provide for a steady progression from the 1975-6 position to that hypothesized for 1980-1 (see p. 146).

This is the budgetary frame of reference for considering options. At the same time it marks the outer limits of the margin for manoeuvre. The effective margin, however, is narrower - both for the budget as a whole and for individual major programmes and sub-programmes. For example, 'some 80 per cent of equipment expenditure is effectively committed up to four years ahead'. This includes funds allotted for 'projects in development, on which large sums have already been spent' and for 'production, maintenance and support of existing weapons systems'.(34) This does not mean that only 20 per cent of the procurement budget is alterable, for projects can be abandoned no matter how much has been spent on them and existing systems can be discarded. The point is that (especially in the early years of an ongoing programme) beyond this 20 per cent one may quickly encounter areas where 'savings' will be offset by budgetory expenditure on current projects and cancellation charges on contracts broken.

On the manpower side similar considerations arise because of redundancy payments for servicemen and civilians discharged. Naturally, these constraints have been taken into account in what follows.

The 'savings' that would be possible on the various major programmes by applying modified notions of 'proper provision' to a revised appreciation of 'essential defence needs' are most conveniently treated on a major programme-by-programme basis.

A. Nuclear Strategic Forces

An immediate decision to withdraw the Polaris submarines from service, close all Polaris-related facilities and activities (at the Clyde Submarine Base, HM Dockyard Rosyth and elsewhere) and halt all work associated with existing plans for 'maintaining the effectiveness' of the force would be unlikely to produce instantaneous 'savings' because of the problems and costs of managing an orderly rundown. But there is no reason to believe that such a rundown would take more than 2-3 years so that all expenditure under this heading might have ceased by 1978-9. A plausible guess at the time-profile of 'savings' would be:

	1976-77	1977-78	1978-79	1979-80	1980-81
£m at 1975-76 Estimates Prices	-	50	50	75	75

(For these 'guesstimates' it would be appropriate to adopt the convenient formula used in the National Institute's 'Economic Review', viz. 'the figures ... are not intended to be more precise than the general statements in the text'.)

B. Navy General Purpose Combat Forces

If the commitment to a balanced fleet were abrogated and it were decided to forego 'ships of high quality' possibilities for 'savings' would include:

(1) Paying-off the Ark Royal, Hermes, Fearless and Intrepid progressively over 3-4 years. The disbandment of the Royal Marine Commandos might appear a logical corollary, but a force could usefully be retained for three roles: Arctic warfare, given ferry and air movement options; to continue to provide detachments in other ships; and for possible offshore security tasks.

(2) Rundown of Patrol submarine (SS) activity with older boats being paid off generally earlier than is currently envisaged and a 'stretching' of the nuclear-powered Fleet Submarine (SSN) building programme. But there is nothing to say that the submarine force must be reduced pari-passu

158

with the paying-off of big surface ships. Indeed a 'new
model' fleet with a heavier weighting of submarines would
be an attractive proposition, given the utility of the SSN
for ASW and as a platform for counter-surface ship weapons
like Harpoon. This is, therefore, a 'savings' possibility
that might or might not commend itself.

(3) Deletion from the programme of provision for the second
and third ASW cruisers and, possibly, some 'stretching' of
the building programme for the new destroyer and frigate
classes. The economics of cancelling the first of class
ASW cruiser (Invincible) are undoubtedly complex. But
this option, the variability of the escort construction
rate and retention/disposal choices for the existing
cruisers (Blake and Tiger) together provide scope for
intra-programme options which would entail expenditures
of £100-£200 millions less than presently planned provision
in 1980-1. Such options would lie within the effective
margin for manoeuvre, representing 25-50 per cent of the
estimated budget for cruisers, destroyers and frigates in
1980-1 (see Table 1).

(4) Reduced provision for afloat support (Royal Fleet
Auxiliaries).

(5) Cancellation of the plans to acquire Sea Harrier - for
the ASW carrier and cruisers; reduced purchases of
helicopters; plus, of course, lower outlays on the running
costs of naval aircraft may be noted in passing that
adoption of a 'smaller shop fleet' philosophy would mean
earlier release of in-service aircraft, including Buccaneer
and Phantom squadrons, for transfer to the Royal Air Force.

In round figures the year-by-year distribution of these
'savings' possibilities might be as shown in the table.

		1976-77	1977-78	1978-79	1979-80	1980-81
£m at 1975-76	(1)	5	10	15	20	20[a]
Estimates Prices	(2)	5	10	30	40	50
	(3)	-	50	100	100	150[a]
	(4)	-	-	-	-	10[a]
	(5)	-	10	30	40	50[a]
	((6)	-	-	-	-	-)

say 275

(The superscript a denotes related sub-programmes. The
final line (6) is included for comparison with Table 1: no
'savings' are credited here because even a smaller ship Navy
would justify retention of some overseas facilities - e.g.
Gibraltar. The total is rounded to the nearest £25 millions:
a range 'guesstimate' would be £250-300 millions.)

C. European Theatre Ground Forces

Some agreed reordering of arrangements for the defence of
north west Europe is not unlikely within the next four or
five years. Neither renewed Congressional pressure for US
troop withdrawals nor further reorganization of the Bundeswehr
can be ruled out. The determination of the Dutch to cut back
on NATO commitments may prompt a general willingness to
contemplate fresh interpretations of flexible response, new
concepts of operations, radical restructuring of forces. There
could be a 'conceptual breakthrough' at Vienna. Within such
a process the United Kingdom might be able to begin troop
withdrawals from West Germany with a view to retaining, say,
three 'new style' divisions (plus corps troops and artillery)
as its contribution to the order of battle in Northern Army
Group (NORTHAG). Alternatively, notice of intention to make
such a withdrawal might be served unilaterally in order to
promote a more general disposition to change structures or
organization. Whatever the circumstances, reduction in the
strength of BAOR is a sine qua non for significant 'savings'
on this programme. The Berlin garrison offers no real scope,
even if politically acceptable terms for modifying the force
level there could be agreed with the French and the Americans.
Nor for that matter do 'Home Forces' - in themselves; only if
there were a smaller Rhine Army would further cuts in the
number of teeth arm units in the United Kingdom make any sort
of sense. Assuming then,

> (1) a phased rundown of BAOR to c. 30,000 by 1980-1, with
> appropriate adjustments in equipment purchases,
> (2) no change in the Berlin garrison,
> (3) cuts in Home Forces to reflect the reduced roulement
> requirement of a smaller Rhine Army,

the order of magnitude of budgetary 'savings' attainable on
this programme would be:

		1976-77	1977-78	1978-79	1979-80	1980-81
£m at 1975-76	(1)	-	50	100	150	200[b]
Estimates Prices	(2)	-	-	-	-	-[b]
	(3)	-	-	25	50	75[b]
						275

(The superscript b denotes related items. 'Savings'
possibilities under (1) would be lower if, as a counterpart
to 'savings' on RAF Germany, it were decided that Rhine Army
should have enhanced missile air defences and/or improved
artillery capabilities.)

D. Other Army Combat Forces

Present budgetary projections to 1980-1 take account of the
intended reductions in Army strengths outside Europe announced

in March 1975.   By 1980-1 it has been assumed that no more
than £25 millions (or thereabouts) will be assigned to this
heading.   By either withdrawal of the residual garrisons
(Cyprus?) or retention only on condition that by 1980-1 host
territory contributions would be wholly covering their
expense, the whole of this provision could be struck from the
Budget.   Phased rundown or gradual increases in contributions
might yield 'savings' in the intervening years as follows:

	1976-77	1977-78	1978-79	1979-80	1980-81
£m at 1975-76					
Estimates Prices	—	5	10	20	25

E. Air Force General Purpose Forces

If it were decided that the 'right' air force for a country
of the United Kingdom's resources and stature were not one
equipped to cover most of the spectrum of mission capabilities
but a force designed to perform effectively in a more limited
range of roles, then major savings would be possible on this
programme in the period to 1980-1.   If the priority roles
were air defence (of the United Kingdom), local air superiority
and close support (in Europe), strategic and maritime
reconnaissance (especially over the northern part of the
EASTLANT area) and provision of tactical airlift for reinforcement or stationed forces, these possibilities would include:

(1) Deletion from the programme of provision for the air
defence variant of the MRCA.   The United Kingdom air
defence task is currently performed by Lightning and,
increasingly, Phantom squadrons (with, of course, ground-
based or airborne radar control).   Phantoms would have to
fulfil the role throughout the 1980s if the interceptor
MRCA were cancelled.   However improvements to them, and
their Sparrow missiles, would be possible.   Budgetary
'savings' on air defence would depend on how extensive an
effort were made in this direction.   In particular, the
question would arise of how far to continue work now
funded on the MRCA programme - for example on the AI
radar and Moving Target Indication - with a view to such
improvement.   In the longer-term, and for the European
environment, the replacement of air defence Lightnings/
Phantoms by a cheaper, lightweight fighter, 'optimized'
for the local air superiority task, becomes an issue.
Given pressures for interoperability (even standardization)
in AFCENT and to establish a 'two-way street' in Atlantic
arms acquisition some British participation in the F.16
programme might not be ruled out.   On the other hand, if
the new Allied Air Forces Central Europe headquarters is
successful in making possible new flexibility in the
employment of air power, some division of labour among
national contingents could become feasible; and in such
circumstances the preferred British contribution might be

in close support and, possibly, battlefield interdiction.   The
option of greater emphasis on enhancing the Army's own ability
to defend itself from air attack, by increased provision for
surface-to-air missiles (SAMs), enters the reckoning too.

(2) Additional expenditure on Jaguar and, perhaps, Harrier
production to reflect the priority of the close support role.
Some capacity for strike/attack is afforded by these aircraft, so there would be no question of complete inability
to undertake certain interdiction missions.   The other
side of this coin, however, is that neither of these types
is the (relatively) 'cheap ... sturdy ... simple, rugged
aircraft ... capable of very high sortie rates' which
might be ideal.   (See p. 154.)   Thus, again in the longer-
term and with interoperability and 'two-way streets' in
mind, a case could be made for including a plane like
Fairchild's A.10 in the 'lower-cost Air Force' of the
1980s.

Outlays over and above those now envisaged under this
Offensive Support heading (2) would be the counterpart,
arising from a reassessment of mission priorities, of
substantial reductions under the Strike/Attack/Reconnaissance
heading, namely,

(3) Cancellation of the 'common' strike reconnaissance
MRCA.   This project's raison d'être virtually disappears
once it is judged (or assumed) that the ability to perform
high-speed, all-weather deep penetration at extreme low
level, culminating in a single-pass weapon delivery on a
point target is not crucial to the Alliance's deterrence
and defence posture on the Central Front.   It has been
shown that there is a basis for such judgment.   (Part of
the argument, however, hinges on the emergence of
alternative means of attacking certain types of target,
for example, RPVs and improved artillery weapons.   Whether
the whole of the 'savings' theoretically attainable under
this heading would be fully realized in relation to the
budget as a whole would depend on what investment in such
means were thought worthwhile.   Cf. C(1) above.)

(4) Reduced provision for tanker aircraft and for transport
squadrons, made possible by changes elsewhere in the programme.
There would be no case for fewer maritime aircraft: indeed
if more intensive surveillance and 'policing' of a 200 mile
exclusive economic zone were judged desirable a case could be
made for acquiring additional but cheaper, simpler patrol
aircraft e.g. Hawker's Coastguarder.

(5) Further closures of operational stations, or reduced
running costs of existing ones, as a consequence of the
shift to a 'lower cost air force' operating if not substantially
fewer at least relatively cheaper, less sophisticated aircraft

with more modest maintenance and support requirements.

(6) Some reduction in the expenses of General Support and Headquarters and of air movement (civil charter) made possible by changes elsewhere in this and other major programmes (e.g. withdrawal from deployments outside Europe, reduction in the size of the Army).

Needless to say it is impossible to quantify these 'savings' possibilities accurately. All that can be inferred about existing budgetary provision is that, for this major programme as a whole, planned outlays probably lie somewhere around c.£850 millions a year throughout the period to 1980-1. What the effective margin for manoeuvre might be depends on the extent of irrevocable expenditure commitments and liabilities to cancellation charges and suchlike. In addition, as has been shown, there are numerous intra-programme options at least some of which would have to be taken up. Moreover, although the weight of the MRCA in procurement projections is not in dispute, it has been moderated by the 'stretching' decided upon during the Defence Review. It must also be borne in mind that the functional costing incorporates personnel, operating and maintenance expenses in addition to capital expenditure. However, if a parsimonious line is taken on intra-programme options; if it is assumed — not unreasonably — that procurement provision, overwhelmingly on the MRCA, still dominates the figures; and if cancellation charges and the like are regarded as precluding net 'savings' only in the immediate future — then plausible guesses can be made about the scale and time profile of 'savings' possible. The former might be, on average, 30-33 per cent of planned spending around 1980; the latter a distribution characterized by negligible opportunities before 1978-9. In round figures, and subject to wider margins of error than in some of the other speculations that have been made, the position might be as shown in the table.

		1976-77	1977-78	1978-79	1979-80	1980-81
£m at 1975-76	(1)					
Estimates Prices	(2)	–	–	180	200	220^c
	(3)					
	(4)				5	10
	(5)			5	10	15^c
	(6)	5	10	15	20	30^c
						275

(The superscript c denotes related sub-programmes. The attribution of early 'savings' to (6) presupposes immediate decisions on MRCA to permit 'support' economies: that of a minor and late 'savings' to (4) is explained by the fact

that (further) transport squadron cuts would not be possible
until the end of the period and - even then - could not be
extensive.)

   A slight digression is necessary at this point to overcome
a presentational difficulty. The foregoing paragraphs
indicate the main 'savings' possibilities in the 'mission
programmes' of the Defence Budget. They have been designated:
 Programme Option Modules . They are building blocks
identified for use in framing coherent Defence Programme
Options to 1980-1.   Further possibilities exist in the
budget's 'support programmes'. But support is responsive to
front-line needs.   It is not practicable, therefore, to
speculate about options in the logistic, administrative and
training 'tail' without prior knowledge of the 'teeth' it
exists to sustain.   Yet if one wishes to elucidate options
for 'savings' of a given size, in the present case $\underline{c}$. £1,000
millions at 1975-6 Estimates prices, it is impossible to
settle this - partly because options are the purpose of the
exercise, partly because it depends on what may be possible
in the support area.   Fortunately there is a way out of this
impasse.   With a contracting defence effort several factors
operate to lower the teeth/tail ratio, as expressed in
financial terms.   They include economies of scale lost in
production/repair facilities; military overheads in training,
R & D administration, welfare support; and increased
expenditure on pensions and gratuities.   The ratio is
currently around 54 : 46 per cent.   Contraction on
the scale to be considered might shift the ratio to 50 : 50
(or worse).   Certainly within a postulated spending reduction
of $\underline{c}$. £1,000 millions - on $\underline{c}$. £4,500 millions - it would not be
realistic to expect 'support' to yield much more than one-third
of the total.   (These last two statements are virtually
equivalent.)   In other words, in considering options for a
defence programme costing $\underline{c}$. £1,000 millions less in 1980-1
something like £325-50 millions would be a reasonable
expectation for the 'support programmes' contribution.
   Before briefly considering the likely sources of such a
contribution it is worth noting the corollary: that $\underline{c}$. £650
millions would have to come from the 'mission programmes'.   In
practical terms this means that one, but only one, of the
'modules' for the most costly major programmes - B, C, and E -
would not have to be taken up.   This is a glimpse of the
central conclusion of this analysis.   Its simplicity may
raise suspicions that the outcome has been contrived.   This
key factor, after all, is that the immediately preceding
discussion leads to the identification of savings possibilities
of $\underline{c}$. £275 millions (in 1980-1) on each of the three high cost
'mission programmes' - one naval, one army, and air force.   If
it were fortuitous it would be remarkable.   In fact a culinary
and cosmetic trick or two have been used, precisely in order
to highlight the central conclusion.   But they do not invalidate that conclusion, which is the logical outcome of the

argument as a whole. It derives from the weight of these
programmes in the functional costing (they account for all but
2 per cent of expenditure on combat forces); from the
balance among them; and from the constraints imposed by,
first, the effective margin for budgetary manoeuvre and,
secondly, the requirement for deliberate 'incrementalism' in
an analysis aimed at generating feasible policy alternatives.

Returning to the mainstream of the discussion, the checklist
of 'savings' possibilities can now be completed by considering
the support area.

F - L 'Support Programmes'

With 'mission programme' options open, as it were, only a
general indication can be given of where 'savings' totalling
around c. £350 millions in 1980-1 might be achieved on these
programmes (F - L in Table 1).

(F) Reserve Forces. A 20 per cent reduction in planned
outlays is the most that could be expected under this
heading. Indeed if substantial reductions were made in
Rhine Army there would be a powerful case for a corresponding expansion of the TAVR; however the financial
cost would be modest because little equipment spending
would be necessary.

(G) R & D not included elsewhere - that is, not attributed
to specific missions. A major cut-back in planned R & D
provision was a feature of the Defence Review. But it
has been shown that a lower cost defence effort necessarily
entails 'opting out' of some areas of military technology
(or taking only a share of a European - or Atlantic -
investment). Reduction in planned provision of c. 25
per cent might therefore be feasible.

(H) Training. Existing budgetary projections undoubtedly
take 'credit' for economies in training that have yet to be
planned let alone effected. But a further 20 per cent
'saving' would be a reasonable expectation.

(I) Production, Repair and Associated Facilities. This
programme comprises the Naval Dockyards (overheads) and the
Repair and Maintenance, Storage and Supply organizations
in the United Kingdom. Assessment of 'savings' possibilities is hazardous. More than one-third of the industrial
and non-industrial civil servants engaged in defence are
employed in these activities, which means that any rundown
would have to be slow. At a guess a £50 millions cut in
the 1980-1 budget target might be manageable.

(K) 'Other Support' Functions is the designation for a
portmanteau functional heading which incorporates provision

for the Whitehall organization, local administration and
communications, meteorological services, Service pensions
and family and personnel services for the armed forces in
the United Kingdom.   In framing the hypothetical costing
for 1980-1 it was assumed that outlays here would rise in
the next few years, increased pension and welfare expenses
more than outweighing modest economies elsewhere.   For
similar reasons the effective margin for manoeuvre on
this programme is probably not much more than 10 per cent
of current projections - say, £125 millions in 1980-1.

(J L) <u>Stocks/Miscellaneous Expenditure and Receipts</u>. At
a guess, in the context of contraction under consideration,
provision under these headings might be halved.

In view of the technique used to gauge the scope for
support 'savings' it would be inappropriate to speculate on
a pattern of development in the intervening period.   Suffice
it to say that few of the 'savings' could be realized at once.
For 1980-1 the assessment translates into very round figures
as follows:-

		1976-77	1977-78	1978-79	1979-80	1980-81
£m at 1975-76	(F)					15
Estimates Prices	(G)					50
	(H)		See text above			75
	(I)					50
	(K)					125
	(J L)					35
						350

The identification of <u>Programme Option Modules</u> is thus
complete.   The full checklist of 'savings' possibilities
that has been built up, on the basis of the rationales for
options elaborated earlier in the paper, is summarized in
Table 2.   The cryptic entries in the 'Remarks' column are
to serve as a reminder of the arguments of substance in the
foregoing pages so that the Table can be used as a point of
reference for final presentation of specific <u>Defence
Programme Options to 1980-81</u>   (35)   But it must be stressed
that it is upon those fuller arguments that the analysis
rests; and also that the figures in the Table 'are not
intended to be more precise than the general statements in
the text'.

Defence Programme Options to 1980-1

The concluding stage of this exercise has now been reached.
The case for developing alternatives to the planned defence
effort - and, in particular, <u>specific</u> alternatives - has
already been made.   Although an 'extensive and thorough'

policy review has recently been completed this does not mean that there can, or should, be a moratorium on debate about 'how much for defence?' At the same time the rhetoric of 'generalized objectives and numbers chosen more or less at random' is unhelpful. Arguments about imprecise alternatives are little better than no debate at all – not least because, invariably, they degenerate into dialogues of the deaf.

It is equally important to concentrate attention on plausible alternatives. This means taking the existing programme and budget as the point of departure – because policy starts here and now. Charting courses in the far oceans of the imagination is an agreeable pastime. But it ignores this reality and so is of little use in serious argument about where governments are heading. In the short run, helmsmanship is at least as important as navigation. It is for this reason that the present analysis proceeded from an overview of current budgetary projections through a deliberate procedure, considering in turn:

- The 'conceptual basis' of the existing programme, the definition of 'essential defence needs' underlying it and the judgments made about 'proper provision' for them.

- The options for change which would arise given a marginal modification of the 'conceptual basis', a more stringent definition of security priorities and a relaxation of some crucial, but vulnerable, assumptions concerning provision.

- The likely financial implications of these individual options, estimated with due regard for any limitations on the scope for budgetary choice.

This procedure has yielded the information summarized in Table 2.

The final objective, however, is development of specific and plausible programme options with a 1980-1 defence budget target in mind: namely, that level of expenditure – assessed as £3,000 millions at 1974 Survey prices, equivalent to c. £3,675 millions at 1975-6 Estimates prices – at which the United Kingdom's defence/GNP proportion would be approximately equal to that of her principal European allies, and which would also represent a 'true' reduction in defence spending

This task can now be addressed directly. The question is: in what ways can the 'building blocks' of Table 2 be manipulated to produce alternative defence programmes embodying estimated expenditure in 1980-1 of not more than c. £3,675 millions at 1975-6 Estimates prices? Or, putting it another way: what combinations of the so-called 'modules' would yield total 'savings' of c. £1,000 millions, on the same price basis?

The short answer has been hinted at already. The essential conclusion reached by the progression of argument followed is

that to bring the defence budget down to the datum level
postulated would require:

(a) Deletion from the present planned programme of the
entire provision for nuclear strategic forces and army
combat forces outside Europe, which would mean,

(A) Rundown of the Polaris force and facilities.

(D) 'No cost to UK arrangements' to be made for
garrisons in dependencies, by withdrawal or full host
territory 'contributions'.

plus

(b) Taking up the full potential for expenditure
reductions available on two of the three remaining
'mission programmes' viz.

(B) Navy General Purpose Combat Forces: by
deliberately embracing a smaller ship fleet
philosophy, involving the paying-off of large
surface units and changes in the new ship
construction programme (certainly including
striking from it provision for the second and
third ASW cruisers).

and/or (C) European Theatre Ground Forces: by reducing
force levels , including a rundown of BAOR by
some 25,000 men to an eventual strength of c. 30,000.

and/or (E) Air Force General Purpose Combat Forces: by
setting out to furnish a lower cost air force in
the 1980s and beyond; that is, a force equipped
to cover less than the full spectrum of tactical
air missions, without the MRCA but with more aircraft
for 'battlefield' roles.

plus

(c) Achieving proportionate reductions in outlays on
'support programmes', which is to say, realizing all the
'savings' in support activities that the general scale of
contraction should allow - on a rough estimate of teeth/
tail relationships during such a process. (These might
amount to around one-sixth of present planned expenditure
by 1980-1.)

That is what would be necessary, in practice, to 'keep
down with the Europeans' (in terms of defence/GNP proportions)
on realistic growth rate assumptions and speculations about
allied defence budget trends - on the assumption, naturally,
that the logic of the procedure adopted in this exercise is
accepted.

168

To express this conclusion more fully, in the form of specific, plausible, but individual Defence Programme Options, all that is necessary is to make explicit the permutations on the three main 'mission programmes'. In a lower-cost defence programme of this order, it has been asserted that the pressure of priorities would effectively preclude continuing provision for the Polaris force and for garrisons outside Europe. But in relation to the three 'heavyweight' Service programmes there would be scope for discretionary choice. It would be possible to:

I   Adopt the Smaller ship Fleet philosophy and make reductions in European Theatre Ground Forces (ETGF) while planning to maintain a 'full spectrum' Royal Air Force, complete (in time) with nearly 400 MRCAs.

OR

II   Elect to move to both a Smaller ship Fleet _and_ a Lower cost Air Force, thus avoiding the necessity for reducing ETGF levels.

OR

III   Preserve the existing naval programme, with its Balanced Fleet philosophy ('ships of high quality'), allowing the burden of adjustment to fall on the Royal Air Force, notably RAF Germany, and on BAOR and United Kingdom Land Forces.

Options I and II would be 'available', so to speak, with or without a somewhat diminished submarine force; Options II and III with a larger or smaller complement of maritime, transport and tanker aircraft.

The composition of each of these principal options — which are implicit in the summary conclusion just presented — is indicated in Table 3, which also shows the breakdown of the £1,000 millions 'saving' in 1980-1 attributable to each.

It will be apparent that there are at least four other options implicit here, namely, those which result from the remaining permutations possible on the 'heavyweight' Service programmes. These do not 'hit' the 1980-1 budget target: one would yield greater 'savings', the rest a good deal less. But they are none the less worth stating, affording as they do a general indication of the broader range of choice.

In principle, if all the 'rationales for options' were thought especially compelling it would be possible to:

IV   Opt for the Smaller ship Fleet _and_ reduced ETGF levels _and_ a lower-cost Air Force — entailing an all-round diminution of military stature.

If, on the other hand, reconsideration of priorities and

'proper provision' were thought possible for only one service's roles, the options would be:

V the Smaller ship Fleet.

OR

VI the reduced ETGF levels.

OR

VII the Lower-cost Air Force.

The implications of these options cannot be elaborated here. It must suffice to say that, if specific alternatives are indeed the best material for well-directed debate, the range of choice is wide enough for plenty of discussion. (36)

One last presentation is required. Although the 1980-1 budget target has been the focus of the exercise, the final aim - as stated earlier - is the identification of Programme Options to 1980-1 which might be regarded as alternatives, for discussion, to existing defence plans. How, then, do the possibilities reviewed compare with the present programme - the 'benchmark' programme? The answer is in Table 4. This shows, for both the principal and subsidiary options, the time profile of expenditure implied by the 'savings' estimates. The expenditure series have been derived by deducting from the 'benchmark' budgetary projections the year-by-year 'savings' attributable to each option. (For 'mission programmes' these are, of course, the possibilities summarized in Table 2. For 'support programmes' a distribution of proportionate reductions has been assessed by simple rule of thumb. (37))
In effect, Table 4 displays the likely budgetary implications of a selected array of policy choices, cast in the form of specific 'alternative futures'. These have been developed from a definition of the United Kingdom's core security interests more stringent than that which underlies the present programme; and following a revision of the assumptions about appropriate provision implicit in that programme. They are submitted, in conjunction with the detailed arguments behind the options, to elucidate the kind of change in security arrangements that expenditure reductions of the given scale would require. If debate about the scale of the future defence effort is to have sharpness and definition - specifically, if lower cost defence programmes, to reduce the British effort to the European norm, are to be realistically appraised - the agenda for discussion should be framed, it is contended, on a basis such as this.
For all practical purposes that completes the exercise. A summary of the main points of the paper and brief, concluding observations follow Table 4.

170

Table 2 PROGRAMME OPTION MODULES
ESTIMATED 'SAVINGS' POSSIBILITIES ON MAJOR PROGRAMMES/SUB PROGRAMMES
£ millions at 1975-76 Estimates Prices

Major Prog.(Sub)	1976 /77 £m	1977 /78 £m	1978 /79 £m	1979 /80 £m	1980 /81 £m	Remarks
A	–	50	50	75	<u>75</u>	Rundown of Force & facilities.
B						
(1)	5	10	15	20	20^a	Rundown of Provision (Ships).
(2)	5	10	30	40	50_a	Stretch SSN Bldg: Rundown SS Ops.
(3)	–	50	100	100	150^a	Cancel Cs. 01?/02/03: Stretch D & F Bldg.
(4)	–	–	–	–	10^a	Reduced Afloat Support Reqt.
(5)	–	10	30	40	50^a	Cancel Sea HARRIER Some Helis.
(6)	–	–	–	–	<u>–</u>	Reduced Costs of Carriers/Amphibs.
					275	
C						
(1)	–	50	100	150	200^b	Rundown to c.30,000: equipment
(2)	–	–	–	–	$–^b$	purchases adjusted.
(3)	–	–	25	50	$\underline{75}^b$	Reduced roulement reqt.
					275	
D	–	5	10	20	<u>25</u>	No Cost to UK by 1980-81.
E						
(1)	–	–	180)	200)	$220^c)$	Cancel MRCA AD Var: Cancel MRCA
(2)			)	)	)	(Common) & prolong JAGUAR & HARRIER
(3)			)	)	)	prod'n ? Arty & SAM purchases (See C(1))? ? F16 or A10 purchases?
(4)	–	–	–	5	10_c	Tankers and Other a/c
(5)	–	–	5	10	15_c	
(6)	5	10	15	20	$\underline{30}_c$	Reduced force level.
					275	
F				(15))		
G				(50))		
H				(75))	(350)	Reductions equal to c. 50% of
I				(50))		'achievement' in mission
K				(125))		programmes.
J,L				(35))		

Table 3 PROGRAMME OPTIONS TO 1980–81
OPTIONS FOR DEFENCE BUDGET TARGET
(Target 'Savings': £1,000 millions at 1975–76 Estimates Prices)

			Def. Budget Saving 1980–81		
			£m		
Option I	A.	Rundown Polaris & Facilities	75		
	B.	Smaller ship Fleet	275		(225)
	C.	Reduced ETGF levels	275		
	D.	No cost to UK arrngts.	25		
		Mission programmes	650		
	F.L.	Proportionate Reductions	c. 350		
				c. 1,000	(925)
Option II	A.	Rundown Polaris & Facilities	75		
	B.	Smaller ship Fleet	275		(225)
	D.	No cost to UK arrngts.	25		
	E.	Lower cost Air Force	275		(265)
		Mission programmes	650		
	F.L.	Proportionate Reductions	c. 350		
				c. 1,000	(900)
Option III	A.	Rundown Polaris & Facilities	75		
	C.	Reduced ETGF levels	275		
	D.	No cost to UK arrngts.	25		
	E.	Lower cost Air Force	275		(265)
		Mission programmes	650		
	F.L.	Proportionate Reductions	c. 350		
				c. 1,000	(985)

Options IV
        V
       VI   See Text
      VII

Note: Figures in parentheses identify the effects of excluding 'non-related
items' i.e. submarine reductions in (B) and tankers and other aircraft
reductions in (E).   See Table 2 and Text pp.159–62.

Table 4   <u>PROGRAMME OPTIONS TO 1980-81</u>
£m at 1975-76 Estimates prices

Programme	1976-77	1977-78	1978-79	1979-80	1980-81
Benchmark	4525	4675	4675	4675	4675
Option I	4515	4425	4150	3925	3675
'Savings'	10	250	525	750	1000
Option II	4510	4475	4075	3775	3675
'Savings'	15	200	600	900	1000
Option III	4520	4525	4150	3775	3675
'Savings'	5	150	525	900	1000
Option IV	4510	4400	3925	3575	3275
'Savings'	15	275	750	1100	1400
Option V	4515	4500	4350	4225	4100
'Savings'	10	175	325	450	575
Option VI	4525	4550	4425	4225	4100
'Savings'	-	125	250	450	575
Option VII	4520	4600	4325	4200	4100
'Savings'	5	75	350	475	575

Notes:

(i)   The 'fully inclusive' options have been used for this
tabulation.   See text p. 169.

(ii)   For Options IV and V/VI/VII suitably adjusted values
have been used for the proportionate reductions in
support: the 'savings' on major programmes (A) and
(D) are included.

(iii)   For the purposes of this Table only a year-by-year
distribution of support reductions has been estimated.
See text p. 170.

Summary and Conclusions

It would be impracticable, and tiresome, to try to precis all the material in this paper. But there is some merit in trying now to isolate the spinal column of argument that runs through it; and there are one or two final observations to be made.

The essential reasoning may be summarized as follows:

(a) Following the 1974 Defence Review the Government, for its own good reasons, adopted a defence programme entailing annual budgetary outlays on broadly the existing scale to 1980 and beyond. This means that the financial weight of security provision will continue for some time to bear more heavily on the United Kingdom than on other NATO powers in Europe.

(b) To equalize 'burdens' over a period of years would require a further reshaping of the defence effort aimed at achieving 'savings' — measured against present plans — of c. £1,000 millions by 1980-1. That would imply a contraction of some 20 per cent compared with the position now.

(c) Deciding whether, and if so exactly how, to embark on such a course requires account to be taken of the external and internal costs and consequences; the existence of potential benefits, arising from the alternative uses of resources, is self-evident. Neither the international political and strategic repercussions nor the domestic employment (and other economic) implications can be adequately assessed in general or in abstract terms however. It is necessary to develop and appraise quite specific policy alternatives: to generate, in other words, (costed) programme options.

(d) The existing defence effort reflects a certain idea of military stature, a structure of security priorities and numerous judgments about what constitutes proper provision for defence needs. Since 'reshaping' begins with what exists, rationales for options are best derived from re-considering these bases. If, but only if, the United Kingdom could accept a somewhat lower place in the pecking order of military power and influence, marginally revise her notions of essential defence needs, and most important, reconsider assumptions about what amounts to 'proper provision', reasonable guidelines for possible reshaping could be established. (The crucial judgments about provision in this connection are those concerning the immutability of ground force levels in Europe and the necessity of having both a 'balanced' navy, with 'ships of high quality', and a 'balanced' air force, equipped to handle the 'full spectrum' of tactical air missions.)

174

(e) The margin for budgetary manoeuvre varies across the range of military functions. But, were the bases reconsidered in this way, 'savings' possibilities would arise in all major mission areas. On revised assumptions provision for the navy's bigger ships, a slice of Rhine Army and the RAF's MRCA might be deleted from the expenditure projections.

(f) At the reduced level of spending necessary to equalize burdens pressure on priorities would rule out maintaining the Polaris force and garrisons in overseas dependencies. To achieve the 'savings' level sought (by 1980-1) would also require two of the three Services to accept a much diminished position. A £3,675 million budget in 1980-1 could accommodate either a fully 'balanced' fleet or an undiminished BAOR or a 'full spectrum' air force, but not two of these and assuredly not all three.

This last is the main conclusion of the essay.

The final observations to be made are these. First, the perennial paradox of defence budgeting shows up in this analysis, as perhaps inevitably it would. Each of the principal programme options implies a major transformation in two of the three Services and in the United Kingdom's general military dispositions. A reduction of the defence effort by some 20 per cent (in financial terms) entails quite fundamental change.

Secondly, the results of the inquiry underline the necessity of 'the options approach'. In the detailed argument of earlier sections of the paper it has been made clear that many alternatives, which are feasible in the sense that choice is not immutably constrained, would be realizable only in a context of general alliance reappraisal – unless the United Kingdom were prepared to ride out profound reverberations. But there are significant differences among the options in the scale and visibility of the repercussions they might provoke, or in the extent and sensitivity of the general reappraisal that might be required to validate them. All the options are disruptive, but some might be more disruptive than others. Similarly in the domestic setting: each of the 'alternative futures' sketched would pose far-reaching adjustment problems for particular industries and regions, for individual servicemen and civilian employees. But there are significant differences among them in the likely incidence of the problems, in time and space and sector. Choice should take account of these.

Third, and last, it will bear repeating that this exercise has been undertaken to elucidate the kind of change significant reshaping might require. There is no prescriptive content. Rather an array of policy choices has been displayed. It is the irony of exercises of this sort that their outcomes are agenda for discussion: the end is a beginning.

September 1975.

Postscript

(May 1977)

to

DEFENCE PROGRAMME OPTIONS TO 1980-1

As originally drafted in the autumn of 1975, the purpose of
this paper was a straightforward one.   It was to illustrate
the nature of the choices which a British government would
face were it decided to move to a lower-cost defence effort
than that embodied in the programme framed after the 1974
Defence Review.   The argument was designed specifically to
elucidate the implications of one particular judgment: that
the United Kingdom's annual provision for defence should, by
1980-1, have been brought down to a level where the defence
budget represented no higher a proportion of Gross National
Product (GNP) than was the norm among her major European
allies.

Events in the last eighteen months have not invalidated
the main themes of the analysis.   Budgetary projections
for defence have been revised downwards following the 1975
and 1976 public expenditure survey 'rounds' (and are undergoing further scrutiny in the current 1977 'round').   But
the official position is that the 1974 Defence Review
decisions remain the basis of the programme.   Formally at
least, the Government intends that the reduced funding which
continuing economic distress is making necessary in the later
1970s will be 'made good' by greater allotments to defence in
the early 1980s.   Recent adjustments affect the phasing of
expenditure but not the underlying evaluation of defence's
place in national priorities or the structure of defence
priorities themselves.   There has been no explicit
abrogation of the broad conclusion which guided the Defence
Review: that by the mid-1980s the United Kingdom's defence/
GNP proportion should stand at around 4.5 per cent, less than
at present but still somewhat higher than her European
allies' allocations.

It follows that the issue of lower-cost alternatives to
present plans is still of interest;   and the implications
of adopting 'equalized defence/GNP proportions' as a target
retain a particular significance.   Furthermore, it remains
broadly true that the scale of adjustment to current intended
spending which would be necessary to bring about 'equalization'
is as outlined in this paper.   And the 'nature' of the
defence programme change which would be necessary to effect
it, if the obvious — but increasingly unsatisfactory — device
of equi-proportionate 'cuts' were rejected, remains roughly
as described in the main text.   That is to say, at the kind
of lower level of expenditure which would be required to

ensure equalized proportions, the United Kingdom could not
cling to current judgments and assumptions about what
constitutes 'proper provision' for her security needs. In
particular maintenance of existing inter- and intra-service
balances would no longer be feasible. It would be necessary,
in fact, to shift to a defence posture based on, for example,

I   reduced naval provision and a diminished ground
    forces' contribution to NATO while maintaining an
    air force equipped for the 'full spectrum' of
    tactical air missions;

OR

II reduced naval provision and a lower-cost air force,
    enabling the European theatre ground force level to
    be maintained;

OR

III an undiminished maritime contribution to the
    Alliance but a reduced commitment of ground and
    air forces (in the United Kingdom and Germany).

And with the sort of reduced defence effort involved preservation of a strategic nuclear retaliatory capability and
retention of remaining overseas garrisons would hardly be
practicable.

In these major respects, then, the paper can stand as
drafted more than eighteen months ago. But developments in
the intervening period do make it appropriate to revise and
qualify certain aspects of the analysis. The aim of this
postscript is to present relevant supplementary argument.

The material consists of three parts. The first of these

(a) notes how the post-1974 Defence Review programme
     has been revised in successive public expenditure
     survey 'rounds',

(b) defines the new benchmark against which options for
     change must now be measured,

(c) examines certain implications of these changes, of
     which the most obvious are that the new benchmark
     programme has a rather different configuration than
     the original and the earliest year now amenable to
     any influence by policy choice is 1978-9.

The second part is devoted to updating the costings of the
programme options. The argument of the paper was elaborated
using data from the 1975-6 Estimates; all the key values
were, therefore, expressed at 1975-6 Estimate prices. Because
of the pace of inflation recently these numbers bear little
relation to those now current in defence policy discussions,

which are based on the expenditure information in the
Statement on the Defence Estimates 1977 (Cmnd 6735), i.e.
on 1977-8 Estimate prices. Some elementary conversions
have been made, putting values on the same present-day
prices basis as the new benchmark programme. The third
and last part of the postscript contains a collection of
loosely-related observations on the specific programme
options developed in the paper proper. They record matters
on which I would now give a rather different emphasis to
the discussion if I were composing the paper afresh.

Programme Revisions and the new Benchmark (1)

The 1974 Defence Review decisions led to adoption of a
defence programme entailing budgetary allocations at a more
or less constant level throughout the later 1970s and early
1980s. At 1975-6 Estimates prices the 'steady state' to be
attained from 1977-8 involved annual outlays of c. £4,675
millions, as originally assessed. (See p.173, Table 4.)
During the 1975 public expenditure survey 'round' the
budget projections underwent a minor upward revision for
technical reasons (based on new information on the likely
timing of outlays). More important, the results of this
scrutiny included decisions to reduce intended provision over
the period 1977-80, bringing down the de facto budget
ceiling for these years to c. £4,530 millions at 1975 Survey
prices (which for all practical purposes equate with 1975-6
Estimate prices).
Expressed at 1976-7 Estimates prices - or, what is
virtually the equivalent, 1976 Survey prices - this 'steady
state' level was c. £5,600 millions. The actual Estimate
for 1976-7 was £5,632 millions. At this juncture (February
1976) the aim was to achieve the levelling-off at the
fractionally lower expenditure plateau from 1977-8. However
the reduction below the post-Defence Review spending profile
was represented as a temporary phenomenon. It was anticipated
that after 1980 the originally-planned expenditure level might
be resumed.
The annual appraisal of public spending undertaken in 1976
brought yet further downward revisions to the target Estimates
for 1977-8 and 1978-9. It was also intimated that in the
1977 'round' the projections for 1979-80 and after would be re-
examined. The 1977-8 and 1978-9 'cuts' - £200 millions and
£230 millions respectively, at 1976 Survey prices - brought
the intended expenditure totals for those years down to
c. £5,400-5,450 millions on this price basis.Translated into
1977-8 Estimates prices / 1977 Survey prices this corresponds
to a level of c. £6,275-6,325 millions. The actual Estimate
for 1977-8 was, in fact, £6,329 millions and the target
Estimate for 1978-9 stands at £6,275 millions   (With
inflation continuing to run at high levels the actual 1978-9

Defence budget will be higher than this, of course.   If no
further paring takes place in the current public expenditure
survey the actual Estimate is likely to exceed £7,250
millions, which is nearly twice the c. £3,700 millions (at
1974 Survey prices) which was the initial Defence Review
datum!)

The Government has been at pains to stress that the 1976
programme revisions, like the previous year's, are solely a
reflection of the short-run economic outlook and of the fact
that recovery from recession has been slower than expected. (2)
Accordingly it has emphasized that the defence
programme continues to rest on the framework of priorities
set in the 1974 Review.   The inference must be that higher
expenditures are envisaged in the later years of the
Ministry of Defence's 10-year Long Term Costings (i.e. for
the early and mid-1980s).   If the intention is to make good
in some way the 'below par' funding that has had to be
accepted, a rising trend from c. £6,550 millions to, say,
£7,200 millions or thereabouts (at 1977 Survey prices) must
be planned for.

What, then, is the pattern of spending for the next few
years which present plans foreshadow?   Expressed at the
1977-8 Estimates prices / 1977 Survey prices which are the
currency of this year's discussions on defence expenditure
statistics, the projections - the new benchmark against
which programme options need to be measured - are as follows,

	1977-8	1978-9	1979-80	Average 1980-1/1983-4
New Benchmark	6,329 (actual)	6,275 (target)	6,550 (under review)	6,800 (provisional)

This profile for projected spending corresponds to that
intimated in a recent Ministry of Defence submission to the
Select Committee on Expenditure's Defence and External
Affairs Sub-Committee (See HC 254 (1976-7), Report, page x.)
The £6,800 millions average budget level for the early 1980s
is consistent with the presumption that the current programme
incorporates target Estimates for the mid-1980s - albeit
highly provisional ones - which lie around £7,200 millions
(at 1977 Survey prices).   (See previous paragraph.) (3)

In Table P.1 below the main consequences of these several
revisions for the defence programme and budget are summarized.
The first line of the table is the benchmark against which the
discussion in the main text was conducted.   (See Table 4 on
p.173.) The final line is the new benchmark against which the
effects of any programme options for the future must now be
measured.

What are the main implications of all this for the analysis
of alternatives to present plans as developed in the paper
itself?   The first point to note is that the configuration of
the new benchmark programme is strikingly different from that

used as a basis for the main discussion. As Table P.1
shows, in place of a sequence of budgetary projections
running level from the later 1970s what is now envisaged
are lower expenditures for a couple of years to be followed
by rising outlays, in real terms, to regain the post-Defence
Review programme values. Whether this means that the
margin for budgetary manoeuvre in the early 1980s should be
considered greater or less than hitherto is an open question.
In my judgment the significant inference is that the kind of
programme change elucidated in my original analysis must now
be regarded as the sort of thing which would be required to
stabilize the defence budget in the early 1980s (rather than
actually to reduce it). The corollary is that the achievement
of 'equalized proportions' in such circumstances would come
from maintaining a 'steady state' defence budget in a setting
of increased growth rather than from the establishment of a
reducing trend in a slow (or no) growth context.

Secondly, and related to this of course, the passage of
time has meant that options for change taking initial effect
in 1976-7 and 1977-8 have passed beyond recall. Indeed,
since the target Estimate for 1978-9 already reflects the 1976
public expenditure survey 'round' revision (the precise
effects of which are only now being worked out), it is possible
that this year's projection too cannot be altered to any
significant extent. Thus the year-by-year incidence of the
'savings' that might accrue from adoption of alternative
programmes would be two or even three years later than
originally assessed. Put another way, even the immediate
exercise of major policy choices would have significant
effect only in the early 1980s; the only worthwhile debate
now concerns Defence Programme Options to 1983-4 (and beyond).

This being so, a further problem arises: the matter of
their incidence apart, are the budgetary 'savings' attributed
to the costed options reviewed in the paper still valid?
Obviously the actual values require up-dating to bring them
to the same prices basis as the new benchmark; and that is
done in the next section of this postscript. What is at
issue here is: has time foreclosed any options or eroded
the 'savings' which certain choices might have permitted?
On this question two points must be made,

(a) A second ASW cruiser has been ordered and is under
     construction on the Tyne; the restructuring of Rhine
     Army is proceeding and with it the disposition of the
     four new-style divisions; the MRCA/Tornado programme
     has entered the production phase in earnest with the
     ordering of a second batch of the aircraft. In each
     of the major mission programme areas, therefore, the
     political and economic pressures against disruptive
     change have been strengthened. And even if plans
     were to be modified, cancellation/compensation costs
     and nugatory expenditure would be higher than
     eighteen months ago.

Table P.1.

PROGRAMME REVISIONS 1975-77
£m on successive prices bases

| Programme | Financial Years | | | | | |
	1978 -79	1979 -80	1980 -81	1981 -82	1982 -83	1983 -84
1. Original Benchmark (1975/76 EP)	4675	4675	4675	(4625)	(4650)	(4675)
		avge =		4650-4675		
after 2. 1975 Revisions						
- techincal	4723	4694	(4700)	-	-	-
- Survey 'cuts'	4530	4530	(4530)	-	-	-
at 1 April 76						
3. Revised Programme (1976-77 EP)	5600	5600	(5600)	(5600)	(6000)	(6200)
after 4. 1976 Revisions						
- technical	5633	5636	5636	-	-	-
- Survey 'cuts'	5403	5636	5636	-	-	-
at 1 April 1977						
5. Revised Programme (1977-78 EP)	6275	6550	(6550	(6550)	(6900)	(7200)
		avge =	c. 6800			
New Benchmark (See Table P.2)		6550	6550	6550	6900	7200

Notes

Figures in parenthesis are my estimates; all other figures in lines
1-5 have a documentary sources, viz.

1. Cmnd 5976, p.23 converted to 1975-76 Estimates prices (to nearest
    £m25)

2. Public Expenditure to 1979-80 Cmnd 6393, pp.16-17 and Cmnd 6432,p.31

3. Cmnd 6432,p.31 converted to 1976-77 Estimates prices (to nearest £m25

4. The Governments Expenditure Plans - Vol.II,Cmnd 6721-II, pp.2-3.

5. HC 254 (1976-77),p.x,converted to 1977-78 Estimates prices
    (to nearest £m25).

The conversion factors between successive price bases are as follows:
        1974-75 EP (74 SP) to 1975-76 EP (75 SP) : 1.23026
        1975-76 EP (75 SP) to 1976-77 EP (76 SP) : 1.23346
        1976-77 EP (76 SP) to 1977-78 EP (77 SP) : 1,16192

(b)   Consistent with the continuing commitment to the
      nominal order of battle defined in the 1974 Review,
      the direct impact of the 1975 and 1976 programme
      revisions has been on 'support' expenses. The
      official line has been that the adjustment to
      expenditure plans involves 'trimming the tail without
      blunting the teeth' and that no diminution of the
      United Kingdom's contribution to NATO's front line
      is involved. The claim is dubious; but that is
      not of immediate concern. What is relevant for
      present purposes is this: because of the attention
      paid to support programmes these past two years it
      is now unlikely that given 'cuts' in planned
      spending on mission programmes would make possible
      the proportionate reductions in support assumed in
      the main text.

Thus, in addition to up-dating the 'savings' values
assigned to the paper's costed options to put them on a
current prices basis, it is necessary to modify somewhat
the actual 'savings' calculations themselves. My assessment
is that the expenditure reduction (in real terms) which would
follow from adoption of the programme options I have considered
is now slightly less than when they were first elaborated over
eighteen months ago.

Option costings: recasting and revaluation

At the beginning of this postscript it was stated that the scale
of adjustment to current intended spending necessary to bring
about equalized defence/GNP proportions remains broadly as outlined in the main text; and that the nature of the programme
changes required to effect such adjustment remains as described.
But how must the actual programme option costings be recast
because of recent programme revisions and their implications?
   To take the assessment of attainable budgetary 'savings'
first. So far as their year-by-year distribution is concerned,
higher cancellation charges and other unavoidable costs reduce
the 'savings' realizable on mission programmes in early years;
and, in practice, it would be unreasonable to expect lower outlays before 1979-80. So far as their size is concerned, reassessment is more difficult. It would serve no useful
purpose - and might create an impression of spurious accuracy -
to calculate afresh 'savings' estimates for the individual
modules from which the principal options are composed. (See
Table 2, p.171.) It seems preferable to make a general allowance for reduced 'savings' possibilities by a 10-20 per cent
mark-down of the values of the time-streams of 'savings'
attributed to the main options (cf. Table, 4, p.173).
   Revised assessments of these time-streams - incorporating
both the distribution and size adjustments - are tabulated below;
the figures are at 1975-6 Estimates prices to facilitate
comparison with the initial calculations.

182

£m at 1975/76 Estimates prices	Year 1 1979-80	Year 2 1980-81	Year 3 1981-82	Year 4 1982-83	Year 5 1983-84
Option I 'Savings'	5	200	375	625	900
Option II 'Savings'	10	150	450	700	900
Option III 'Savings'	—	100	375	700	900

Needless to say these values are rough-and-ready estimates. They should be regarded as the centres of ranges rather than as 'point' estimates. (It should also be noted that exception could be taken to the assumption that even immediate decisions for programme change could not be given effect until 1979-80.)

To permit measurement against the new benchmark — as established in the previous section and set out in Table P.1 on p.181 - it is necessary to convert the 'savings' to the same prices basis (1977-8 Estimates/1977 Survey prices). There is no reason to suppose that the general conversion factors which enable the defence programme and budget as a whole to be translated from one prices basis to another should not be applied to these assessments. The calculation is, therefore, a straightforward one. The results are given in Table P.2 which is an exact counterpart to Table 4 of the paper proper (see p. 173). That is to say, it sets out for each of the principal options (I, II and III) the time profiles of expenditure implied by the 'savings' estimates. As for the earlier table the series have been derived by deducting from the (new) benchmark budgetary projection the (recast and revalued) yearly 'savings' attributable to each option.

Table P.2

PROGRAMME OPTIONS TO 1983-84
£m at 1977-78 Estimates / 1977 Survey prices

Programme	1979-80	1980-81	1981-82	1982-83	1983-84
New Benchmark (From Table P.1)	6550	6550	6550	6900	7200
Option I 'Savings'	6543 7	6250 300	6000 550	6000 900	5900 1300
Option II 'Savings'	6535 15	6325 225	5900 650	5900 1000	5900 1300
Option III 'Savings'	6550 —	6400 150	6000 550	5900 1000	5900 1300

Note

The conversion factor 1975-76 EP (75 SP) to 1977-78 (77 SP) which has been used to revalue the 'savings' estimates is c. 1.433: figures here are rounded.

Among other things this presentation demonstrates strikingly what later implementation of the programme options elaborated in this paper would entail. In order to 'compensate' for the temporarily low funding provision for 1977-80 the Government's present plans envisage rising budgets in the early 1980s and not the more or less level trajectory originally plotted for post-Defence Review expenditures. Exercising options for change along the lines developed here would thus have the effect of stabilizing the defence budget in the early 1980s - at a level which I estimate at around £6,000 millions (at 1977 Survey prices) - rather than lowering it in the later 1970s (which would have happened had they been adopted from, say, 1976-7). As noted earlier the corollary is: achievement of the equalized proportions target, which is the specific focus of this exercise, would emerge from stable outlays in the growth setting which the early 1980s are expected to provide, instead of from falling expenditures in the low (or no) growth situation of the later 1970s.

This being so, two other observations are worth making. First, an alternative reading of Table P.2 is: if, in the early 1980s governments decide that they are not prepared to allot to defence sufficient resources to consummate existing plans but instead elect to 'bring defence expenditure under control' at the existing level of current spending, it will require changes of the kind elucidated in the main paper to achieve such a goal. At the moment, for their own good reasons, neither Ministers nor officials are willing to acknowledge the possibility, of course. Be that as it may, on each recent occasion when governments have actually had to choose whether or not to endorse defence programmes calling for significant yearly increases in expenditure (in real terms) i.e. in 1964 and again in 1973 and 1974, they have in fact opted for 'defence reviews' and for the imposition of a de facto budget ceiling. (4) Moreover, as I have argued else- where, examination of the United Kingdom's experience since 1945 reveals a clear preference, on the part of both Conservative and Labour governments, for allotting the dividends of growth to private uses and civil public uses. Certainly over the last 20-25 years the scale of defence provision, in real terms, has been remarkably constant taking one year with another. (5)

In a sense the second point is simply a gloss on the first. It is also very much a personal observation. Having regard to this recent historical record of budgeting for defence my feeling is that, when economic growth does get underway in the United Kingdom, governments will not

184

in the event choose to return to the defence expenditure path
sketched after the 1974 Review. The urge to relax checks on
consumption, sustain investment and restore the mid-1970s'
cuts in civil public expenditure programmes will outweigh
any inclination to fund rising defence budgets. Ministers
will again be attracted by the idea of holding military
spending at its existing level; and some figure like
£6,000-£6,500 millions (at 1977 prices) will become a datum
for the future programme as '£2,000 millions (at 1964 prices)'
and '£3,800 millions (at 1974 prices)' have been in the
recent past. What this means is that the present exercise
in developing programme options, though initiated simply to
show the kind of change which 'equalizing proportions' might
require and intended to be neither prescriptive nor predictive,
may well turn out to have been a timely undertaking. For it
illustrates the tough choices which future British governments
may have to face anyway, whether they think that 'keeping
down with the Europeans' in terms of defence/GNP proportions
is a sensible guideline for settling the size of the defence
effort or not.

   If this reading of the situation is correct, or even half-
correct, it is all the more important to ask: are the options
analysed in the paper sound in themselves? Do my original
substantive arguments - about rationales for a smaller ship
fleet, a lower-cost air force and so on - require revision
and updating as well as the budgetary calculations?

Option Rationales: Shifts of Emphasis

Needless to say there are several points where I would now
shift the emphasis in my original exposition. In the
remaining paragraphs of this postscript I review them
briefly. At the outset however I would stress that the
essential features of the analysis remain unaltered. In
particular, on re-examining the argument after a lapse of
more than eighteen months, I see no reason to qualify
certain basic propositions, namely,

   (a) that discussion of possible lower-cost defence
        efforts in terms of specific, costed options is
        preferable to debate on generalized objectives
        or figures chosen more or less at random,

   (b) that the current programme and budget is the only
        worthwhile point of departure for the examination
        of alternatives,

   (c) that the ideal, and indeed the only logical, way
        to identify those specific programme options which
        are worth serious analytical attention is to look
        critically at the conceptual basis of existing
        provision for defence (including the judgments and
        assumptions underlying it),

and, finally,

(d) that, for the United Kingdom at the present juncture,
     the options which invite elaboration are programmes
     which explicitly acknowledge the nation's reduced
     circumstances.

Nor do I think it necessary to modify the main thesis
developed from these foundations.   I still believe that the
pertinent questions for British defence include whether nuclear
strategic forces and remaining garrisons outside Europe are
'essential' to the national order of battle in the 1980s;
whether European theatre ground force levels should be
considered immutable;   and whether it is feasible (or desirable) to regard preservation of the structural 'balances' of
the individual Services as a planning imperative. (6)

My supplementary observations are therefore related to the
texture of the analysis rather than its structure;   for the
most part they concern matters of nuance and detail, that is
all.   The main points are summarized below under five
headings.

1. Nuclear Strategic Forces

It is not really admissible to gloss over the question of
the future of nuclear strategic forces, as I did originally,
with the bald assertion that in a defence budget some 16-20
per cent below the level planned pressure on priorities
would simply squeeze out continuing provision for a retaliatory
capability.   (See pp. 149 and 175.) Indeed, because of growing
evidence that the new United States' Administration may choose
to side-step many of the questions posed by the Soviet Union's
intermediate range ballistic missile (IRBM) forces should
they threaten to obstruct progress to a SALT II agreement,
the case for retention in the military balance of some
European counterweight to these forces is now stronger than
at any time in the recent past.   This is not the place for a
report on the burgeoning debate on 'Euro-deterrence'.   Nor
is it possible to discuss the proposition that Europeans need
to become engaged in existing (or new) arms control negotiations in which nuclear capacities lying outside the ambit of
current exchanges could be considered. (7)   Suffice it to
say that, while these matters are at issue, identifying that
approach to the future of the nuclear strategic forces which
would best serve the United Kingdom's security interests is
anything but simple.   It follows that the implication of my
original argument - that any lower-cost defence effort based
on a stringent definition of 'needs' would have to exclude
completely provision for the kind of capability the Polaris
force represents - is misleading, to say the least.

2. European Theatre Forces

In my discussion of European force levels, leading to the
conclusion that existing dispositions in the Allied Forces
Central Europe (AFCENT) area should not be regarded as
immutable, I made passing reference to 'inefficiencies
associated with inappropriate stress on staying power' and
'failure to exploit technological opportunities favouring
the defence'. (See p.161). I would now be inclined to lay
greater emphasis on these points. During the last eighteen
months or so criticism of both these aspects of NATO doctrine
and concepts of operations has been mounting. One of the
most biting indictments of the inadequacy of the Alliance's
capacity to withstand initial attack has come from Senator
Sam Nunn; and his case is compelling. (8) Both American
and European analysts have developed important ideas on how
techniques of precision and discrimination might be exploited
to enhance deterrence via increases in the effectiveness (and
credibility) of conventional defence capabilities and new
emphasis on 'varying response with circumstances'. Many of
their arguments are persuasive. (9) All this lends weight
to the thesis that, were it not for the resistance of those
committed by tradition and instinct to preserving their
expertise in familiar areas, it would probably be possible to
devise for the AFCENT area a defensive posture no less
effective than the present one but which could be sustained
with _fewer_ troops and at _lower_ cost.

3. 'Balanced Forces' : Tactical Air

In that part of the analysis of the 'balanced forces' question
dealing with air force mission priorities I noted that
committing resources to aircraft for deep strike/interdiction
missions (including offensive counter-air operations) was
beginning to look less and less sensible, especially when at
the expense of provision for support for ground forces in the
meeting engagement or penetration battle. (See pp. 153-4).
Highly controversial though the topic is, I believe that the
main thrust of that discussion to be sound. But I would not
now argue that 'close air support and local air superiority
are the priority missions' (p. 154). I would say instead
that capabilities for battlefield interdiction and for air
combat over the battlefield, with local superiority as an
aspiration, are what matter most. Both soldiers and airmen
are becoming more and more dubious about the feasibility of
intimate air support of ground troops in the battle area
itself. Air power can best help friendly forces, the
argument runs, by performing shallow or 'battlefield' interdiction to impair the protagonist's ability to sustain or
replace the formations he has already committed to the
engagement. Local air superiority is no less desirable
than hitherto; armies ask that air forces 'keep the enemy's

air off our backs' as well as preventing or delaying the
appearance of the next echelon of ground troops.   But it
may be more difficult to achieve.   However, because of the
costs and difficulties of offensive counter-air (i.e. destroying or harassing the opponent's aircraft on their airfields),
it seems that in the future essentially defensive air combat
may be the most effective, perhaps the only feasible, way of
opposing an enemy's air capabilities.   None of this affects
my principal conclusion about tactical air power for the
selection of programme options: that it is now out-of-date
to assume that the only air forces worth having are 'balanced'
forces, capable of undertaking the full spectrum of traditional
missions.   But the shift of emphasis in the argument is none
the less important for the consideration of appropriate air
force structures for the 1980s and beyond.   The current
programme and budget incorporates provision for an aircraft,
the Tornado, which is optimized for a mission that is
becoming less and less important (on this reasoning);   and
there is no provision for a 'true' lightweight fighter of
the type for which air warfare developments seem to indicate
a clear need. (10)

4.   'Balanced Forces' :   The Fleet

I believe the main thrust of my argument about naval forces
to be sound too.   That is to say, I would still challenge
the fundamental proposition that only a fleet composed on
the basis of existing assumptions about 'balance' can
adequately serve the maritime security interests of the
United Kingdom (and the Alliance).   However I would wish to
qualify to some extent the notion implicit in my subsequent
analysis, that the alternative force structure which most
readily commends itself is one without 'ships of high
quality' and with greater emphasis on smaller surface ships
and submarines.   (See pp. 155-6.)   If I were writing this
part of the paper afresh I would stress that there are other
options.   In particular, I would try to elucidate the
argument occasionally heard in naval circles that, if the
shape of the existing fleet were put to the question, a
stronger case could be made for the ASW cruiser programme
than, for example, some of the new escort programmes.   In
other words, the option of a future configuration for the
Royal Navy based on some 'ships of high quality' but fewer
destroyers/frigates seems to me to deserve fuller consideration than I have given it. (11)

5.   Other Matters

Obviously there are several other matters on which, after
reflection and with the benefit of hindsight, I would not
now express myself in quite the same way as in the original

text.   For example,

(a)   The scope for support economies in British Forces
Germany, through elimination of single-Service
provision of essentially common community services,
is a question which should receive greater attention
in speculation about lower-cost programme options
than I have given it.

(b)   A corollary of the argument in 3 above about
tactical air power is that, ironically, the
rationale for the air defence variant of Tornado
may be stronger than that for the strike/reconnais-
sance version (cf. pp. 161-2).

(c)   A further corollary is that the issues surrounding
definition of AST 403 (the so-called Jaguar/Harrier
replacement) merit fuller ventilation in any analysis
of air force programme options for the 1980s and
beyond.

(d)   A conspicuous omission from the discussion of naval
forces is consideration of what might constitute
appropriate provision for the policing of the United
Kingdom's extended fisheries limits and offshore
constabulary tasks generally; the notion that these
duties should be entrusted not to the Royal Navy but
to some civil coastguard agency has been widely
canvassed and choosing such an option might have
important implications for the 'numbers' question in
relation to (for example) the frigate and mine
counter-measures vessel programmes.

(e)   The whole issue of what scope there might be for
'savings' in the defence budget via the more active
pursuit of interoperability of equipment with NATO
partners (and even standardization), the development
of the 'two way street' in Atlantic arms procurement
and similar initiatives was treated too perfunctorily
in the paper itself.

This list could be extended to include fully twice as many
items.   But no useful purpose would be served by doing so
other than to confirm what one knows already: that no self-
respecting analyst, reviewing a piece of work after a lapse
of several months, considers that his original effort requires
no amendment.
To sum up: there are several aspects of my substantive
argument about programme options where, on reflection, I
detect a need for some shift of emphasis or for the addition
of qualifying remarks.   But, as stated earlier, the essential
features of the analysis do not seem to me to have been overtaken by events.

Conclusion

The purpose of this postscript's parent paper, written in 1975, was to elucidate the kind of change which would be necessary if, by the end of the 1970s, the United Kingdom's defence/GNP proportion were to stand no higher than that of her European allies. Its key conclusion was that to reach this goal (by 1980-1) would require radical surgery on two of the three services. The defence budget would have to fall to a level which could accommodate provision for _either_ a balanced fleet _or_ existing ground force levels in Europe _or_ a 'full spectrum' tactical air force, but not two of these and assuredly not all three.

The supplementary material in these additional pages has shown that,

(a) following recent revisions, defence budget projections foreshadow _lower_ expenditure than previously scheduled in the late 1970s, and compensatory _rising_ outlays in the early 1980s; thus,

(b) implementing the sort of programme change developed in the main paper — on a now inevitably later time-scale and yielding somewhat smaller 'savings' — would simply _stabilize_ defence spending in the 1980s at around £6,000 millions (at 1977 Survey prices); or, put the other way, any government wishing to contain the growth of defence expenditure in the early 1980s would have to choose among options like those elaborated; although

(c) the actual configuration of some of the programme options would bear re-examination if the full analysis were to be conducted afresh.

Of these points the second is clearly the most intriguing. Whereas two years ago change along the lines discussed would have reduced defence spending, from 1980 change will be necessary if things are to remain the same.

May, 1977

Notes

(1) That is, at the price level used for the 1974 Public Expenditure Survey — the routine, annual appraisal of all public spending programmes with which the Defence Review was co-ordinated.

(2) See the observations (including the present writer's own evidence) in Second Report from the Expenditure Committee, Session 1974-5, The Defence Review Proposals , HC 259 (1974-5). (Hereafter HC 259 (1974-5).)

(3) Other measures of burden yield yet different 'rankings' of European states. In a recent examination of France, West Germany and the United Kingdom I concluded that these countries 'bear broadly similar defense "burdens" albeit in different ways'. See D. Greenwood, 'The Defense Efforts of France, West Germany and the United Kingdom', in Horton F.B. et al.(eds), Comparative Defense Policy (Baltimore and London: Johns Hopkins U.P., 1974), pp. 340-62.

(4) HC 259 (1974-5), Report, p. ix and Evidence pp.3 (para 8) and 13 (Q's 10 and 11).

(5) See HC (1974-5), Evidence, pp.58-9.

(6) More generally, the case for debating defence (and other) policy alternatives in terms of 'costed options' has been powerfully argued in the literature on defence budgeting: see the citations in the author's Budgeting for Defence (London: RUSI, 1972), pp. 85-7. (And it has been endorsed by a contemporary Defence Minister, with the observation that 'to debate generalized objectives or figures chosen more or less at random is to avoid the basic issue'. (F. Judd, in 'The Security of Offshore Resources', RUSI Journal for Defence Studies 120 (3 September 1975), p.4).

(7) Throughout this section I am dealing with defence expenditure on the national (Defence Budget) definition and not the broader NATO definition. In 1974-5 the latter was 4 per cent higher than the former. The argument is not affected.

(8) Cmnd. 5976, Ch.1, para 25 (d) (p.10): Twelfth Report from the Expenditure Committee, 1972-3, Nuclear Weapon Programme HC (1972-3), Report, p. vi, Evidence, p.28 (Q.653).

(9) Cmnd. 5976, Ch. 1, para 79 (pp. 34-5) and Ch. 11, para 10 (pp. 41-2).

(10) Cmnd 5976, Ch. 1 paras. 25(a) (p.9) and 47-54 (pp. 16-18); Ch. II paras. 15-24 (pp. 45-6); Ch. VII paras. 30-5 (pp. 87-8); also Ch. 1 para 27(c) (p.11).

(11) Cmnd 5976 Ch. 1 paras. 28 & 29 and 33-44 (pp. 13-15).

(12) Cmnd 5976 Ch. 1 paras. 55-9 (pp. 18-19) and Ch. VII paras. 10-17 (pp. 84-5).

(13) Cmnd 5976 Ch. 1 paras. 60-2 (pp. 19-20) and Ch. VI, Pt.1 passim.

(14) See D. Greenwood, Budgeting for Defence pp.54-5.

(15) Ibid. pp.56-8.

(16) The point applies to all public expenditure programme, of course - not just to defence.

(17) In what follows I have drawn on my own recent contribu- tions to 'the process': 'The Defence Review in Perspective', Survival XVII, 5, September/October, 1975, pp. 223-9, and 'Sights Lowered: the United Kingdom's Defence Effort 1975-84' Royal Air Forces Quarterly, 15, 3, Autumn 1975, pp. 187-97.

(18) Cmnd 5976, Ch. 1, para. 25 (pp. 9-10).

(19) Cmnd 5976, Ch. 1, para 34 (p. 14).

(20) Survival , XVII, 5, September/October 1975, pp. 227-8.

(21) Protocol II on Forces of Western European Union (Paris, 23 October 1954), Article 1.b. (See NATO Facts and Figures 1971 edition, p. 316.) Emphasis added.

(22) Quoted phrases from Mason, R. 'Britain's Security Interests' in Survival XVII, 5, September/October 1975 , p.218.

(23) WEU Assembly Document 663 (2 April 1975), Rational Deployment of Forces on the Central Front.

(24) See, for instance, S. Canby, The Alliance and Europe IV - Military Doctrine and Technology , Adelphi Paper No. 109 (London: IISS, 1975).

(25) For example, K. Hunt, The Alliance and Europe II - Defence with Fewer Men , Adelphi Paper No. 98, (London: IISS, 1973) esp. p.20 et seq.

(26) Canby, _The Alliance and Europe IV_ , p.14.   This is a
      matter to be taken up later in considering the 'balanced
      forces' principle.

(27) See K. Hunt, _The Alliance and Europe II_ , pp. 9-10.

(28) This paragraph summarizes some of the main arguments in
      S. Canby, _The Alliance and Europe IV_ , pp. 36-41.
      (See also J. Erickson, 'Some Developments in Soviet
      Tactical Aviation (Frontovaya Aviatsiya)', _RUSI
      Journal_ , 120, 3,September 1975 , pp. 70-4.)

(29) W.D. White, _US Tactical Air Power: Missions, Forces
      and Costs_ (Washington: Brookings Institution, 1974),
      p. 103 and _passim_. See also 'Spending More, Getting
      Less' _Forbes Magazine_ (15 April 1975), pp. 22-32.

(30) S. Canby, _The Alliance and Europe IV_ , p.40.

(31) Ibid, p.41 (emphasis added).

(32) Cmnd. 5976 Ch. 1 para 25,b (p.9).

(33) Ibid. (p.10).

(34) HC 259 (1974-5) p.58.

(35) The 'packages' referred to at p.144 above.

(36) Some explanation is required of the failure to
      identify, as a further subsidiary option, equi-
      proportionate expenditure reductions on the three
      main 'mission programmes' or equal absolute reductions
      tailored to yield 'savings' of _c_. £550 millions (2 x
      275) in 1980-1.   There are two reasons for this.
      First, as a matter of principle, the law of equal
      misery is an express of the 'balanced forces' notion
      discussed earlier (pp. 13-14) in its inter-service
      guise.   It is difficult to believe that the preferred
      distribution of less funds would be proportionately
      less all round.   Secondly, as a practical matter,
      whereas 'paring' of army forces levels though trouble-
      some may be feasible, in the naval and air force
      programmes for the next few years there are discrete
      blocks of expenditure - on new ships or the MRCA, for
      instance - which must be taken 'all or nothing' to make
      any sense at all.

(37) No support reductions were credited for 1976-7.   For
      the remaining years 'savings' equal to one-third (1977-8
      and 1978-9), one-half (1979-80) and 50-5 per cent
      (1980-1) of those assessed for 'mission programmes' were
      included, rounding in each case to bring the total 'savings'
      - and thus the expenditure series also - to the nearest £25
      millions.

Notes in Postscript

(1) The material in this section is based on the Statements on the Defence Estimates 1975, 1976 and 1977 (Cmnds 5976, 6432, 6735) and the Second Report from the Expenditure Committee, Session 1976–77, Cumulative Effects of Cuts in Defence Expenditure , House of Commons Paper (HC) 254 of 1976-7 (cited hereafter as HC254 (1976- 7)). Any account of programme revisions must pick its way through the confusion of alternative price bases. The marginal notes are intended to mark the prices basis on which the main figures in the paragraph are expressed, which correspond to the values in terms of which discussion was conducted at the time in question. (The abbreviations are 75/76 EP for 1975-6 Estimates prices (and so on) and 75 SP for Survey prices (and so on).)

(2) Cmnd 6735, para 107.

(3) Note also that, in the discussions following President Carter's call for a general commitment to increased defence expenditure in NATO, the Secretary of State for Defence has subscribed to the general undertaking to aim for real increases of 3 per cent a year while pointing out that the United Kingdom would not be able to step up its spending until 'after 1979'. (See The Economist (21 May 1977), p. 71.)

(4) See D. Greenwood, 'The Defence Review in Perspective', Survival , XVII, 5,Sep/Oct 1975, pp. 223-9.

(5) See D. Greenwood, 'Constraints and Choices in the Transformation of Britain's Defence Effort since 1945', British Journal of International Studies, 2 (1976) pp. 27-40, especially Table 2. The data in this table show that, in the twenty years 1956-75 the five-year moving average of defence expenditure stayed within a range of nine percentage points around a stable trend line.

(6) See main text pp. 146-50.

(7) See, for example, R. Burt, 'Technology and East-West Arms Control' International Affairs , 53, 1,January 1977, especially pp. 65-8 and the extracts from Lord Kennet's speech in the House of Lords, 1 December 1976, in Survival XX, 3, May/June 1977, pp. 125-7.

(8) For a useful summary of his views see S. Nunn, 'Deterring War in Europe: Some Basic Assumptions Need Revising' NATO Review (February 1977), pp. 4-7.

(9)   See, for example, J.J. Holst and U. Nerlich (eds.),
      Beyond Nuclear Deterrence: New Aims, New Arms (New
      York: Crane Russak, 1977).

(10) The conclusions about the problems of close air support
     reported here arise from several conversations with
     military and air force officers and also reflect the
     lessons drawn by Israeli air commanders from the experience
     of the Yom Kippur War of 1973.   For an interesting
     discussion of 'offensive v. defensive counter-air' issues
     see the articles by Hine and Latter on The Royal Air
     Forces Quarterly, Winter 1976.

(11) Among other things this means that the rationale for
     retention of the Sea Harrier programme is another matter
     to which I would give greater attention if I were
     conducting the analysis afresh.

3 NOTE ON THE BUDGET APPROACH AND A PROGRAMME WHICH RETAINS THE UK NUCLEAR STRATEGIC FORCE

by Paul Cockle

The Capacity Versus Programme Budgeting Approach

This paper provides an amendment to David Greenwood's paper 'Defence Programme Options to 1980-1' which is based on a programme budgeting approach. It has been argued in this Study Group that this methodology is inferior to the capacity approach which, it is claimed, can make 'genuine' cuts in defence spending. I merely wish to register my view that this is not an argument I accept and that the capacity approach provides an irrational basis for determining defence requirements.

Advocates of the capacity approach argue that defence spending is determined principally by the need to sustain the productive capacity of the arms industry. If indeed that were true, it would be an irrational way for a country to determine its defence policy, since requirements would be falsely determined by the supply of weapons. In my view, however, it would be equally irrational to design a defence policy based on cutting back industrial capacity, for that approach would still be predicated on the same false grounds. The capacity approach does not therefore offer a rational framework for determining a defence policy.

The programme budgeting approach, however, does provide such a framework, for policy goals are articulated into programme expenditures. If cuts are to be made in the defence budget, savings can be set against policy options foregone. Decision making is aided by placing preconceived needs alongside their cost; to consider only one, without the other, is to play roulette with public funds.

Retaining the Polaris Fleet

I argued, at the meeting that considered David Greenwood's paper, that to include the phasing out of the Polaris fleet in all the programme options overly restricts political choice. It is often argued that the fleet should be allowed to run out its natural life since its low operating costs are well worth the benefits it conveys. The period of its 'natural' life is often assumed to be 20 years of service, which means the first would be due for replacement in 1986 and the last in 1990. However, it is claimed that the last boat has a slightly longer life of 25 years and it is quite possible to maintain these boats over a longer period than 20 years, if necessary

The arguments for keeping the fleet are numerous but not all of equal weight. Some argue that Britain's nuclear strategic forces can be perceived by the Superpowers as a trigger which might bring them into disastrous conflict and that its existence therefore constrains them from developing rules of engagement for a war in Europe which might preserve them but destroy Western Europe.

In another vein it has been suggested that the capacity to retaliate - it is accepted there is no question of an independent first strike - confers two grisly options which might deter provocation. The first entails contemplating suicide. Britain could, if the situation arises, threaten to retaliate for any Warsaw Pact military action which is held to threaten vital national interests. The second has been dubbed the blow from the grave. Britain could threaten to retaliate only for a strategic nuclear strike by the USSR on our own soil. Yet another argument emphasizes, whether one approves or not, that nuclear strategic weapons confer a status which may be turned to diplomatic advantage. Then, there remains the recurring theme of an all-European nuclear strategic force. Without entering into the various pressures that could push Europe towards this course, it would be far cheaper for Britain to offer four old boats built with yesterday's sacrifices than make a current contribution to the proposed force. Finally, nuclear issues (SALT, NPT, etc.) are important in their own right and the possession of nuclear arms affords not only the technological knowledge to deal expertly with problems that arise but also insures against total exclusion from Superpower deliberations which might produce undesirable accords.

Since it is not necessary to judge the issue, all that need be said is that it is possible to make a case for retaining the Polaris fleet and to do this within the suggested budgetary restraints, would involve cutting into conventional force expenditures. David Greenwood, working to a fairly high margin of error of £25m, estimated the cost of the Nuclear Strategic Forces in 1980 to be only £75m. If this sum were spread evenly over the three major conventional force programmes. Navy General Purpose Forces (NGPF), European Theatre Ground Forces (ETGF) and Air Force General Purpose Combat Forces (AFGPCF), the cuts required in each would fall within the margin of error. Therefore, whether it is necessary to even change David Greenwood's figures for these programmes is a moot point. However, they could be distributed over years and sub-programmes in the manner shown in Table 1, p.200.

As the sums for each programme are so small, they are not likely to result in the complete scrapping of a major item and are more likely to be met by stretching a little further those programmes suggested by David Greenwood. Thus, within the NGPF programme, the SSN building programme might be stretched fractionally more, as might the destroyer and frigate programme. The total saving by 1980-1 would be £300m instead of £275m. The ETGF cuts would have to be marginally deeper involving a sharper rundown of men to around 27,500 men with appropriate adjustment to equipment purchases as well as cuts in the Home Forces to reflect the reduced roulement requirement. In the case of AFGPCF, the complexity of the adjustments required produce far greater difficulties, and ascribing further cuts of such small size to particular measures is a wasted exercise.

It will be assumed, therefore, that the cuts will come from the first three sub-programmes. Support costs will remain the same as those proposed by David Greenwood.

Table 1 suggests a modification to the Programme Option Modules of Table 2 in David Greenwood's paper (see p. 201). These modifications do not substantially change the conclusions of that paper. All that has been done here is to allow the option of keeping the Polaris fleet at the expense of some conventional capability. In fact, it is possible to have any of Options I to III whilst retaining Polaris (see David Greenwood: Table 3, attached). The only difference would be that elements B, C and E would take values of £300m and not £275m.

TABLE I : Programme Option Modules Modified to Retain
the Polaris Fleet £ Million 1975-76 Estimates Prices

MAJOR PROG	1976-77	1977-78	1978-79	1979-80	1980-81	
Polaris	-	-	-	-	-	
NGPF (1)	5	10	15	20	20	
(2)	5	10	30	40	55	
(3)	-	55	110	120	170	
(4)	-	-	-	-	10	
(5)	-	10	30	40	50	
(6)	-	-	-	-	-	£300m
ETGF (1)	-	50	100	150	220	
(2)	-	-	-	-	-	
(3)	-	-	25	50	80	£300m
Other Army (1)	-	5	10	20	25	£25m
AFGP (1)	(-	-	190	220	245)	
CF (2)	(				)	
(3)	(				)	
(4)	-	-	-	5	10	
(5)			5	10	15	
(6)	5	10	15	20	30	£300m

Support Programmes as Before
(as for David Greenwood,
Tables 2 and 3)

Table 2   PROGRAMME OPTION MODULES
          ESTIMATED 'SAVINGS' POSSIBILITIES ON MAJOR
          PROGRAMMES SUB-PROGRAMMES
          £ millions at 1975-76 Estimates Prices

Major Prog.(Sub)		1976/77 £m	1977/78 £m	1978/79 £m	1979/80 £m	1980/81 £m	REMARKS
A		-	50	50	75	<u>75</u>	Rundown of Force & Facilities
B	(1)	5	10	15	20	20[a]	Rundown of Provision (Ships)
	(2)	5	10	30	40	50	Stretch SSN Bldg: Rundown SS Ops
	(3)	-	50	100	100	150[a]	Cancel Cs. 01?/02/03: Stretch B & F Bldg.
	(4)	-	-	-	-	10[a]	Reduced Afloat Support Reqt.
	(5)	-	10	30	40	50[a]	Cancel SEA HARRIER Some Helis. Reduced=
	(6)	-	-	-	-	-	Costs of Carriers/Amphibs
						<u>275</u>	
C	(1)	-	50	100	150	200[b]	Rundown to c. 30,000: equipment purchases adjusted
	(2)	-	-	-	-	-	adjusted
	(3)	-	-	25	50	75[b]	Reduced roulement reqt
						<u>275</u>	
D		-	5	10	20	<u>25</u>	No Cost to UK by 1980-81
E	(1)	-	-	180)	200)	220[c])	Cancel MRCA AD Var: Cancel MRCA (Common) & prolong JAGUAR & HARRIER prod'n ? Arty & SAM purchases (see C(1))? F16 or A10 purchases ?
	(2)			)	)	)	
	(3)			)	)	)	
	(4)	-	-	-	5	10	Tankers and Other a/c
	(5)	-	-	5	10	15[c]	
	(6)	5	10	15	20	30[c]	Reduced force level
						<u>275</u>	
F					(15))		
G					(50))		
H					(75))		
I					(50))}(350)		Reductions equal to c. 50% of 'achievement' in mission progs.
K					(125))		
J,L					(35))		

Table 3 PROGRAMME OPTIONS TO 1980-81
OPTIONS FOR DEFENCE BUDGET TARGET
(Target 'Savings': £1000 millions at 1975-76 Estimated Prices)

Def. Budget Saving
1980-81
£m

Option I	A.	Rundown POLARIS & Facilities	75	
	B.	Smaller ship Fleet	275	(225)
	C.	Reduced ETGF levels	275	
	D.	No cost to UK arrngts	25	
		Mission programme	650	
	F.L.	Proportionate Reductions	c. 350	
			c. 1000	(925

Option II	A.	Rundown POLARIS & Facilities	75	
	B.	Smaller ship Fleet	275	(225
	D.	No cost to UK arrngts	25	
	E.	Lower cost Air Force	275	(265
		Mission programme	650	
	F.L.	Proportionate Reductions	c. 350	
			c. 1000	(900

Option III	A.	Rundown POLARIS & Facilities	75	
	C.	Reduced ETGF levels	275	
	D.	No cost to UK arrngts	25	
	E.	Lower cost Air Force	275	(26
		Mission programme	650	
	F.L.	Proportionate Reductions	c. 350	
			c. 1000	(98

Others	IV		
	V	See Text	
	VI		
	VII		

Note Figures in parentheses identify the effects of excluding 'non-
related' items i.e. Submarine reductions in (B) and tankers and
other aircraft reductions in (E). See Table 2

4 DEFENCE PROGRAMME OPTIONS TO 1981: POLITICO-STRATEGIC AND ECONOMIC IMPLICATIONS

by Alan Lee Williams

In his paper 'Defence Programme Options to 1980-1' David Greenwood has identified a series of Programme Option Modules from which, he suggests, a defence programme can be assembled costing £1,000 million (at 1975-6 Estimates Prices) less by 1980-1 than the programme now adopted by the Government. In constructing these modules he has carried out a brief analysis of the strategic implications of the programme changes but has, intentionally, not carried out a thorough exploration of the employment and other economic consequences of such a reduction; nor has he fully analysed the politico-strategic cost or identified any potential benefits from alternative resource uses.

Given that the Government's own defence review, reducing the programme by about £5,800 million (£4,700 million updated to 1975-6 Estimates Prices) over nine years involves far reaching changes in the pattern of deployment of defence resources, it is immediately apparent that further reductions, totalling more than £5,500 million over eight years, must have even greater politico-strategic and economic implications. Fuller analysis is therefore needed in order to complement Mr Greenwood's paper. The Study Group will need to consider how best to carry out this task; I have set out below a preliminary assessment of the subjects we shall need to examine in detail.

The Defence Burden and the Threat

As Mr Greenwood states, there is no balance which is self-evidently right, between the needs of defence and the real resource costs of meeting them. The last Defence Review was intended to reduce our share of GNP, devoted to defence, to about 4½ per cent - a figure more nearly equal to that of our European allies; it also resulted in a pattern of expenditure in which perhaps 95 per cent of our effort is devoted to NATO (the keystone of UK defence). Before we decide to recommend a programme reduced still further, we need to be sure that the resulting sum total of NATO military expenditure provides an adequate and effective deterrent to the Soviet threat and to examine carefully the proposition that the resources allocated to defence by the several members of the Alliance must represent an exactly equal burden on their economics. Since this consideration is outside our terms of reference, I mention this only in passing.

We shall also need to examine the fact that the USSR, with a GNP no more than half of the USA, spends rather more on arms in real terms than the USA. Coupled with Mr Brezhnev's reaffirmation that there will be no let up in the ideological struggle, despite recent moves toward detente, we need to be sure that NATO's cohesion would be maintained in the event of a substantial reduction in the UK's contribution.

Option A. Nuclear Strategic Forces

Abandonment of Polaris would leave the French as the only
European influence on the strategic nuclear balance of
power.   There would be total reliance on the US to provide
the essential third element in the triad of conventional,
tactical nuclear and strategic nuclear forces, on which NATO's
strategy of deterrence is based.   It is unrealistic, not
least because of the provisions of the Nuclear Non-Proliferation
Treaty, to expect other European powers, individually or
collectively, to acquire a similar capability.   Given the
limitation placed on anti-ballistic missile systems, Polaris
should remain an effective part of NATO's strategic nuclear
armoury at very little cost; there may be an argument for
giving it up in the context of multilateral disarmament, but
there is no argument of cost-effectiveness for unilateral
scrapping.

Option B. Navy General Purpose Combat Forces

To assess the politico-strategic implications of adopting
Option B an analysis has to be made of the role envisaged
for the big ships of the Royal Navy.   As Mr Greenwood
points out, the 1975 Defence White Paper does not provide
firm guidance on this matter; he suspects perhaps that
the quality enshrined in these ships is pursued for its own
sake.
     Before we can come to a conclusion on this, we shall need
to examine the Soviet maritime threat.   Their Navy is
steadily increasing in quality, with an increasing emphasis on
nuclear powered submarines; much of it would be deployed in
the Eastern Atlantic and Channel where the British Navy makes
the main contribution to the Alliance.   Would a 'light ship'
Navy have any effect against a 'quality' opposition?
Further analysis is needed of the likely nature of a future
war at sea, not least to determine the role of the ASW
cruiser and the Maritime Harrier.   I think it would be
unrealistic to plan on the assumption that the USA will
provide the quality and we will provide only the small ship
support; as a result NATO's maritime affairs would be totally
dominated by the USA.

Option C. European Theatre Ground Forces

The kernel of Option C is the proposal to reduce BAOR by
25,000 men over 5 years to 30,000 as part of a NATO-agreed
reordering of arrangements for the defence of North-West
Europe.   The front line units of BAOR are deployed in a
vulnerable sector, against which the Warsaw Pact could
launch the main weight of its attack.   In the absence of a
mutual and balanced programme of force reductions, British

reductions of the order postulated would either require other
NATO allies to provide forces to fill the gaps created, or
entail accepting that in the British area of responsibility
there was no longer a credible conventional defence.   Do we
think that either of these alternatives is realistic?

A radical reshaping of NATO's ground forces would require
the adoption of new technologies and/or increased reliance
on reinforcements before hostilities began.   Much more
analysis is needed before we can be sure that these could
provide the basis for an acceptable alternative strategy,
otherwise we should have to revert to the nuclear tripwire.

Option D.   Other Army Combat Forces

Britain's commitments outside NATO do not consume a large
proportion of the Defence Budget, although at the beginning
of 1975 British troops were deployed in many different parts
of the world.   The 'Other Army Combat Forces' programme is
not the only element in this deployment;   yet a Defence
Review decision to withdraw from all non-NATO commitments,
including the Mediterranean garrisons, would have saved at
most £150 million a year at 1974 prices.

As it is, the Defence Review decisions will result in
withdrawal from Malaysia, Singapore, Gan, Mauritius, and
Malta, with reductions in force levels in Masirah and
Cyprus, and a higher contribution from the Government of
Hong Kong.   In the present situation in the Mediterranean,
withdrawal from Cyprus is surely not a realistic option;
equally I doubt whether it would make sense to contemplate
withdrawal from Hong Kong if the Colony will not meet the
full cost of its defence.

Option E.   Air Force General Purpose Forces

The large reductions proposed in expenditure on air forces
are based on querying the need for a 'balanced' air force
equipped 'to cover most of the spectrum of mission capabilities';   the size and shape of the RAF would be limited by
having it undertake a smaller range of roles than at present.

The cheap and simple solution for the provision of aircraft
needs close scrutiny.   While there are advantages in either the
economy achieved or the greater numbers possible, the tasks
that must be faced, which range from the detection of nuclear
submarines in the Atlantic to combat with some of the most
advanced modern fighters and attack aircraft in the world,
simply cannot be achieved without cost.   The search all the
time is for a compromise between the two requirements, cost
and capability.

Would it be possible for the RAF at this stage to 'opt
out' of some of its roles?   This would require specialization
in roles by different NATO member countries.   However, in the

light of the increasing Warsaw Pact threat, we cannot assume
that the members of NATO would agree to specialize in this
way unless the resulting financial savings were used to
increase the effective level of overall defence rather than
simply to reduce the defence efforts of individual members
nations.   Air support is necessarily a combination of
capabilities.   The decision on how and where it will be
used must surely depend on the strategic, technical and
tactical situation at the time.

Options F-L

In constructing the hypothetical costing for 1980-1, to
provide the base line for the programme option modules,
Mr Greenwood suggests that the teeth/tail ration will shift
towards the tail - possibly to a 50/50 split.   This seems a
not unreasonable assumption but we shall need to examine the
manpower implications of Options A-E before we can be sure
that savings f £350 million can be found from programmes F-L
without totally undermining the fighting effectiveness of our
residual combat forces.

Manpower and Industrial Implications

It should be possible to assess the manpower implications
of a reduction in defence expenditure of £1,000 million a
year.   There will not only be the direct redundancies (and
loss of recruiting) arising from the force cuts in options
A-E;   there will also be substantial reductions in military
support and in civilian employees.
     The effect on defence industries will also need to be
examined - not only as a result of reductions in direct
purchases but also from a loss of defence exports.   Shipbuilding could be hit very hard if option B were adopted
at a time of world-wide slump in ship orders;   Vickers,
Yarrows, Swan Hunter and Cammell Laird would all suffer.
Closure of a Naval Dockyard could also be necessary.
     Delection of the MRCA would have serious implications
for the nationalized aerospace industry, particularly if its
replacements were bought off the shelf overseas.   The
conventional armaments industry would also have to shed
labour.   There are also the indirect effects of these cuts
on sub-contractors and local trades and services around the
factories affected by defence cuts.
     In total therefore, the scale of resulting unemployment,
could be quite serious.   Re-deployment, especially of some
of the more specialized skilled men and of those in areas
of high current unemployment, may not be easily accomplished.
There is a real danger that they will merely add to the pool
of unemployed but, more hopefully, they will come on the
labour market at a time of economic recovery;   in that case

significant investment would be needed, which would reduce
the effect of public expenditure savings from further defence
cuts   Analysis is needed of the scale of this problem
before we can come to any rational decision, and these
preliminary figures make it quite clear that rational
analysis requires a simultaneous consideration of strategic
and employment repercussions.

5 STRATEGIC AND POLITICAL IMPLICATIONS OF REDUCED DEFENCE PROGRAMMES

by Dan Smith

The task of this paper is to consider political and strategic consequences of implementing the reduced defence programmes presented by David Greenwood (1) as illustrative possibilities of the kind of reductions needed to achieve a cut in military spending of £1,000m a year (at 1974 prices).

It must be said that this is an inherently unsatisfactory task.   There are strong economic reasons for cutting defence, and from an economic point of view the level of reductions can be determined according to criteria based on the needs of the economy, alternative uses for resources, and so on.   But the cuts, and force reductions consequent upon them, are bound to be arbitrary from a strategic point of view.   To be fair, they are no more arbitrary than the base line from which reductions start, or than the as yet unstated increases in military spending aspired to by the Conservatives.   The Study Group is not alone in its sinfulness.   Indeed, some degree of arbitrariness is perhaps inevitable, and the Study Group has heard very strong arguments that the shape of military spending is determined less by real strategic need than by the needs of industry and service sectional interest.

But if the Labour Party is to free Britain's national security policies from the distortions of the present, then it must at some time get down to the task of deciding what security needs it perceives Britain as having, within the context of Labour's domestic and foreign policies, and how they can be best met.   Defence policy could then be, as it should be, considered alongside policy on arms control and disarmament, international relations and Britain's world role as well as alongside industrial and economic policy, in a logical and coherent fashion.

Nevertheless, the present exercise is worthwhile in helping to determine the implications of Labour's present policy.   The reduced programmes considered in this paper can be summarized as follows:

(a)   phasing out Polaris;

(b)   paying off Ark Royal,   Hermes, Fearless, and Intrepid, cancelling the second and third ASW Cruisers and Sea Harrier, running down patrol submarine activity, and 'stretching' building programmes for Fleet submarines, destroyers and frigates;

(c)   reduction of the British Army on the Rhine to about 30,000 men, with appropriate equipment adjustments, and a proportional cut in Home Forces to reflect a reduced requirement for rotating men to and from BAOR;

(d)   cancellations of both variants of MRCA, reductions in tankers and air transport strengths, and increased expenditure on Harrier and Jaguar.

Other programme reductions not considered here, but
included in Greenwood's paper, are phasing out provision
for ground forces outside Europe, and proportionate
reductions in support programmes.

Greenwood calculates that to meet the target for overall
cuts, it would be necessary to implement programme I plus two
of the other three (plus phasing out ground forces outside
Europe together with proportionate support reductions).
Paul Cockle has argued that the target could be met without
needing to include programme I in the reductions (2),
although there is also some doubt as to whether 'I plus 2'
is enough to meet the target. (3)   A study by Kenneth Hunt
opened with the judgment that

> The present level of force is unlikely to be available
> in the years to come, and defence structures, concepts
> and deployments will have to be altered to match reduced
> resources and take advantage of new technology. (4)

The same point of departure is used here.   It is assumed
that reducing military spending is an _imperative_, and we
therefore have to find sensible ways of doing it.   The only
fundamental changed assumption in this paper, therefore, is
that Britain's role in NATO cannot continue to be as expensive
as in the past

In places, the programmes presented by David Greenwood are
modified.   Like the programmes, these modifications are 'non-
prescriptive'.   The intention of the paper is not to recommend
an alternative strategy for Britain, but to discuss sensible
ways in which our military spending could be reduced.   To do
this, it is necessary to start with a digression away from the
programmes, to discuss the background against which such
reductions might be made, in terms of the military balance and
technological developments in weaponry.

The Numbers Game

Most debate on defence policy takes place against the background of the military balance.   Before yet one more person
comes out to play in the numbersgame, it is worth saying that
this is proving to be a more and more unsatisfactory way of
calculating the comparative effectiveness of opposing military
forces.   Yet it is on numbers that much of the public
justification for defence policy depends.

In evidence to a House of Commons committee, Rear Admiral
Berger, Assistant Chief of the Naval Staff said,

> We would not wish to give you the impression in our answers
> to all your questions that we think other than that the
> Soviet maritime threat is a very formidable one indeed, and
> it is growing.   It is growing in quantity, and probably in
> quality, faster than NATO. (5)

211

This is a remarkably unhelpful comment. While one cannot
take issue with his assertion that the Soviet Navy is large,
one can take issue with the implication of Soviet preponderance; one can ask what the Soviet Navy threatens, and one can
straightforwardly disagree with his final sentence.

Table 1 indicates a rough numerical comparison of NATO and
Warsaw Pact naval strengths, while Table 2 compares recent
naval constructions rates. In crucial categories numerical
superiority belongs to NATO. The overall numerical parity
is no guide to comparative quality and effectiveness about
which the American Defense Department's judgment is:

Soviet naval peacetime presence increased sharply in the
late 1960s but now appears to have stabilized at a
level below that of the overall US presence; however,
in certain areas such as the Mediterranean, the Soviet
Union continues to deploy more forces than the US. When
the peacetime fleets of allies on both sides are tallied,
it is clear that the US and its allies deploy naval forces
in peacetime which are superior to those deployed by the
Soviet Union and its allies. (6)

Table 1: Naval General Purpose, NATO (including France) and
the Warsaw Pact. (Excluding reserves and ships
under construction)

	NATO	WARSAW PACT
Aircraft Carriers	17	0
Cruisers	45	35
Destroyers	172	107
Frigates	212	113
Corvettes	73	196
Attack submarine (nuclear-and diesel-powered)	202	225
Cruise Missile submarines	0	68
Amphibious Warfare Vessels*	110	95
Replenishment, support & transport vessels**	220	100
Ocean & Coastal Mine Hunters & Sweepers	284	401

*   Excludes landing craft etc.

**   Excludes tugs, tenders, other small support vessels, and
merchant marine shipping expected to have military role
if called for.

Both the asterisked categories' figures were rounded to
nearest 5.

Source: Jane's Fighting Ships 1975/76

There has been significantly more major naval construction
in NATO countries in the past decade, which would seem to
support the view that NATO's numerically roughly equal forces
are superior and more modern than the Pact's.   Only in the
category of submarines has there been more construction in
Warsaw Pact countries over the past 10 years, and over the
past 5 NATO's submarine construction rate has outstripped
the Pact's by 66 per cent.   To this picture can be added
a recent report in The Times which reported, on the basis
of information supplied by the Pentagon, that the US has
built twice as many 'large combat ships' as the USSR in the
past 15 years. (7)   The US Defense Department's Annual
Report announced the intention of expanding the American
fleet over the next decade (8).   In other words the
changes in the Soviet Navy has left it less powerful than
the American Navy which is now to be expanded.

Table 2:   Major Naval GPF construction, NATO (including
France) and the Warsaw Pact

	NATO		WARSAW PACT	
	1965–74	1970–74	1965–74	1970–74
Aircraft Carriers	3	1	0	0
Attack & Cruise Missile Submarines	80	51	81	30
Surface Combatants over 1,000 tons	148	76	84	23

Source: IISS Military Balance 1975/76

NATO nations have built, and will continue to build more
major ships than the Warsaw Pact.   Rear Admiral Berger was
wrong, and gave inaccurate advice to the House of Commons.
The role of the Soviet Navy is more limited than normally
implied, and more limited than the US Navy's.   The assessment
made by Barry Blechman three years ago has now gained general
currency:

Generally, and with the exception of strategic submarines,
the Soviet Navy does not appear to be designed to project
the Soviet Union's power into distant oceans but to defend
the security and the interests of the USSR - by preventing
attacks on its homeland and by limiting the role of the

213

United States and other western powers in regions close to
Soviet shores, notably the Middle East. (9)

It is a very different kind of navy from the American one.
Comparing the opposing navies the Pentagon notes that NATO
allies 'emphasize the missions of sea control for defense of
(sea lines of communication), projection of power ashore for
use in wartime, and naval presence to control crises in
peacetime', whereas the Soviet emphasis is on 'sea denial
and defense against the US capability to project power, by
carrier air or amphibious operations, onto the European and
Asian land mass', together with a growing diplomatic use of
naval presence (10). The introduction of aircraft into the
Soviet Navy may alter this judgment (though the alteration
is not automatically necessary), but the analyses from both
Blechman and the Pentagon make clear that the Soviet Navy's
changed role since the 1950s has been in response to western
sea power, and the threat it can be seen to pose, and still
excludes the role of power projection.

What, then, has all the fuss been about? Admiral Holloway,
US Navy Chief of Staff, let the answer slip in his posture
statement:

> My judgement is that, today, we retain a slim margin of
> superiority with respect to the Soviet threat in those
> scenarios involving our most vital national interests,
> yet thirty years ago we were in a position of overwhelming
> superiority. What has happened to the balance of naval
> power? (11)

The concern is the loss of overwhelming superiority, and
with it the freedom of action is provided. Michael Klare's
conclusion in a recent article goes right to the heart of the
matter:

> By gaining 'new visibility in areas of the world which
> have traditionally been within the Western sphere of
> influence' (to use Admiral Rectanus' words), Moscow has
> called into question the invulnerability of Western
> fleets there and thus their utility as instruments of
> coercion and influence. Thus, it is not Western
> shipping that is threatened by Soviet naval deployments,
> but Washington's strategy for continued Western hegemony
> in remote Third World areas. (12) [Emphasis in original].

This judgment appears to be supported by Admiral Holloway's
concern, as well as by the differences in Soviet and American
naval roles. In this sense the Soviet Navy need not be
considered a threat to Britain's vital interests, and
certainly not to those interests as one hopes the Labour
Party perceives them.

The 1976 White Paper on defence states categorically that,
'In conventional ground and air forces, the imbalance in

central Europe has moved further in the Warsaw Pact's favour' (13). This judgment would appear to be challenged by Table 3, which is only a partial comparison of ground and air forces in Europe, but which suggests that, if anything the numerical balance has moved slightly NATO's way. A study by the Pentagon in 1973 indicated that, in a European battle, tactical air forces of each side would be 'about even at 5,000 to 6,000 each';(14) this must cast some doubt on the figures in Table 3 for both 1973 and 1975, which may be attributable to excluding American, British and French air defence forces. The same report stated that the worldwide air balance worked out at about 10,000 to 11,000 aircraft each, and, apart from the White Paper's assertion, there is no evidence to suggest that the balance has swung either way.

In ground forces, it is the disparity in tank numbers which is the most common cause for alarm about the military balance. It is at this point that a comparison of forces based purely on numbers gets to its most deceptive and confusing. As Les Aspin has argued, 'American production (of tanks) has been low for years because the US Army was satisfied that it had enough tanks'. (15) If you decide to have fewer tanks, it's a bit unfair to complain when the other side has more.

The qualitative superiority of NATO tanks has often been argued as a counterweight to their numerical inferiority. But even including this factor is not very helpful. If NATO is not planning a massed tank attack in central Europe then it does not need as many tanks as the Warsaw Pact has.

If NATO's anti-tank weapons are at all effective then it matters less what the tank balance is, quantitatively or qualitatively. A military force's adequacy should be judged according to the needs and missions of the country or alliance in question, not according to some other country's or alliance's needs and missions. If the discussion of tanks could concentrate on the job they have to do, and the best ways to stop them from doing it, it would be on a much sounder basis.

Probably the most striking feature of the Soviet Army is its sheer size, which is a boon to those who wish to paint a picture of Soviet hordes 'sweeping through Europe'. With 1.8 million men it ranks only second in size to the Chinese army. In his study of the Soviet Army, however, Jeffrey Record pointed out that this is not an unreasonable size 'given the size of the ground forces of the nation's principal adversaries'. (16) NATO ground forces, according to Record, in Europe number about 2 million personnel (and this figure excludes US units stationed at home but earmarked for European contingencies), which, together with the Chinese total, adds up to around 4.5 million. (17) From this one can see that the Soviets have considerable excuse for alarm at the military balance, and alarmism is probably as rife in the Kremlin as anywhere It is, after all, quite easy to get alarmed if you are properly selective about the figures.

Table 3: Numerical comparisons of certain forces in Europe 1973 and 1975, NATO (excluding France) and the Warsaw Pact

a) All Europe	NATO 1973	NATO 1975	WARSAW PACT 1973	WARSAW PACT 1975	Ratio NATO : Warsaw Pact 1973	Ratio NATO : Warsaw Pact 1975
Tanks	8,650	10,500	23,200	26,250	1: 2.7	1 : 2.5
Soldiers/Marines (000s)	1,130	1,200	1,200	1,240	1: 1.1	1 : 1.03
Tactical Aircraft* (i)	2,756	2,909	5,495	4,955	1: 2	1 : 1.7
Light bombers	146	158	280	255	1: 1.9	1 : 1.6
Fighter/ground attack	1,560	1,700	1,525	1,525	1.02:1	1.1 : 1
Interceptors (i)	625	625	3,050	2,625	1: 4.9	1 : 4.2
Reconnaissance	425	425	640	550	1: 1.5	1 : 1.3
b) North & Central Europe						
Tanks	6,500	7,000	17,000	19,000	1: 2.6	1 : 2.7
Soldiers/Marines (000s)	600	625	900	895	1: 1.5	1 : 1.4
Tactical Aircraft (i)	1,890	2,050	4,300	4,025	1: 2.3	1 : 2
Light bombers	140	150	250	225	1: 1.8	1 : 1.5
Fighter/ground attack	1,100	1,250	1,400	1,325	1: 1.3	1 : 1.1
Interceptors (i)	350	350	2,100	2,000	1: 6	1 : 5.7
Reconnaissance	300	300	550	475	1: 1.8	1 : 1.6

(i) See text for comments on the possible inaccuracy of these figures.

Sources: IISS Military Balance 1973/74 and 1975/76.

Total NATO manpower is 5.1 million personnel, compared to 4.6 million for the Warsaw Pact. (18) However, these figures include a total of 3,575,000 for the USSR, whereas the Pentagon is now estimating the Soviet total at 4,800,000. Les Aspin has thrown a cog in the works of this remarkably high Soviet total estimated by the Pentagon, revising it downwards on the basis of information supplied to him by the Defense Intelligence Agency, as shown in Table 4. It has to be said that NATO's manpower figures could probably also be adjusted up, down, and sideways in a similar fashion, and that the DIA has contested one or two of Aspin's adjustments. On the other hand, it is true that NATO uses large numbers of civilians in tasks for which the Warsaw Pact uses soldiers, and the exercise also usefully demonstrates, once again, the dubiousness of the numbers game.

Table 4: Adjusted comparison of Soviet and American military
manpower (000s)

Soviet total	4,800	American total	2,127
minus Internal Security &		minus Chaplains &	
Border Guards (430)		Construction (3)	
minus Construction Troops (250)		minus Far East Troops	
minus Military working on farms		(115)	
and railroads (150)			
minus Political Officers (70)			
minus Civil Defence (20)			
minus Supply, Storage and			
R & D Troops (170)			
minus Coast Guard (60)			
minus Sino-Soviet Border Troops			
(500)			
minus Czech Garrison (55)			
minus Other Civilians in			
Uniform (300)	2,795		
minus Extra Air Defence			
Troops* (475)			
minus Extra strategic Force			
Troops* (275)			
	2,045		2,009

Figures in brackets indicated numbers estimated for each mission.

*
In these two missions the USSR maintains more men than the USA;
Rep Aspin's contention is that these extra men are no extra
threat to the USA and he has therefore 'adjusted them out' in
the comparison.

Source: Defence Intelligence Agency and Representative Les
Aspin (Dem Wis) reported in the New York Times, 24
April 1976.

In the attempt to compare the two military forces, one
method used has been comparison of military budget. Almost
the entire case put forward by James Schlesinger for
increased American military spending was constructed on the
hypothesis that the USSR was outspending the USA and this was
in itself dangerous ; (19) looking at the budget instead of
what it buys is an approach to strategy discussed nearer the
end of this paper. But even if budget comparison were a
useful criterion in comparing forces, the way it is done is
very dubious. Under pressure the CIA this year has revised
upwards its estimate of Soviet military spending concluding

217

that the Soviet defence spending has been 35 per cent more than the USA's since 1970, and 40 per cent more in 1975. (20) The method used is, in the words of Andrew Marshall, a leading critic of the former CIA estimates, to estimate 'what the US would have to spend in order to replicate within the existing US context the physical dimensions and operational capabilities of the Soviet military'. (21)   In other words, how much would it cost the American government to 'buy' the Soviet military establishment, at American prices?

Whatever the answer to the question is, and whatever conclusions can be drawn from the answer, appear to have been invalidated by Les Aspin who seems to make a hobby of messing up easy assumptions and conclusions in such matters.   He argues that if the calculation is done the other way round (i.e., working out how much it would cost the Soviet government to 'buy' the American military establishment, in roubles and at Soviet prices), then the result is the exact opposite — the USA appears the bigger spender. (22)   As he points out, the exercise involves formidable problems in assembling the data, and the article based on the calculations did not tabulate them, but the reason he gives for these conflicting results seems sound:

> Whereas the dollar comparison is weighted by the fact that it prices the manpower-intensive Soviet forces in terms of high US personnel costs, the rouble comparison is weighted by the fact that it prices the technology-intensive US forces in terms of the high costs of Soviet technology. (23)

Different things have different relative prices in different societies.   In addition, he points out that, since the American arms inventory includes items much more sophisticated than the Soviets have (in the field of computers and advanced electronics), a calculation such as this is bound to underestimate the cost to the USSR of 'replicating' 'the physical dimensions and operational capabilities' of the American military.   As he rightly concludes, 'so the answer to the question, "who is spending more on defence", depends on the price system used'. (24)

Most of the conventional comparisons of military force between NATO and the Warsaw Pact are very dubious, serving to confuse rather than clarify.   No clear picture can be provided and firmly held to.   So much depends upon estimation and interpretation of disputed data, that only the most limited conclusions are really possible.

However, the summary of the problems in this section may suggest that the naval balance is not alarming;   that the central European conventional balance has moved NATO's way; that the Soviet Army is not overlarge;   that worldwide NATO is the more powerful alliance;   that either the USA or the USSR spends more on defence, depending on which way you look at it, but that comparison is no guide to military effectiveness; that the tank balance depends not on quantity, nor indeed on quality, but on an assessment of missions (this also applies to aircraft). The intention of this section has been to clear the air a little to provide the breathing space for the 'relaxed assumptions' on which the reduced programmes were based.

New Technology of Warfare

The second background factor is the veritable revolution in military technology that has occurred over the past decade. Any attempt to think or plan ahead, even over a relatively short period, in military matters cannot ignore these developments. They are especially relevant to this study because, if assumptions are to be relaxed, we need to consider the extent to which they should be changing anyway. How can tactics and perhaps strategy adapt to the new technology; how can it be employed to advantage?

In general, the military environment must change as a result of these developments. Precision Guided Munitions (PGM) - especially Anti-Tank Guided Weapons (ATGW), Surface-to-Air-Missiles (SAM), and Anti-Ship Missiles (ASM) - challenge the utility of many of the most valued, and costly, weapons systems now in service, production, or development. Along with PGM have come developments in surveillance, detection and communications which many have described as creating an 'Electronic Battlefield'.

These weapons and developments have been surfacing over a considerable period of time. For instance, the Condor ASM which is now in limited production in the USA grew out of an operational requirement established in 1962; development was begun in 1965, although numerous problems delayed initial production until 1973/74, since when there have been further modifications to it. (25) In many ways these advances in military technology grew out of the Vietnam war. The in-effectiveness of, especially, the more exotic surveillance gadgetry in that war have led many to question their ultimate usefulness; even here, though, it appears that some of these developments could be utilized in other wars. (26) Wider awareness of both PGM and the other kinds of developments began when General Westmoreland presented his vision of the future in 1969: 'I see battle fields on which we can destroy anything we can locate through instant communications and the almost instantaneous application of highly lethal firepower.' (27)

Over the years some staggering claims have been made for PGM. Norman Augustine, Assistant Secretary of the US Army for R & D, has asserted that, 'On today's battlefield a fundamental change has taken place: What can be seen can be hit. What can be hit can be destroyed.' (28)

An only slightly less ambitious claim was made by James Digby of the Rand Corporation when he wrote, 'With PGM seeing a target can <u>usually</u> lead to its destruction' (29) (emphasis added), and dated that that situation twenty years into the future .   A major US army programme, Terminal Homing, introduced in 1972, is described as, 'a technology programme out of which may grow definitive projects to provide a first-shot, first-hit capability for both rockets and cannons'. (30)

Whatever else may be said to qualify these claims, expectations are clearly extremely high.   The intention of the Terminal Homing Programme is even more ambitious than the claim made by Augustine (since he did not specify how many attempts were needed to hit the target).

The first real public demonstration of the effectiveness of the new weaponry was in the Arab-Israeli October 1973 war. Egypt and Syria between them lost as many tanks as the USA has in Europe.   Israeli tank units were for a time neutralized by Egyptian ATGW of Soviet make.   Until the point when the Egyptian army re-opened the offensive, apparently to relieve the hard pressed Syrians by drawing off Israeli strength, Egyptian defences held firm against all Israeli counter-attacks. (31)   Almost all the recent interest in and enthusiasm for PGM stems from the experience of that war.

As yet, however, as might be expected with weapons in a relatively early stage of development, PGM have many problems. Notably their performance in low or nil visibility will nowhere near match their fair-weather performance, and this is especially relevant if one is attempting to draw lessons from the Yom Kippur desert war and apply them to the very different European conditions.   Other major problems include the use of smoke-screens to defeat manual ATGW, Electronic Counter Measures (ECM) against both SAM and semi-automatic ATGW, and the use of heavy 'suppression fire' against operators of SAM and ATGW. (32)

However, it has been pointed out that ECM against SAM will be expensive and difficult to mount, and by no means assured of success;   in an 'ECM race' it is more likely that SAM will win than manned aircraft, especially since cost restraints will become very relevant. (33)   The judgment made by Major-General Farrar-Hockley, that 'Future electronic countermeasures and counter-countermeasures will doubtless prefer fluctuating favours on both air defence and air attack', (34) may turn out to be inaccurate.   With suppression of SAM also very difficult, especially for missions demanding deep penetration of enemy space, it is likely that air defence will maintain an adequate and effective superiority.

Equally, ATGW operators can work on a 'shoot and scoot' basis especially with semi-automatic ATGW which have a 'fire and forget' capability — once fired, the weapon seeks out its target while the operators select another target or get out as fast as possible — which will make them very elusive targets. And artillery engaged in suppression fire can, of course, be 'counter-suppressed'.   In the continuing struggle, on the one hand, to suppress SAM and ATGW, and, on the other, to protect them, artillery range will be of prime importance, which suggests increasing adoption of terminally guided rocket artillery for suppression and counter-suppression.

It is possible to draw a number of conclusions and lessons from all this. Firstly, as General Weyland, the US Army Chief of Staff, put it rather laconically, 'The battlefield was never a safe place, and modern weapons systems have made it more deadly than ever.' (35) Major-General Farrar-Hockley has suggested that weapons like <u>Milan</u>, a medium range ATGW, would 'restore to the infantry what they have lost in terms of the present anti-tank gun falling behind in range and hitting power compared to the tank'; he pointed out that, in these terms, by 1950 the battle tank was 'streaking away', but that its advantage is now being eroded. (36) Steven Canby has pointed out that, together with night vision and sensor devices, individual companies now have 'a greater anti-tank capacity than a whole mechanised battalion before 1970'. (37) In turn this means that 'small units can be very powerful when equipped with PGM' and that 'it will become much less desirable to concentrate a great deal of military value in one place or vehicle'. (38) Increasingly this suggests the deployment of small highly mobile units with a very high weapon-to-man ratio, and increasingly it challenges the continuing validity of tanks, not only by blunting their attacking power, but also because they will no longer be needed to protect the infantry.

The role of tactical air power also comes under question, as Canby has argued:

> (The Arab-Israeli war) showed that, in a hostile air-defence environment, tactical air power cannot operate without extremely expensive losses. Israel was finally able to destroy the Egyptian air-defence system in one area only, and that by ground forces. This presents a paradox: the purpose of NATO air power is to support ground forces; but if ground forces have to defeat the enemy's ground forces first, what is tactical air power for, except for neutralising enemy air power? (39)

A loss rate of over 2 per cent per sortie flown has always been thought of as prohibitively high, and it has generally been thought possible to exceed that even before the intro-duction of guided SAM; with PGM the loss rate will be even higher. Both the Warsaw Pact and NATO will be able to create extremely hostile air defence environments, and it may be that even the role of neutralizing enemy air power will eventually slip out of the hands of tactical air power. Such a development, even if it is likely, must be dated well into the future, but not so far ahead that it is irrelevant for procurement decisions now being taken. Concentrating a lot of value into one vehicle will not make sense either on the ground or in the air. If there is any future for tactical air power it may be with cheap and rugged aircraft with the very limited role of 'helping the army in situations in which it has been thrown off balance'. (40)

These considerations may help explain why the tank and aircraft numerical balances were dismissed in the previous section.   It is no longer the case that the best way to stop a tank (or an aircraft) is with another tank (or air-craft).   Compared to tanks and other armoured vehicles, to all kinds of aircraft and helicopters, these missiles are easy to carry, require little training to use, and appear to be simple and reliable in operation, characteristics which, as has been widely recognized, 'add up to formidable effectiveness against an opponent superior in tanks and aircraft'. (41)   This judgment is obviously very relevant both to consideration of the present central European military situation, and to the effect on it of the kind of reductions we are considering.

Many people have concluded that the tactics made possible by PGM will help the defence more than the offence, particu-larly because of the increased expense of attacking vehicles such as tanks.   How long this will hold true, if at all, is uncertain.   Important qualifications upon this view have been provided by Richard Burt, but he held the view that ground will be more difficult to seize, adding, 'it will be almost as difficult to recover'. (42)   If the defending side needs to go on the counter-attack it will encounter exactly the same problems it has just been posing to the enemy.   The conclusion from this is that the defence **must** be strengthened, because counter-attack is less effective, just as offence in general is less effective.

The same problems for high value targets and concentrations will exist at sea.   Aircraft and small ships armed with PGM could provide a major challenge to even the best protected aircraft carriers, large cruisers etc.   On 18 April 1976 the US Navy carried out a successful test firing of a guided missile from a hovercraft moving at 'more than 60 knots' against a target five miles away.   The 100 ton vessel has achieved speeds of 82.3 knots, and Navy spokesmen were reported as claiming that an enlarged version could be ocean-going and capable of long cruises. (43)

Obviously there will be numerous hurdles to surmount before this development has advanced to the stage of deploy-ment, but it may well be a sign of the coming times.   Armed with ASM, quite possibly capable of hitting targets over the horizon, potentially with its own SAM air defences, it will be a lethal and elusive ship.   It may be thought particularly appropriate for the Royal Navy, which is largely deployed within relatively easy reach of land, obviating the necessity for long cruises, in areas where land-based aircraft could provide target acquisition (although the use of Remotely Piloted Vehicle (RPV) launched from the ship might also be considered).

It is, of course, impossible to say precisely where these developments are leading.   But it is possible to say that tank offensives will be less effective because of PGM, and

that the utility of tactical air power is also diminished.
Counter measures against various types of PGM may be possible,
but those who point out the problems for PGM must also
recognize the problems of counter measures, especially
against PGM deployed defensively. Smallness and speed look
like being tremendous advantages, together with good concealment. High cost equipment may just not be cost effective.
The prospect raised is increased defensive strength with
reduced manpower and equipment costs.

And while one can conclude together with the American
Director of Defense Research and Engineering that, 'the
nation which exploits them more imaginatively and deploys
them more aggressively will gain enormous advantage' (44)
one should add that these developments also serve to raise
questions about the validity of existing weapons, technologies,
and tactical doctrines, which may no longer have an effective
role.

Reduced Programmes

I - Cancellation of Polaris

This programme simply involves phasing out all provision for
the Polaris strategic nuclear force. Such a step would not
amount to unilateral nuclear disarmament, as it leaves
tactical nuclear weapons with both army and RAF, but it would
still be significant.

Military arguments for the retention of Polaris have
always been rather vague. Arguments that the four submarines,
of which never more than two and often only one are at sea at
any time, constitute an independent deterrent have never
carried much weight. Whatever uncertainties exist about
nuclear deterrence, and there are many, it is certain that a
threat by Britain to use strategic nuclear weapons against a
nuclear-armed opponent is tantamount to an offer to commit
suicide. It is equally hard to argue that Polaris makes any
meaningful contribution, for either good or bad, to the
strategic nuclear balance. Britain's 192 warheads, of which
a maximum of 96 can be fired at any one time, look rather
pathetic beside the American and Soviet totals of approximately
8,580 and approximately 3,260 warheads respectively. (45)
The degree of overkill which the British warheads can add to
the American overkill has no strategic significance whatsoever.

There is, indeed, some doubt about Polaris' role. The
British government continues to refer to Polaris as 'a
contribution to NATO's strategic deterrent', (46) but the
Pentagon appears to disagree; in the Annual Report for FY
1977 Polaris is included in the list of NATO's 'theater
nuclear system' and there is a discussion of their use in a
'theater wide nuclear war'. (47) One would not have thought
that Polaris' warheads were accurate enough for use on or near
the battlefield, in which case the Pentagon may be referring

to their use against cities outside the USSR, although this would be a novel definition of 'theater nuclear systems'. The distinction between tactical and strategic weapons is always difficult, and is of little interest to the people against whom they are used, but it is significant that the Pentagon quite categorically disagrees with the British government on Polaris' role.

In any case, the effect of abandoning Polaris is more likely to be political than military. The political case for retention of Polaris has always been more comprehensible, though not stronger, than the military case. It has been argued that Britain is given a special status by possessing strategic nuclear weapons, although evidently it is not a status with which the Pentagon endows Britain. The spectacle of Britain tagging weakly along behind the USA, ineffectually pretending to be a great power because it has nuclear weapons, is not, one would have thought, likely to enhance British status with anybody. It is a reflection of outdated pretensions, easily seen through. Precisely what status, and in whose eyes, would be lost is impossible to define. It is arguable that by dropping the sham Britain would gain status. Ultimately, however, status depends on numerous other factors, and there is no reason in this context why Britain should not scrap Polaris.

It has also been argued that the Polaris force has the effect of tying the USA in to the NATO alliance. This argument depends on the assumption that the USA is seeking to leave the alliance. Used more subtly, the argument goes on to suggest that, in time of need, the USA would be prevented from abandoning Britain and Europe to a Soviet dominated fate (which it might do rather than risk all-out nuclear war) by the Polaris force, which could be used independently to force the issue. One might prefer, however, that the issue were not forced, Polaris not used, and Britain not annihilated in the inevitable response. In any case this argument depends on the dubious assumption that American spares for Polaris are a more significant indicator of commitment to Europe than American troops in Europe. And if the conclusion of the argument is that abandonment of Polaris risks shattering the NATO alliance, then that is plainly farcical and unworthy of serious attention. The role of Polaris within the alliance is minimal.

On the other hand abandonment of Polaris might be used by a government in an attempt to stimulate wider and faster moves towards arms reduction and disarmament. The attempt might fail, but if Polaris is to be abandoned because of the need to reduce military spending, and if there are no _adverse_ political or military effects, then the attempt should be made to wring all possible _advantage_ from the decision.

The Labour Manifesto has committed the Labour government to abstaining from development of 'the next generation' of strategic weapons; the government has consistently made clear its acceptance of this commitment, which it has

224

explicitly stated includes not MIRVing Polaris missiles or purchasing Poseidon.   In other words, the prospect of a Britain without strategic nuclear weapons is regarded quite sanguinely, as it should be.   There is no reason why the date for the realization of that prospect should not be brought forward, and there might be much to gain.

II - Naval General Purpose Forces

This programme comprises five major identifiable segments:
1) to pay off <u>Ark Royal,</u> the Navy's sole remaining aircraft carrier, which is anyway due for disposal in 1978/79 (48) - there are no consequences from this apart from those that the government's decision shows it is prepared to accept;
2) to pay off <u>Fearless</u> and <u>Intrepid</u>, the Navy's two assault ships leaving the Navy without an amphibious warfare capability, and raising a question mark over the future of the Marines (although Greenwood points out that there would still be possible roles for the Marines (49));
3) to run down patrol submarines activity and 'stretch' building programmes for the new Fleet submarines;
4) to cancel the second and third ASW cruisers and <u>Sea Harrier</u>, and pay off <u>Hermes</u> - to follow the logic through the discussion below assumes cancellation of <u>Invincible</u>, the first ASW Cruiser now under construction;
5) possibly 'stretching' construction programmes for the new frigates and destroyers, and/or paying off <u>Blake</u> and <u>Tiger</u>, the Navy's two cruisers.
    Taken overall, this adds up to a very significant change in the structure of the Royal Navy.   If all options were taken up it would leave the RN without ships larger than light cruisers, and the RN would become, relatively speaking a 'small ship Navy'.   In addition one mission, amphibious warfare, would have been entirely lost.   The question, therefore, is whether the lost mission is any great loss, and whether the other changes diminish the ability to accomplish the remaining missions.
    The bulk of RN activity is already concentrated in the Eastern Atlantic and Channel areas.   Worldwide cruising ability is not demanded of the RN, and the RN tends to operate within fairly easy reach of the home bases, as was made very clear when Mr Jaffray, Assistant Under Secretary of State at the Ministry of Defence, commented to the Expenditure Committee that under most circumstances the proposed ASW Cruiser task force could operate under shore-based air cover. (50)   The <u>Guardian</u>'s comment on this programme's proposals that they would create a Navy which 'could hardly exert control at long range', (51) a comment which may be echoed quite widely, yet which is hardly relevant given the present scope of naval activities.
    In this area, 'the main threat ... comes from the large and growing Soviet submarine fleet, an ever increasing proportion of which is nuclear-powered.   Hence our naval forces are largely devoted to anti-submarine warfare, while retaining a significant general purpose capability'. (52)

There is no explanation as to why, if the Soviet submarine threat has been identified as the cause for concern, other general purpose capability is retained. Power projection ashore, a role central to the US Navy, is already only peripheral to the RN. Of the two assault ships, one is on active duty, and Hermes, now converted to an ASW Cruiser, retains a secondary role for amphibious warfare. This hardly adds up to an effective or 'significant' capability for power projection In any case, whom do we plan to invade in what conceivable scenario within the scope of present RN operations? Keeping just two assault ships may well, as Greenwood suggests, 'owe as much to reluctance to pay off "ships of quality" as to operational imperatives for their retention'. (53) Since the capability is already so limited, there seems little to be lost in scrapping it altogether. The role of the Marines should then be examined very closely; while the main rationale for maintaining the force might appear to have been eliminated there is no inherent reason why it should be disbanded. This is certainly a matter for further consideration given that there are tasks which the force could still usefully perform.

From the White Paper, it is clear that British maritime effort is largely devoted to sea control and sea denial; in other words, if a war starts in Europe, Britain's role will be to fight the 'Battle of the Atlantic' all over again, a point which has not been lost on the Conservatives who now frequently demand that the MoD compare the present tally of Soviet submarines with the numbers possessed by Germany in 1939. This role very much goes hand in hand with the concept of a long war in Europe; while the concept is a dubious one it is not necessary to question it in order to raise doubts about the effectiveness of the ASW Cruisers.

Among other things, these ships will be large targets. The task forces of which they will be the core will be large and unnecessary concentrations of military value. The Cruisers themselves will be protected from the very submarines they are supposed to be hunting and destroying by ASW frigates and destroyers. The government has argued that the ASW Cruisers will be no more vulnerable than other surface ships; to justify the extra cost they would have to be a lot less vulnerable, but in any case the claim is inaccurate – they will be more vulnerable because they are bigger, and since they will also act as command and control centres they will be very inviting targets. Discussing rationales for the reduced programmes, Greenwood suggested that 'In general the most effective (ASW) systems in future seem likely to be land-based aircraft and hunter-killer submarines'. (54) (Emphasis in original.)

To which one might add that smaller ships will also be more effective than the ASW Cruisers, whose ASW capability was reported by the Navy minister to be not greater than a frigate's. (55) Smaller ships could be deployed over

larger areas with equal effectiveness and, obviating the need
for a task force, will also be smaller targets. Paying off
Hermes, halting construction of Invincible, and cancelling
further ASW Cruiser construction will not diminish the RN's
ASW capability, and may enhance it.

If the ASW Cruisers are struck out, then there is obviously
no rationale for Sea Harrier. Once described as a 'desirable'
addition to the ASW Cruiser task force, it has now been
promoted to being 'essential' a fact upon which the House of
Commons Expenditure Committee commented acidly. (56) What
may not have been properly realized is that Sea Harrier is
intended to have a very limited role, described to the
Expenditure Committee by Mr Jaffray:

> The Harrier does not pretend to be a first-class
> interceptor like the Phantom; it does not pretend to
> be a first-class strike aircraft like the Buccaneer;
> but what it does pretend to do is to have a very good
> capability, and an immediate reaction capability,
> against the Soviet long-range maritime patrol aircraft,
> the Bear, which . . . is so crucial for the targetting
> and setting up of the attack from long-range Soviet
> missiles, whether these are air-launched, or surface-
> ship-launched. (57)

So, although he went on to describe it as part of the task
force's air cover, it will not actually be used to protect
the force from aircraft or surface ships.

Although the White Paper gives a slightly different
description of Sea Harrier's mission, (58) this version is
the same as that given by the Navy Minister in a recent
debate. (59)   The only intention is to use Sea Harrier
against the Soviet Bear, in the belief and the hope that
Bear will not be replaced by anything 'very different or
superior throughout the 1980s', (60) which is a little inconsistent since in virtually every other field of weapons
MoD consistently predicts stunning imminent Soviet
improvements.   Questioned about this apparent inflexibility,
Mr Jaffray replied, 'The Harrier will be versatile within
the limits of its capability', (61) which sounds impressive,
but could equally well be said of bows and arrows, and is no
answer to the criticism that a lot of money is being spent on
a very limited ability which may well have only temporary
utility.   Given these limitations it does not seem that much
would be lost if it were cancelled; certainly, the capabilities of Sea Harrier are not enough to justify retaining ASW
Cruisers on that basis alone.

If the general principle of a navy made up of relatively
smaller ships, primarily directed for ASW operations, is
accepted, then Blake and Tiger must have increasingly dubious
roles.   Originally conceived of as orthodox cruisers during
the Second World War (which is when work on them started),
they do not appear to add any significant strength to the

RN's capability, either for ASW or against surface ships, even with their advanced gunnery. Taking up the option to pay these two vessels off towards the end of the 70s might eliminate the budgetry need to slow down construction of new frigates and destroyers. On the other hand, if these construction programmes are affected, it can be argued that with cancellation of the ASW Cruisers reducing the escort requirement, and with the same true as a result of abandoning the amphibious warfare mission, less frigates and destroyers are needed. The bulk of the frigates and destroyers are fairly modern, so immediate replacement is not needed.

Running down patrol submarine activity and delaying Fleet submarine construction might not be necessary, in a budgetry sense, if the frigate and destroyer programmes were slowed. On the other hand there is no very clear reason why patrol activity should not be run down and construction of Fleet submarines slowed. The 1975 Defence White Paper explained quite clearly that the intention behind the maritime strategy was to prevent the naval balance shifting 'so far in favour of the Warsaw Pact that it had an evident ability in a period of tension to isolate Europe by sea' because of the effects this would have on 'Allied confidence and cohesion'. (62) This was put another way by Mr Jaffray — 'But basically we have to ask ourselves how much is enough to deter rather than to win a conventional battle.' (63)   Or, as Greenwood puts it, 'The important thing is to be there.' (64) (Emphasis in original.)   These principles leave plenty of room for manoeuvre, and focus attention on the thought that the RN's present role is more as a morale-booster for NATO than anything else. A direct worsening of the maritime situation is unlikely to result from taking up the submarine option in this programme.

In conclusion it appears that this programme could be implemented without drastic consequences. Paying off Ark Royal and the two assault ships merely jettisons peripheral missions, and, in the former, involves no change of plan. The ASW mission can be accomplished at least as effectively and at reduced cost if Hermes is paid off, and Sea Harrier, Invincible and the other two ASW Cruisers are cancelled. Paying off Blake and Tiger would then leave a choice between delays in construction of frigates and destroyers, or delays and a run down in the submarine force. While it would be possible to strike a balance between these two options (i.e. lesser delays and a smaller run down), it could justifiably be argued that the submarine programme should remain un- affected, on the basis of greater effectiveness in both ASW and anti-surface ship operations. This does not, of course, mean doing without destroyers and frigates; it merely means introducing new ships more slowly than now planned into an adequately effective force.

No new mission concepts are necessary, but what would be needed is a change of 'self-image' by the Royal Navy, which will have to get used to doing without very large ships and invasion plans.

III - European Theatre Ground Forces

This programme hinges on the reduction of BAOR from 55,000
men to about 30,000, with appropriate adjustments in equipment and in Home Forces.   Two major criticisms of this, or
any other suggestion for BAOR reductions are that they lower
the nuclear threshold and harm allied confidence and cohesion;
both criticisms are discussed below.

   The BAOR reduction amounts to a 45 per cent cut.   Although
a major reduction of British forces, it is less dramatic
within the overall NATO ground forces context.   It corresponds to a 3.5 per cent reduction in North and Central
Europe, 1.8 per cent reduction of NATO ground forces in the
whole of Europe, and a 3 per cent reduction of NATO ground
forces within the region covered by the MFR talks in Vienna.
(If French ground forces were also counted, the percentages
would be lower.)   Proportionate equipment reductions might
mean, for example, reducing the BAOR tank force from 650 to
350, corresponding to reductions of 4.3 per cent, 2.8 per
cent and 4.7 per cent of NATO tank strengths in the same
areas, again excluding French forces. (65)

   The effect of this on combat strength needs to be re-
assessed on the basis of the present reorganization of BAOR,
including the elimination of the Brigade level of command,
and thought likely to 'increase the number of company-sized
combat teams by over a quarter'. (66)   Thus if the 45 per
cent reduction were carried out proportionately on combat
and support units, one might expect a reduction in combat
personnel of <u>30 per cent</u> compared to present levels (pre-
reorganization).

   This figure is not immutable.   Canby has suggested a
fairly basic re-structuring of NATO ground forces logistic
arrangements;   he suggests that by concentrating 'logistic
assets' centrally, instead of maintaining them separately
at each level of command, it would be possible to halve the
size of NATO's 'divisional slice' (i.e. personnel in a
division plus a share of non-division support personnel),
while maintaining about the same 'foxhole' combat strength.
(67)

   Restructuring like this might have less effect on BAOR,
whose divisional slice is relatively small compared to the
Americans', but further structural reforms might lessen
the effect of the reductions on combat strength, making it
perhaps 25 per cent lower than present levels.

   The programme might be further altered by assuming that
manpower and equipment reductions do not need to be proportional, or by cutting more sharply in more expensive equipment to make manpower reductions smaller.   However, the
present BAOR reorganization includes 'improving the weapon-
to-man ratio'; (68)   the extra room for manoeuvre to diminish
the size of the combat reduction compared to present levels

might therefore be only marginal.

At any event, with this programme we are talking about a BAOR with something like 75 per cent of its present combat strength, still a major reduction, but not as dramatic as might first appear.

A British Government seeking to implement this programme would have two main options: other NATO nations could be asked to fill breaches they could identify as resulting from the programme, in ways that they thought fit; alternatively, altered conditions might promote altered strategic or tactical doctrines.

While it might be felt that a British Government had some justification in asking other NATO nations to step in because it could no longer afford present force levels, this does not seem a very satisfactory course, if only because the allies might wish to bargain over it, possibly extracting an unwelcome quid pro quo. As already shown, the effect of this BAOR reduction on total NATO ground force levels is very small; its importance to NATO is more likely to be as a possible indicator of declining British interest in the central front - in a sense, its symbolic value rather than actual effect. This, of course, raises a basic strategic issue - the importance to Britain of the central front. A decision that its importance had declined would be of fundamental significance in NATO, with implications nobody should ignore; on the other hand, it might be seen to be important for Britain to recognize that changed priority, and act accordingly, despite the profound effects on the Alliance's co-ordination and relations. It would be perfectly possible, for example, with or without enforced reductions, to argue that the real role for the British Army is to be found in Northern Ireland, and in acting as a reserve force for the European front in crisis or in war. In such a case one might then argue for going beyond the scope of this programme, reducing BAOR still further but without Home Force reductions, maintaining a system of 'dual basing' with equipment ready for troops if they had to be sent to Europe. At this point the discussion begins to stray outside the terms of reference for this paper, but it is important to recognize that it would be quite possible to make a basic reassessment of Britain's strategic priorities, disregarding for the moment their desirability on a number of different levels.

If such a reassessment were not undertaken, and if implementation of this programme did not imply changed strategic priorities, the British Government could then either ask the allies to lend a hand in the BAOR region in Europe, or undertake a tactical re-ordering, with less profound but nonetheless significant consequences.

In seeking this possibility the developments in technology described earlier might be of some help. General Merglen has already been quoted, drawing the lesson from the 1973 Arab-Israeli war that new ATGW could provide formidable defence against an enemy numerically superior in tanks, (69) and Digby's comment about the increased power of small units might also be borne in mind (70). Following a similar track, Kenneth Hunt has commented, 'What is sought here is primarily weapon systems that will improve the defence <u>even if the enemy has them</u> (such as barrier systems), but also those that tend to reduce the handicap of having fewer men.' (71) (Emphasis added.)

It would be possible, for example, to consider deploying BAOR for defence, with reinforcements from other NATO partners expected to take on any counter-attack role that were necessary. Given that it will be easier to hold on to ground, but harder to regain lost ground (72) it could be argued that limited ground forces in the front line should not be bled to provide a counter-attack capability that might be inadequate, and might only be needed because defensive capabilities had been needlessly reduced.

The role of NATO ETGF is described in the 1976 Defence White Paper in quite limited terms: 'NATO's conventional forces deter, and would defend against, conventional attacks. They deter because they could repel a limited incursion upon NATO territory, and because they would provide a <u>stalwart first defence</u> against a major conventional attack on Western Europe.' (73) (Emphasis added.)

If this is indeed their role, then the suggestion that BAOR be deployed purely for defence does not seem so terribly revolutionary, and one can then take up and fully accept Canby's judgment, which may cause a shiver in the spines of some army units but nevertheless appears very valid: 'Conventional war should therefore be concerned with thwarting the attacker and preventing these forces from occupying ground. Combay and support forces (and their underlying technology) which are not geared to this objective or are needlessly expensive and redundant.' (74)

Once a possible change of role for BAOR is considered, and in the light of this view, a heavy question-mark hangs over the role of tanks. Their role as a counter-attack force would be invalidated because the mission had been abandoned by BAOR, and they would not be needed to protect infantry; one can only conclude that BAOR could then do without them altogether.

An article in the <u>Observer</u> recently suggested that tanks have only a short period of useful 'life' left, and asked, 'But is the Army, and particularly its tank regiments, prepared to accept that? Or will the Tank become tomorrow's equivalent of the well-bred cavalry horses on which British military leaders placed their faith in the 1920s?' (75)

History is replete with examples of once-useful tactics and weapons being employed with disastrous effects, with the effect of new weapons being unrealized by military chiefs. This is the much referred-to syndrome of 'preparing for the last war', an understandable failing, especially in services which place so much emphasis on tradition, but a serious one. It is therefore worth considering whether or not BAOR's role

might, in order the better to provide 'a stalwart first
defence', be switched away from that of a fully balanced
ground force, to the more limited one of defence and
attrition.   This would not mean, of course, that BAOR
would become an immobile defensive force; such a proposal
would be foolish.   Alternatives to that include providing
a very mobile defence, based on small units and utilizing
their increased power, together with their comparative
speed of movement and ease of concealment.   While the force
would not be able to follow through a major counter-attack,
capability for light fast raids could be an important attribute.
At any rate, the possibility is posed and could be considered.

While this possibility has less fundamental significance
than a basic strategic reassessment, it would nevertheless be
an important new departure.   It would probably meet with a
strong counter-attack from NATO allies, even though it does
not necessitate abandoning the concept of forward defence,
and would increase defensive strength in the BAOR region
compared to the present situation.   It could be the occasion
for a major tactical disagreement with serious results for
NATO's tactical cohesion.

Once again, it would be for the British government to weigh
priorities.   A possible decision would be to press on with
the changes in the belief that they benefited NATO's defences,
reflected the altering of strategic and tactical conditions by
technological developments, and would anyway be adopted by
other NATO countries in due course.   It could also be argued
that the tactical disagreements resulting from such a change
would be no worse than present tactical disagreements (Britain
and the Federal Republic of Germany have completely different
tank warfare doctrines, for example), and that there should
therefore be no ill-effects on cohesion arising from this
specific change, and that, in any case, there would be no
desire to see a tactical disagreement needlessly become a
major political and strategic row.

These points are taken up further on in the paper.   But
it can be said now that if the government were to decide that
the military effects of these or other tactical changes were
beneficial, as it could rationally do, then it might also be
able to press on with the changes despite disagreements with
allies.

Of course, a decision to do away with tanks would provoke
a very hostile reaction from many quarters within Britain.
Even a suggestion here that this could be a possibility for
consideration may evoke strong reactions.   These could partly
arise from the view that it is tanks which 'make' an army,
and partly from the fact that the Warsaw Pact has so many
tanks.   It cannot be said too emphatically that forces
should be structured according to one's own assessment of
needs, priorities, missions, and ability — not according to
the other side's.   If Soviet literature draws strongly on
World War II experience to justify the continuing emphasis
on dense armour and numerical preponderance, (76) that is no

reason for NATO allies to follow suit. In the past, tanks
have protected infantry – a task for which they are not now
needed; they have combatted other tanks – which ATGW can
do better; they have been the basis of offensives – which
NATO is not considering; they have carried out counter-
attacks – a role on which less emphasis might be placed.
Given all this, it can sensibly be argued that BAOR's limited
resources can be more effectively utilized by placing a much
lower priority on counter-attack, and accordingly scrapping
tanks.

As a last comment on this possibility, it should be recognized that if the new technology is to be applied to its full
effect, it will not be by grafting it onto old doctrines (as
now appears to be happening), but because its implications for
tactics, strategy, and for structure have been fully realized
and acted upon. If the need to reduce military spending
clears the air for consideration of how best to do that, then
so much the better.

A further consideration must intervene here. Reductions such
as this programme entails effectively mean fewer men available
to be sent to Northern Ireland. While the government's policy
is to reduce army forces there, and while most people hope it
can be carried through, the possibility that it will not work
must be considered. It could be that the reductions in this
programme would need to be modified in the light of that;
certainly it would be daft to reduce the manpower of the
British Army without consideration of its major active role.
Short of the kind of reassessment discussed earlier, it might
be possible to increase the provision for air transport, so
that if BAOR were temporarily drawn on for some contingency
in Northern Ireland (or elsewhere in Britain, come to that),
the speedy return of the troops to BAOR remained possible.
On the other hand, of course, a major reassessment of the
utility of the troops in Northern Ireland might go hand in
hand with the reductions in this programme.

Implementing this programme would probably require an
increased contribution to what is now the BAOR region by
other allies, or a fundamental reconsideration of the role
of the army in the European front, or tactical changes;
these appear to be the choices, even though the programme's
manpower reduction is marginal in the context of all NATO
ground forces in Europe, because the significance of the
reductions is not purely military but also, especially for
our allies, political. Limiting BAOR's role might provide a
way of mitigating the most profound consequences of reductions; although this would undoubtedly result in tactical
disagreement, it might be no worse than the disagreements NATO
is already engaged in. Quite certainly, NATO allies would
not find it easy to accept such reductions or such tactical
changes; whether or not they are proceeded with would depend
upon the priority attached to the various factors.

233

IV - Air Force General Purpose Forces

This programme is based on cancellation of MRCA, reduced provision for tankers and air transport, and increased expenditure on Harrier and Jaguar. Arguments about this programme will revolve around MRCA.

It is worth starting, however, by pointing out that in the light of the discussions above, air transport reductions might be undesirable. Certainly, if the decision were to provide a means of moving soldiers quickly in case of crisis, between Northern Ireland and BAOR, it would be foolish to reduce air transport capacity below identifiable lower limits. At the same time, requisitioning of civil aviation might be a method of providing some air transport capacity (after all, British Rail ferries are already earmarked for sea transport). While it is correct, working from the budget, to include air transport in this programme, it really needs to be considered alongside the preceding one, and decisions on it should be taken in the light of decisions on ETGF.

Cancellations of MRCA would result in an eventual withering away of deep penetration roles for British tactical air power, in maintaining Harrier and Jaguar for close support and battle-field interdiction (together with Buccaneer) in using Jaguar in the reconnaissance role which it is already taking over from Phantom, (77) and in using Phantom in the interceptor role for air defence, supplemented by SAM.

There are several doubts about the continuing value of deep penetration roles — strike and deep interdiction (raids on enemy airfields and preventing at source movements of men and equipment). As one important analysis put it:

Not only must the attack aircraft be able to fly a long distance carrying a munition payload sufficient to make the trip worthwhile, it must also be prepared to cope with a variety of increasingly sophisticated area defences en route and to deliver its payload accurately against point targets around which enemy air defences are apt to be particularly intense and well organized (more so than in a fluid battle environment). Moreover, barring the use of nuclear weapons (which, it is important to note, the same kind of aircraft is equally capable of delivering), this feat must be accomplished again and again at a low long-run rate of attrition in the attacking force. (78)

As has already been noted, an attrition rate of 2 per cent (i.e. 20 aircraft lost for every 1,000 sorties flown) has always been regarded as prohibitively high, enough to force precipitate cancellation or curtailment of the offensive. (79) This rate, surpassable for some time, will be well exceeded by guided SAM. The most likely counter is replacing conventional weapons with nuclear ones, or using long-range missiles, or both; this is far from a happy thought, but it does suggest that deep penetration roles for manned aircraft are a dead duck. Loss of the two missions under this heading through cancellation of MRCA is no loss — it merely saves money.

Many have suggested that close support is the main tactical air mission, together with the derived mission of providing air superiority. Canby suggests that 'terminal guidance, by correcting the artillery's weakness, removes the basic rationale underlying much of the close air-support mission', (80) and this could naturally lower the value of air superiority as well. Even if this is true, it has already been noted that tactical air might have a role in helping a defending army recover when it has been thrown off balance. Even if the ground forces' defensive capability can be strengthened, it is clearly desirable to have as much insurance on that capability as possible; if it is more difficult to regain lost territory, and therefore less reliance can be placed on counter-attack, tactical air may have a useful role, operating in an environment where SAM defences, though still formidable, would be harder to organize. The battlefield interdiction role — hindering the movement of men and equipment close to the battle area — has been suggested as requiring the flexibility that only manned aircraft can provide, because of the highly fluid situation. But this role has always been best accomplished by destroying road and rail networks, which are not mobile targets, and therefore do not need such flexibility — medium range missiles might do the job just as well, and more cheaply.

All of this suggests a requirement for a small 'crisis period' tactical air force, ideally made up of cheaper and specialized aircraft. Increased expenditure on _Harrier_ and _Jaguar_ may therefore not be necessary. In any case, it is not certain that even this limited utility for tactical air power will be of long duration. Further improvements in stand-off technology, the development of RPV's for target acquisition, and 'sowing' the battlefield with sensors, either beforehand or by scattering them from RPV's and drones, might all lead to the elimination of the close air support mission in the late 80s or 90s. But for the next decade, at least, close support aircraft may be useful and important in a fairly limited role, and should not be scrapped since the RAF has types in service capable of fulfilling that role. At the same time, and although these are not the 'cheap and rugged' aircraft which would be best suited for this role, these considerations would rule out the option mentioned by Greenwood (81) of purchasing the American A-10, or any other aircraft, for this role.

If this removes the rationale for procuring the so-called 'common' MRCA, it does exactly the same for the Air Defence Variant (ADV). Procuring the ADV, but not going ahead with the 'common' version would result in an interceptor whose unit production and development price was around £10 million at autumn 1975 prices. There are, in addition, grounds for doubt about the success of converting the 'common' version into the ADV, doubts which centre on the performance of the engine, which, at the time of writing, is reportedly far short of what is required.

This reflects a common problem in placing emphasis on versatility in aircraft design, or indeed in the design of any general purpose machine. In the continuing debate over multi-purpose aircraft versus specialized aircraft William White's judgment appears more and more relevant:

> Firstly, while it can perform several different operations, it is generally less proficient at doing any single one than a machine designed specifically for that operation and no other would be.
> Second, a machine intended to perform several operations must have the performance qualities needed to satisfy the most demanding one, and when it is used where lesser performance would suffice, efficiency suffers from over design. (82)

An earlier paper prepared for the Study Group discussed the way in which these general principles applied to and are demonstrated by the case of MRCA.(83) But the major argument against both versions of MRCA is, as has been said about TSR-2, 'It was designed in one type of strategic environment, but was expected to survive in quite another.' (84)

This still leaves the role of interceptor, or more accurately, the mission of air defence, which must be fulfilled by something. The programme as envisaged by David Greenwood includes using Phantom for that mission. This would, however, be seen only as a stop-gap.

The Ministry of Defence argues that an ideal air defence system for Britain must include both interceptor aircraft and SAM. (85) Given Britain's geographical situation, it may be difficult to envisage an air defence system composed solely of SAM, even though it is this method of air defence which has so increased the odds against successful performance of deep penetration missions, particularly if aircraft striking against Britain were armed with stand-off missiles which could eliminate almost entirely the need to penetrate British air space. But the problems posed by stand-off technology to British air defence are no more likely to be solved by manned interceptors than by SAM. A 'one-shot sure-hit' capability would be required, together with an extremely fast reaction time. The MRCA interceptor force would have a strength of 165 aircraft, probably augmented in the early stages of its deployment by some Phantoms. The ability of this size of force to mount continuous patrol, in adequate numbers, even during a relatively limited period of time in a crisis, should not be over-rated; the additional air defence capability of manned interceptors on top of SAM may be only marginal, but the extra cost will be vast. Therefore, it could be suggested that while deployment of Phantom for air defence temporarily might make sense, efforts should be devoted to SAM

236

development, including use of sea-based SAM, rather than to
procurement of a replacement interceptor.   This will not be
a 100 per cent satisfactory solution to the problems of
British air defence nor will manned interceptors provide a
completely satisfactory solution.   Ultimately Britain may
have to accept a degree of geographically imposed insecurity
in this regard.

In general, this programme appears to recognize realities,
especially the growing problems about deep penetration missions,
and attempt to adapt to them rather than demand major unforced
strategic changes.   Indeed, it looks like a more realistic
and cost effective programme than presently pursued by Britain.
The provision for air transport could be modified in the light
of decisions over ETGF, and it might not be necessary to take
up the option for increasing expenditure on Harrier and Jaguar.
Strategic effects of implementing this programme are by no
means unacceptable.   The RAF would, like the Royal Naval if
programme II were implemented, have to become used to a
different status, but it will probably have to get used to a
lower status in the years to come anyway.

The Nuclear Threshold

Discussing reductions of US forces in Europe, John Newhouse
concluded, 'The United States would be most unwilling to exercise the nuclear option until after non-nuclear combat had
reached a high threshold.   Among the perceived effects of a
substantial withdrawal of US troops from Europe would be a
lowering of that threshold.' (86)

This is an argument which can be used against any proposed
reduction of conventional forces, ground or air, in Europe.
The argument is that reducing conventional forces means they
will be no more than a 'tripwire' for nuclear weapons in time
of war, which diminishes deterrence by leaving fewer alternatives, and less time to steel the will for reluctant use of
tactical nuclear weapons.   While this is a serious criticism
it does have a number of weaknesses.

Behind arguments about the nuclear threshold, and evident in
the quotation from Newhouse, there often seems to be a view
that there is a bad time and a good time to use nuclear
weapons.   In fact, there is no good time to use them;

> Studies and war games ... showed that high casualty
> rates and a great amount of collateral damage were
> likely to result from a tactical nuclear war in Europe.
> Even under the most favourable assumptions, it appeared
> that between 2 and 20 million Europeans would be killed,
> with widespread damage to the economy of the affected
> area and a high risk of 100 million dead if the war
> escalated to attacks on cities. (87) (Emphasis added.)

Defending western Europe with tactical nuclear weapons would
destroy large parts, if not all, of the 'defended' countries.
By concentrating on _when_ tactical nuclear weapons should be
used, the nuclear threshold argument blurs the question of
whether they should be used.

If their use is initiated, it is extremely unlikely that
the resulting nuclear exchange will remain limited:

> Is it, for example, conceivable that an aggressor,
> having achieved his immediate aims thanks to a prepon-
> derance of conventional forces on the spot, would stand
> meekly by and watch his forces being wiped out by
> tactical nuclear bombardment when he had it in his
> power to inflict unacceptable damage by the use of
> strategic nuclear weapons? (88)

There is a clear distinction between conventional and
nuclear weapons, the so-called firebreak; once that bridge
has been crossed, there is no further clear distinction all
the way up to the thermonuclear holocaust. The 1976 Defence
White Paper states that tactical nuclear weapons 'represent a
link between NATO's conventional and strategic nuclear forces'
(89) - that is precisely what is wrong with them; strategic
nuclear war is tied more closely to conventional war. This
is done to aid deterrence, but at great cost to defence. And
if it is indeed the case that thus aids deterrence, then what
happens to Newhouse's and others' case that effective
deterrence demands a high nuclear threshold? It may be
true that a high threshold aids deterrence, or it may be true
that a low threshold and a clear link between conventional and
strategic forces aid deterrence, but one would appear to rule
out the other. In any case it will be extremely difficult
for critics of these programmes to use the argument that the
nuclear threshold will be lowered, at a time when it has just
been announced that NATO has nuclear depth charges. (90)

These concerns have been answered by those who argue that
it is possible to have limited use of nuclear weapons, even
at a strategic level, and that use of one category does not
necessitate use of the next one up. Of course, this is
possible, but there would be numerous pressures forcing
escalation once the firebreak had been crossed. Most
obvious is the tendency, once committed to a policy, to
push it a little bit further to achieve success, and then a
bit further again, even when an objective observer can see
the policy is self-defeating. Less obvious, but equally
important for battlefield use of nuclear weapons, is the
fact that the radioactivity of nuclear explosions blanks
out or distorts radio communications over a wide area for
a considerable period; lost communications means inaccurate
targeting, which in turn necessitates the use of larger
weapons.

One use suggested for tactical nuclear weapons is the
famed 'shot across the bows' - use of a small number of

nuclear weapons, perhaps against a major Warsaw Pact city,
to show resolve and determination, and force the enemy to
think again. It would take a remarkably stiff Soviet
upper lip not to respond in kind, to show equal resolve and
determination and force NATO to think again. Alain
Enthoven has commented that, in eight years of studying
the problem in the 60s, he saw no convincing scenario in
which use would make sense. (91)
    Even though tactical nuclear weapons would be useless
for defence, and unlikely to have the desired effect as
a 'short across the bows', it can still be argued that
they act as a deterrent; this argument takes the form of
asserting that they introduce uncertainty into the
situation, leaving the Pact in doubt as to what response
they would face to any aggression. This is a novel
departure for the theory of deterrence, at any level,
which has normally been thought to depend upon creating
absolute certainty as to the risks involved in any course
of action. Perhaps more important is the comment by
Jeffrey Record that, 'It is equally likely that in an acute
crisis (uncertainty) would serve to reduce the prospect of
non-nuclear combat by encouraging a pre-emptive Pact nuclear
strike.' (92)
    In conclusion, these weapons just should not be there,
and the British government should make every effort to
achieve a satisfactory method of removing them. They would
be disastrous for defence, make the most dubious contribution
to deterrence, increase the likelihood of nuclear war in a
crisis period, and do not make sense in any of the scenarios
which have been constructed to rationalize their use. They
are a menace to the countries they are supposed to be
defending. Whether or not Britain reduces its military
spending it should be trying to have them removed from
Europe. The argument about the nuclear threshold is
completely invalid, diverting attention away from the real
problems about nuclear weapons in Europe, which exist regardless of these programmes. A final point is worth making:
the nuclear threshold is not some mystical point which moves
up and down the graph of combat according to natural laws —
it is the point where NATO takes the decision to use nuclear
weapons; the nuclear threshold is low or high depending upon
that decision. Given the possibilities of strengthening
defensive power, even if BAOR is reduced, there is no
necessity for NATO to use nuclear weapons earlier rather than
later. Best of all, NATO should decide not to use them at
all.

Nato's Confidence and Cohesion

A second criticism of any reduction in any conventional
forces is that it will necessarily lower NATO's confidence,
and weaken its cohesion. Mr Jaffray put it very plainly to

the Expenditure Committee:

> I think it is really not possible to demonstrate at what
> level NATO's forces would fail to deter the Soviet Union.
> It is not a black and white situation at all.   All that
> one can say, I think, is that the more conventional forces
> that NATO has, the more credible NATO's strategy will be
> in a situation of nuclear balance or nuclear equipoise,
> and the greater the confidence that the Alliance will
> have in that strategy — and, of course, the converse
> would be equally true. (93)

Mr Gilmour, the Opposition Defence spokesman, expanded
this to say that cutting defence 'damages our position in
the eyes of NATO and damages NATO itself' and that 'cuts
of defence expenditure in NATO have a domino effect'. (94)
His assertion that cuts, or since he was referring to
present government policy, a refusal to increase spending,
have a domino effect is hardly plausible — the FRG, Italy,
and the USA are all increasing their military spending, and
France is also planning to.   The other arguments deserve
more consideration.

Mr Jaffray's speech is far from hard and fast.   As he
says, it is impossible to demonstrate conclusively the point
where deterrence fails.   While this does not mean that one
can reduce forces without regard for the consequences, it is
a warning against attempting to draw a clear line as the
minimum.   Mr James of the MOD attempted to draw a clear line
for the Expenditure Committee with regard to air defence,
and then found himself qualifying his position:

> The size of force that we have described in the memorandum
> ... is the minimum that we need to have an effective
> deterrent posture in air defence ... How long and in what
> circumstances that force could fight, I think even my
> professional colleagues may find it very difficult to
> answer.   What, of course, is a question none of us can
> answer is how long it might be required to fight. (95)

He is virtually saying, 'This is the minimum, but we have
no criteria on which we base that judgment.'
The difficulties of assessing this kind of criterion have
led to the approach that the defence budget should be as big
as possible at all times, an approach apparently held by Mr
Gilmour.   Thus, to the age-old question of 'How much is
enough?' comes the answer, 'As much as you can get.'   This
is an approach ridden with errors, two of the major ones being
amply demonstrated by Alan Lee Williams in his memorandum to
the Study Group: 'Before we decide to recommend a programme
reduced still further, we need to be sure that the resulting
sum total of NATO expenditure provides an adequate and effective
deterrent to the Soviet threat.' (96) (Emphasis added.)

240

Before going any further it is worth commenting that the sum total of NATO military expenditure will be barely affected by the size of cuts we are discussing. More interesting is the view that it is the scale of spending which provides the deterrent (rather than what is represented by that spending). This may just have been a slip, but it does reflect a very common error — to be more impressed by a large budget than by what it buys. And this leads to the second error — to equate cost with capability. This is no basis on which to proceed. If allied confidence is to be shattered because NATO governments follow this approach, then it is time they were re-educated.

In the absence of this approach, there are still many over-simplifications. A great deal of alarm was recently caused by a 'worst case' analysis of a massive Warsaw Pact attack catching the West unawares (97). While consideration of the worst possible is often worthwhile, it should not be taken too far. As Enthoven has explained, a massive surprise attack <u>simply could not occur:</u>

> While the Pact might attempt <u>either</u> a concealed mobiliza-tion followed by a limited-scale attack, <u>or</u> deliberate mobilization followed by a maximum-scale attack, they could not have a concealed mobilization followed by a maximum scale attack. Warsaw Pact preparations to attack NATO on a large scale would severely disrupt their economies and probably have to be publically announced. (98) (Emphasis in original.)

Even if they were not publicly announced, the preparations would be very very obvious.

In fact, partly because of this outright aggression is not expected by most intelligent people, including the government. (99) The threat now being guarded against is the use of military force for political ends, to influence NATO countries' foreign or even domestic policies. There has been a lot of discussion as to how this might happen. Dr R.J. Vincent made a brave attempt to analyse and identify something specific as the basis of this perception of threat, but concluded that attention should be directed towards 'the general impact which Soviet military power has on its environ-ment' rather than specific uses of it; as he argued, 'The concern here is directly with the view of the <u>beholder</u> of military power, and only <u>in-directly</u> with the utility of force to the <u>holder</u> of it.' (100) (Emphasis in original.) If the beholder's eye were troubled this might result in concessions in a crisis.

Now, here we have a real mess. If NATO lacks confidence, it may make concessions in a crisis. Why should it do that? — presumably because it fears the consequences if it does not. What consequences? — actual use of military force. But NATO does not expect the Warsaw Pact to use its military force directly — it does not expect invasion — so it does not <u>need</u>

to concede.   So it should not lack confidence.   So it will not have to concede.

If this logic is faulty, then what is needed to prove its faultiness is a specific demonstration of the consequences which NATO has to fear; if there are no such consequences, it need not have that fear.

In any case, it has been argued that the reduction in these programmes need not diminish the military effectiveness of Britain's NATO contributions.   There is no reason why NATO's confidence should be harmed by them.   Naturally, were such reductions carried out, there would be plenty of complaints, but it is hardly conceivable that at the end of the day NATO will officially decide it has lost confidence as a result and is now prepared to give in.

Indeed, what is at stake in these reductions is not so much the future of NATO, as NATO's trust in the future of Britain's contribution to NATO.   There might be a general fear that Britain was about to follow France's example and withdraw its forces from the integrated command.   Perhaps, however, a worse fear should be Britain's inability to fulfil its paper commitments because of its economic difficulties. Real and forcible opposition to these reductions by NATO might ultimately be self-defeating, and a government seeking to handle these reductions intelligently and minimize the resulting discord would do well to make that point quite strongly.

The military cohesion of NATO would be especially affected by ground force reductions, whether or not tactical changes were initiated, and by Britain taking an intelligent stance over tactical nuclear weapons.   It does not make for easy co-operation, if one member of the alliance begins to take a different course over crucial tactical and strategic questions.   However, it is not really desirable for NATO to cohere around bad policies.   It would have to be decided whether NATO cohesion were an absolute priority, to be achieved at all costs, regardless of where the concensus finally emerged, or whether NATO should strive for the maximum possible cohesion, but not at the cost of other important British priorities.

Negotiating cohesion at the Vienna MFR talks would undoubtedly be weakened by unilateral British reductions within the MFR region.   But, while it is right and important for European nations to seek 'not only to strengthen inhibitions regarding the use of force but also to reduce others' ability to use it', (101) this aim should not necessarily over-ride all other policy considerations. An objective assessment of the three-year old Vienna talks can only conclude that they have produced nothing, and do not look like producing anything meaningful for some time. NATO negotiators in Vienna will undoubtedly be annoyed, and it may well affect the Warsaw Pact commitment, to the talks, but it cannot really be argued that it will hinder progress when there has been almost none.

242

NATO cohesion might also be marred to the extent that progress on standardization of equipment and collaboration in procurement were held up. Given the very profound political and strategic implications of this process, there are many who might not be displeased if progress were delayed for a while. Without significant public discussion of the policy, the British government appears calmly to be heading toward standardized defence policies in NATO Europe together with increased collaboration in development and production in military industry, along a path which could lead to the creation of multinational European arms firms and an integrated EEC weapon procurement agency, along the lines recommended in a report to the EEC Commission last year.(102) If this is indeed the direction of government policy, consciously or unconsciously, then its implications and possible consequences far outweigh those of a reduction in British military spending. These are not technical questions – they are fundamental points of policy; what the policy is and means should be considered first, before any further progress in this direction.

Any reduction in military spending in NATO will undoubtedly arouse hostility in NATO. This will doubtless rather isolate Britain, and lower trust in Britain's continuing commitment to the Alliance, and especially the central front. While this will adversely affect NATO's essential cohesion, there is no military reason why NATO's confidence in its military effectiveness need be diminished, although one cannot doubt that claims will be made that it is so affected. One result of this will be to diminish British status inside the alliance.

These political consequences are important and deserve consideration. But they should be balanced against other priorities, particularly the perceived economic advantages of reducing military spending and redeploying resources. In weighing up the various factors, we should attempt to distinguish between the real strategic consequences of reducing forces, and the response in the eye of the beholder. A hostile response may be based on a misjudgment of the real effects; a British government need not be put off its policy by that response and could decide to implement the policy and ride out the shock waves, which would doubtless diminish to nothing after a few years.

It has to be said that a reduction in British military spending of £1,000 million per year at 1974 prices is not going to cause the collapse of NATO. It will not significant- ly, if carried out via reductions like the programmes considered here, affect NATO's overall force levels or military strengths. Depending upon the priorities as assessed by the government, it would be possible to proceed with it.

243

Conclusions

It appears that programmes I (Polaris), II (Naval GPF) and IV (RAF) could be implemented without adverse strategic consequences. Missions lost are ones in which Britain already has only a peripheral interest, or ones which will be ineffective, and the important question is how best to carry out the remaining missions. Programmes presented by David Greenwood have been modified in places, with the intention of tying up a couple of loose ends, and normally with the effects of producing increased savings.

Programme III (ETGF) is more problematic. While the BAOR reductions are actually less dramatic than might at first appear, some far reaching changes might be necessary to implement this programme. Barring a complete strategic reassessment, fairly basic tactical changes could be made within the context of present NATO strategy, without diminishing NATO's defensive effectiveness.

The strategic consequences of all these programmes therefore seem acceptable. Strong political opposition within NATO should be expected, and this would almost certainly weaken NATO's cohesion. On the other hand, it does not seem likely that these political effects will really damage NATO's military credibility as it now stands – in other words, there is unlikely to be strategic 'feedback' from political consequences. A government implementing these or similar reductions will have to do it in the face of strong opposition, which sensible handling and argument might mitigate somewhat, but which will nevertheless be important.

In the end, the crucial issue is one of priorities. Those who accept the value of cutting military spending will probably argue that the price Britain will pay within NATO is not too high for the economic benefits which will accrue; those who deny the value of those cuts will probably argue that too little will be gained at too high a price. So we might ultimately come to a discussion of these competing priorities, whether, for example, complete NATO cohesion is worth more than aiding industrial performance through re-allocation of resources away from military industry. A discussion of the more technical military issues can help identify what priorities are at stake, and how far military priorities are at stake, and may help disarm some strong but tendentious criticisms of military reductions. The final decision about defence, however, is a political one.

Richardson Institute, London
May 1976

Notes

In these notes the term 'Adelphi' refers to a paper in the
Adelphi Series, published by the Institute of Strategic
Studies, London; the term 'Brookings' refers to a publication
of the Brookings Institution, Washington DC.

1. Study Group Paper, No.2: David Greenwood, Defence
   Programme Options to 1980-1.

2. Study Group Paper, No.3: Paul Cockle, A Programme Option
   which retains the UK Strategic Nuclear Forces.

3. This is because Greenwood's calculations were based on
   1975 prices, while the target cut was assessed by the
   Study Group in 1974 prices.

4. Kenneth Hunt, Defence With Fewer Men, Adelphi 98 (Summer
   1973), p.1.

5. Second Report from the Expenditure Committee Session
   1975-6 Defence (hereafter referred to as Second Report)
   Minutes pp.76 and 77, Q233.

6. Donald H. Rumsfeld, Annual Defence Department Report FY
   1977 (hereafter referred to as Defence Report), pp.
   155-6, section IV/C/2.

7. 'Naval Balance remains on side of US' in The Times
   (4 May 1976).

8. Defence Report, p.157, Section IV/C/2.

9. Barry M. Blechman, The Changing Soviet Navy (Brookings,
   1973), p.26.

10. Defence Report, p.154, Section IV/C/2.

11. Admiral James L. Holloway, FY1977 Military Posture of
    the United States Navy, p.9.

12. Michael Klare, 'Superpower Rivalry at Sea', Foreign
    Policy (Winter 1975/6).

13. Statement on the Defence Estimates 1976, Cmnd 6432, Ch.I,
    para 20.

14. 'Study Insists NATO Can Defend Itself', The Washington
    Post (7 June 1973).

15. Les Aspin, 'Budget Time at the Pentagon', The Nation
    (3 April 1976).

16. Jeffrey Record, *Sizing up the Soviet Army* (Brookings 1975), p.9.

17. The disparity between Record's figures and the ones in Table 3, taken from the IISS, is probably accounted for by Record including French totals, British Home Forces, and troops in Portugal.

18. *The Military Balance 1975/6* (IISS, London, 1975).

19. See, for example, his first major article after leaving the Pentagon, James Schlesinger, 'A Testing Time for America', *Fortune* (February 1976).

20. 'CIA Estimates that Soviet Spent 40% More Than US on Defence in the Last Year', *The New York Times* (28 February 1976).

21. Reprinted in 'Estimating Soviet Defence Spending', *Survival* (March/April 1976).

22. Les Aspin 'How to look at the Soviet-American Balance', *Foreign Policy* (Spring 1976).

23. Ibid.

24. Ibid.

25. Defence Market Survey *Market Intelligence Report* (1974), 'Condor'.

26. For a full discussion of the genesis and drawbacks of the 'Electronic Battlefield', see Michael Klare, *War Without End*, (New York, Random House, 1972), Ch.7.

27. General William C. Westmoreland, Address at the Annual Luncheon of the Association of the United States Army, 14 October 1969, reprinted in the *Congressional Record* (16 October 1969).

28. Norman R. Augustine, 'One Plane, One Tank, One Ship, Trend for the Future', *Defence Management Journal* (April 1975)

29. James Digby, *Precision Guided Munitions*, Adelphi 118 (1975), p.4.

30. Defence Market Survey *Market Intelligence Report* (1975), 'Terminal Homing'.

31. See Elizabeth Monroe and A.H. Farrar-Hockley, *The Arab-Israeli War, October 1973 - Background and Events* Adelphi 111 (Winter 1974/75), pp.26-7.

32. See Richard Burt, 'New Weapons Technologies and European Security', _Orbis_ (Summer 1975).

33. Steven Canby, _Military Doctrine and Technology_, Adelphi 109 (1974/5), see p.37.

34. Monroe and Farrar-Hockley, _The Arab-Israeli War_, p.31.

35. General Fred C. Weyand, _Statement on the Posture of the United States Army FY 1977_, to the Senate Armed Services Committee (3 February 1976), p.14.

36. _Second Report_ Minutes p.150, Q589.

37. Canby, _Military Doctrine_, p.24.

38. Digby, _Precision Guided Munitions_, p.4.

39. Canby, _Military Doctrine_, p.4.

40. Ibid.

41. General A. Merglen, 'Military Lessons of the October War' in _The Middle East and the International System - I The Impact of the 1973 War_, Adelphi 114 (Spring 1975); the papers which make up this publication were presented at a Seminar in September 1974.

42. Burt, 'New Weapons Technologies'.

43. 'Breakthrough in sea war tactics', _The Guardian_, and 'US Vessel Fires Missile at High Speed', _The International Herald Tribune_, both of 19 April 1976.

44. Malcolm R. Currie, _Overview Statement_ on the Department of Defence Programme of Research, Development, Test and Evaluation FY 1977 (10 February 1976), p.14.

45. Figures were taken from _Jane's Weapon Systems 1976_ and _The Military Balance 1975/6_, and reconciled by author's estimate.

46. Cmnd 6432, Ch.I, para.42.

47. _Defence Report_, p.99 Section III/A, and p.106 Section III/D/2.

48. See _Jane's Fighting Ships 1975/76._

49. Greenwood, _Defence Programme Options_, p.25.

50. _Second Report_, Minutes, p.74, Q222.

247

51. David Fairhall, 'The Chop Logic of Cutting Defence', The Guardian (14 May 1976).

52. Cmnd 6432, Ch II, para 7.

53. Greenwood, Defence Programme Options, p.23.

54. Ibid.

55. See Pat Duffy, Hansard (12 May 1976), col. 477. Since the paper was written it has been suggested that a technical analysis of the ASW Cruiser reveals that it has a greater ASW capacity than a frigate; the argument that it is so appears quite strong, and I apologize for passing on without comment what now seems possibly to have been an error on the part of the minister.

56. Second Report, p.xxi.

57. Second Report, minutes, p.74, Q222.

58. Cmnd 6432, Ch III, para 2.

59. See Pat Duffy, Hansard (12 May 1976).

60. Second Report, Minutes, p.75, Q223.

61. Ibid.

62. Statement on the Defence Estimates 1975, Cmnd 5976, Ch. I, para 25b.

63. Second Report, Minutes, p.70, Q198.

64. Greenwood, Defence Programme Options, p.22.

65. Based on figures in The Military Balance 1975/76.

66. Cmnd 6432, Ch II, para 13.

67. Canby, Military Doctrine, p.18.

68. Cmnd 6432, Ch II, para 13. Naturally, the completion of this reorganization will change the base line of combat strength from which reductions begin, and the reduction in combat strength will accordingly be somewhat sharper.

69. See note 41.

70. See note 29.

71. Hunt, Defence with Fewer Men, p.14.

248

72. See note 42.

73. Cmnd 6432, Ch.I, para 29

74. Canby, _Military Doctrine_, p.14.

75. Andrew Wilson, 'What NATO should really worry about', _The Observer,_ (21 March 1976).

76. See Record, _Sizing up the Soviet Army_, p.41.

77. Cmnd 6432, Ch II, para 24.

78. William D. White, _US Tactical Air Power_ (Brookings 1974), p.68-9.

79. Ibid.

80. Canby, _Military Doctrine,_ p.39.

81. See Greenwood, _Defence Programme Options_, p.29.

82. White, _US Tactical Air Power_, p.56.

83. Study Group Paper No.14, Dan Smith, _Tornado — Cancellation, Conversion and Diversification in the Aerospace Industry._

84. Alan and Geoffrey Lee Williams, _Crisis in European Defence_ (Charles Knight, London,1974), p.60.

85. _Second Report_, Minutes pp.111-17, Memorandum SCOE 77/1 presented by Ministry of Defence; see paras 8-12 and 19.

86. John Newhouse and others, _US Troops in Europe_ (Brookings 1971), p.153.

87. Alain C. Enthoven and K. Wayne-Smith, _How much is Enough? Shaping the Defence Programme 1961-1969_ (Harper and Row, New York, 1971), p.128.

88. Fitzroy Maclean, 'Nuclear Deterrence and Conventional Force', _Brassey's Annual 1970_ (Praeger, New York, 1970), p.5.

89. Cmnd 6432, Ch.I, para 30.

90. 'Britain has nuclear sea bombs', _The Sunday Times_ (2 May 1976).

91. Alain C. Enthoven, 'US Troops in Europe: How Many? Doing What?' _Foreign Affairs_ (April 1975).

92. Jeffrey Record with Thomas Anderson, _US Nuclear Weapons in Europe: Issues and Alternatives_ (Brookings, 1974), p.18.

93. Second Report, Minutes, p.71, Q198.

94. Ian Gilmour, Hansard (31 March 1976), col.1352.

95. Second Report, Minutes p.137, Q527.

96. Study Group Paper No.4. Alan Lee Williams, Defence
     Programme Options to 1981: Politico-Strategic and
     Economic Implications.

97. Lord Chalfont, 'The West must act to defend itself
     while it still has the chance', The Times (15 March
     1976).

98. Alain C. Enthoven,'US Troops in Europe'.

99. Cmnd 6432, Ch I, para 25.

100. R.J. Vincent, Military Power and Political Influence:
     The Soviet Union and Western Europe, Adelphi 119
     (Autumn 1975), pp.18-19.

101. See J.I. Coffey, New Approaches to Arms Reduction in
     Europe, Adelphi 105 (Summer 1974), p.3.

102. Reported in The Financial Times (10 October 1975).

B. MILITARY EXPENDITURE AND THE ECONOMY

6 NOTE ON THE MILITARY EXPENDITURE AND NATIONAL PRODUCT: UK AND CERTAIN OTHER COUNTRIES

By Frank Blackaby

1. Introduction

The question set was as follows: 'What cut in UK military expenditure, from its level in 1976, would be needed to bring the UK percentage of military expenditure in gross domestic product down to the average figure for our main European Allies, by 1981?   Assume that our main European Allies are France, Italy and West Germany.   Express the size of the cut (a) as a percentage of UK military expenditure in 1976 (b) in £s million, at 1976 prices.' This paper sets out to answer this question and nothing more.   It does not consider whether this is a good question or a bad question, or whether other approaches to the whole matter would be better.

In order to answer this question, assumptions have to be made (a) about the likely movement of the share of military expenditure in gross domestic product for France, Germany and Italy between 1976 and 1981, and (b) the rate of increase of the UK's gross domestic product (in real terms) between 1976 and 1981.

Of course there is no way of knowing for certain what will happen between 1976 and 1981.   However, when there are clear trends in the past 10-15 years, a reasonable assumption is that these trends will continue, amended if there is any reason to amend them.   These 'central' assumptions are set out in the next section; they are the 'best guess' at what will happen.

In section 4, the 'central' assumptions are varied, to see what difference it makes.

For those who simply want the answer given by the 'central' assumptions: for the UK percentage of military expenditure in gross domestic product to come down to that of our main European allies by 1981, military expenditure would have to be cut by between a quarter and a third, from its level in 1976.   That is a cut of about £1,700 million, at 1976 prices, on the NATO definition of UK expenditure.

2. Assumptions About Military Expenditure and National Product

There are a bewildering number of different figures around for military expenditure as a percentage of national product.   For military expenditure, there are national figures, and figures on NATO definitions.   For national, or domestic product, there are different concepts which can change the levels of the percentages (though they do not make much difference either to the trends through time or to the general pattern of relativities between countries).   Some comparisons use gross domestic product at market prices; others use gross domestic product at factor cost, which has the further complication that, for gross domestic product just as for military expenditure

253

there are national figures which differ from the internationally standardized figures published by the UN.

The principle followed in this paper is, first, to
ensure as far as possible that the same concepts are
used for all countries; and secondly, to prefer internationally standardized figures to national ones. For
gross domestic product, the UN standardized series for
gross domestic product at market prices is to be preferred.
However, the calculations would come out about the same
if gross domestic product at factor cost had been taken
instead. (1) For military expenditure, NATO definitions
are used.

Because of these definitional problems, it follows that
figures derived from other sources — which quite often do
not say which definition of gross domestic product has been
used — will probably not be compatible with the set of
figures given here.

Chart 1 shows the trends from 1960 to 1979 in military
expenditure as a percentage of gross domestic product for
the four countries considered in this paper. The figures
for 1976 — and to a lesser extent for 1975 — are liable to
revision. Note also that the period 1974-6 was a bad
period for national output in all countries, with a
significant fall in output in 1975. As a consequence,
military expenditure, as a percentage of GDP, was off-
trend in this period. It is possible to make an adjustment for this, and estimate military expenditure as a
percentage of 'full-employment' gross domestic product for
the three countries. The figures are shown in the table.

Military expenditure as a percentage of:

		1974	1975	1976
Germany:	GDP actual	3.6	3.6	3.5
	GDP full employment	3.5	3.0	3.1
France:	GDP actual	3.7	3.9	3.9
	GDP full employment	3.7	3.7	3.7
Italy:	GDP actual	2.9	2.8	2.6
	GDP full employment	2.9	2.6	2.4

In considering future trends, it is useful to have
these adjusted figures in mind.

The question to be answered at this stage is: 'What
share of GDP is it reasonable to assume that France,
Germany and Italy will be devoting to military expenditure
in 1981?' In 1976, the actual 'weighted' share for the
three countries taken together was 3.4 per cent. (2)

There are various ways of making a reasonable guess at
these 1981 figures. One method would be to fit a straight-
line trend to the 'full-employment-adjusted' share from
1960 to 1976. This is not an unreasonable assumption:

indeed, it is always risky to assume that the future is different from the past unless there is a very good reason for thinking this. An extrapolation of the straight-line trends would give a 'weighted' percentage for the three countries in 1981 of 2.8 per cent.

However, it may be objected that a definite decision has been taken to alter the past trend. So an alternative estimate is to assume that military expenditure in all three countries rises at 3 per cent a year in real terms from now on, and that they return to growth rates of national product which are a little lower than their trend growth rates. This gives a share of military expenditure in GDP in 1981 of 3.2 per cent for the three countries. This is also the figure that would be obtained by assuming that the 'full-employment-adjusted' shares of military expenditure in GDP were the same in 1981 as in 1976. This is also 3.2 per cent.

For purposes of this exercise, we have taken as the central assumption the figure given above, producing a 1981 figure of 3.2 per cent; that is, that those countries together will be devoting 3.2 per cent of their aggregate GDP to military expenditure in 1981. The figure is biased towards the high side. It virtually ignores the long-term falling trend.

3. The Remainder of the Calculations

It remains to make an assumption about the growth of UK gross domestic product between 1976 and 1981; to apply the 3.2 percentage to the 1981 figure; and to compare this figure with 1976 military expenditure. (All these calculations are at 1976 prices.)

The long-term trend in the growth of UK GDP, at constant prices, is 2.8 per cent: that is the figure from 1954 to 1973, thus discounting the recession since 1973. The 'control assumption' put in for 1976-81 is the significantly higher figure of 3.5 per cent. We have assumed that, whereas for Italy, Germany and France further growth rates are slightly below their trend rates from now on, for the UK the figure is significantly above the trend rate, because of North Sea oil. The calculations cannot, therefore, be accused of bias, to get a high 'military-expenditure-cut' figure. Once again, the bias in the 'central assumption' about growth-rates is one which will tend to reduce the figure of the cut in military expenditure needed.

Applying a 3.5 per cent growth-rate to the 1976 estimate of UK GDP market prices (on UN definitions) (3) gives a figure for 1981 of £140 billion (at 1976 prices). 3.2 per cent of that figure is £4,480 million. This compares with the 1976 figure of military expenditure of £6,188 million. So the cut needed is around £1,700 million at 1976 prices, which is a cut of between a quarter and a third.

4. The Consequences of Alternative Assumptions

How sensitive is this calculation to the assumptions made? The table which follows varies the two crucial assumptions - the UK growth rate, and the share of military expenditure in GDP in France, Germany and Italy in 1981. For the UK growth rate, a high figure of 4.5 per cent a year is taken, and a low figure of 2.8 per cent a year. For the share of military expenditure in GDP for the three continental countries, the high figure is one of no fall at all from the actual figure in 1976 - 3.4 per cent. The low figure is the continuation of historical trends, leading to the 1981 estimate of 2.8 per cent.

On all these variations, a substantial cut in UK military expenditure should be required to approximate to the Continental share of GDP in 1981. The lowest possible cut is 20 per cent, the highest is 39 per cent.

	High Continental percentage in 1981 (3.4)		'Central' Continental percentage in 1981 (3.2)		Low Continental percentage in 1981 (2.8)	
	Implied cuts in military expenditure					
	£(1976) mm	Per cent	£(1976) mm	Per cent	£(1976) mm	Per cent
Fast UK growth-rate (4.25 per cent a year)	1253	20	1543	25	2124	34
Central UK growth-rate (3.5 per cent a year)	1428	23	1708	28	2268	37
Slow UK growth-rate (2.8 per cent a year)	1586	26	1857	30	2398	39

It is, therefore, beyond a reasonable doubt that, to bring the UK share of GDP devoted to military expenditure down to the average of Germany, France and Italy by 1981, a very substantial cut in UK military expenditure is needed. Over a wide range of assumptions, that conclusion stands.

Annexe — Are Adjustments Needed for Conscription?

1. In my view, there is no need to adjust these figures for the fact that the Continental countries have conscription, and we do not, for the following reasons (spelt out below):

   (a) By 1981 it is very possible that either conscription will have been reduced in length, or abolished; and there is no evidence that this leads to a rise in the share of military expenditure in GDP.

   (b) In any case, in making a comparison one must not ignore the fact that, although the professional soldier costs more, he is — according to military judgment — also worth more.

   (c) Even if one were to make an adjustment for conscription, I do not see any good reason for doubting the Ministry of Defence calculations that the adjustment would be a small one.

   (d) In the 'central' projections, allowance is in any case made for a rather slower fall than past trends show for the share of military expenditure in GDP in Germany and France.

   These various points are set out in more detail below.

2. If the presence or absence of conscription did in fact make a great deal of difference to these percentages, then one would expect to see a kink in series of figures at the point when conscription was abolished, or when the period of conscription was reduced. Again, if the presence or absence of conscription was very important, changes in the conscription system should show up particularly in the share of military pay and allowances in total military expenditure.

3. It is in fact difficult to trace any of these effects in the figures. This is probably because, when conscription is abolished or reduced, countries have tended to reduce the numbers in their armed forces; and indeed it is quite reasonable that they should, as the arguments in para 6 below suggest. For example, between 1960 and 1965 military conscription was abolished in the UK; nonetheless, the share of military pay and allowances in total military expenditure actually fell slightly. (4) In France, over the period covered by chart 1, the effective period of conscription was cut from two years to one year; nonetheless the share of military expenditure in gross domestic product trended steadily downwards. In the United States,

257

the transition to an all-volunteer army was between 1969 and
1973.   During this period, the share of military expenditure
in gross domestic product fell sharply.   The share of
military pay and allowances in total military expenditure in
United States rose a little between these two years — from
27.4 per cent in 1969 to 31.1 per cent in 1973.   However,
this was a much smaller rise than in Western Germany, with
its conscript army;   in Germany, the comparable percentages,
for military pay and allowances as a percentage of total
military expenditure, were 35.8 per cent in 1969 and 41.0
per cent in 1973.

4.   There is a further relevant point here: 'Most countries
of Western Europe, although they have not made the complete
transition to an all-volunteer force, have changed their
attitude towards conscription in another way: the concept of
pure conscription — where the cost of the conscript is
little more than his subsistence requirements — has gradually
been abandoned in favour of higher pay and better standards
generally.' (5)

5.   For these reasons, it is unwise to assume that if France
and Germany shifted to an all-volunteer army between now
and 1981 it would mean a significant rise in their figures
for military expenditure as a percentage of gross domestic
product.   It has not done this in any other case;   so
there seems no reason why it should do so in their case.
They would almost certainly offset the additional cost of
an all-volunteer army by cuts in the numbers of military
personnel.

6.   Secondly, behind any upward adjustment in the military
expenditure estimates for countries with conscription
lies the assumption that a conscript soldier is worth
as much as a professional soldier.   Countries with
professional armies certainly pay them more;   but the
military authorities in those countries argue that they
get military value for money.   They make the point that
with the increasing complexity of modern weapons, there
is also an increasing number of military functions where
conscripts only become competent in the last month or
two of their service.   The purpose of an army is to
fight wars;   and a modern war in Europe is likely to be
brief, with little time for calling up ex-conscripts to
the colours.   A professional army is likely to be much
more militarily effective than a conscript army.   After
all, a conscript army is full of people who do not like
the military life;   a professional army is not.   There
is the further point that in a conscript army, compared
to an all-volunteer army, a much larger proportion of the
professional soldiers are pre-empted for purposes of
training.

7.   Finally, even if one were to make some allowances for

conscription, the Ministry of Defence estimate is that it would be a small allowance. Their evidence to the Defence and External Affairs Sub-Committee of the Expenditure Committee reads as follows: (6)

> We have looked at this from the point of trying to calculate what is the hidden additional burden to the economy of France and Germany represented by the fact that they have very large conscript forces to whom, especially in the case of France, they pay very low salaries. We have tried to establish an economic price of the soldier in each country by comparing the wages paid to the conscripts with the average wage paid to persons of the same age group in the country concerned. The conclusion that we have reached is that broadly speaking, in order to measure the additional burden on the economies of Germany and France resulting from the fact that they have taken conscripts out of their economies into their armed forces, you should add on something of the order of a quarter of 1 per cent in the case of Germany and a half of 1 per cent in the case of France.

Note: I am grateful for comments on a first draft of this paper to Paul Cockle and Mary Kaldor; they are not, however, responsible for the conclusions drawn.

<u>Notes</u>

1. This is because military expenditure itself is given at
   market prices, and it should therefore be compared with
   a gross domestic product series which is also at market
   prices.   The UN terminology is 'gross domestic product
   at purchasers values'.

2. The GDPs for France and Germany were given roughly equal
   weights.   Italy was given a weight of 65 per cent of
   France and Germany.

3. Note that on UN definitions GDP at market prices is
   approximately 2.5 per cent below the figure given in
   UK national accounts (<u>Economic Trends</u>, April 1977).

4. See <u>World Armaments and Disarmament</u>,   SIPRI Yearbook
   1974, tables 7.3 and 7.4 on pages 132 and 133.

5. Ibid, p.133.

6. <u>Second Report from the Expenditure Committee Session
   1974-5</u>,   The Defence Review Reports.   H C 259, pp.35-6.

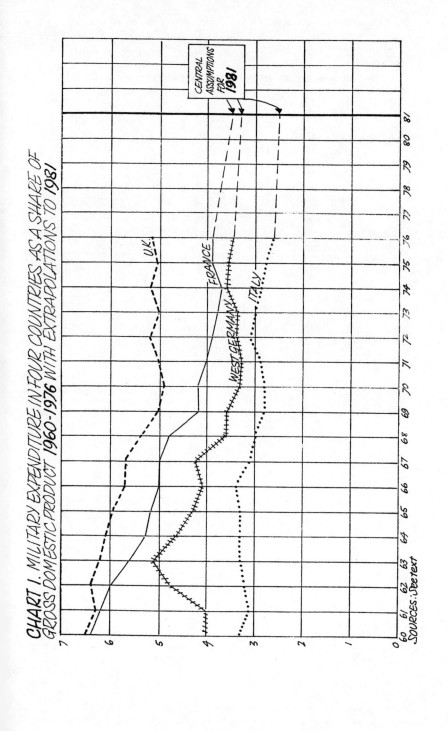

CHART 1. MILITARY EXPENDITURE IN FOUR COUNTRIES AS A SHARE OF GROSS DOMESTIC PRODUCT 1960-1976 WITH EXTRAPOLATIONS TO 1981

CENTRAL ASSUMPTIONS FOR 1981

U.K.

FRANCE

WEST GERMANY

ITALY

SOURCES: See Text

7 THE RESOURCE COST OF MILITARY EXPENDITURE

by Ron Smith

1. The Issues

In 1973, current expenditure on goods and services for military defence, in the UK, was £3,316 million, 5.3 per cent of GDP at factor cost, and rather more than was spent on either education or the National Health Service. One is tempted to ask whether the benefits generated by this expenditure outweigh the costs, or whether the resources should be used elsewhere. Although this is a natural question to ask, it is not obvious that it can have a quantitative answer. In this paper I primarily wish to consider the problems raised by the measurement of costs; though I shall make a few comments on the question of benefits, the measurement of which raises even more formidable difficulties.

I do not intend to consider the related question of distribution, which arises because the costs and benefits of military expenditure accrue to different groups. Nonetheless, this issue is important, because, in practice, decisions about arms procurement and the share of output devoted to military expenditure are rarely made on the basis of the interests of particular groups. For instance, in a consideration of procurement decision making in the USA; Kurth, 1973, identifies four main types of influence: (a) strategic or geo-political considerations; (b) bureau-cratic self-interest and inertia within government and military; (c) electoral calculations by politicians (what are called 'pork barrel' considerations in the USA); and (d) the needs of the economic system and the interests of the defence contractors. In this paper, however, the intention is to consider the aggregate economic consequences of the decisions actually made, rather than to examine why those decisions were made, and which groups benefited from them.

In the literature on defence budgeting, the economic concept of opportunity cost has been widely advocated as the appropriate measure. For instance, Greenwood (1972, p.7) says: 'The concept of opportunity cost, directing attention, as it does to the foregone alternative, the other opportunity, is a reminder that spending entails allocation and choice among competing opportunities.'

Later (p.36) he argues that there has been a failure by the government to take account of the opportunity cost of one line of spending in terms of another.

As an example he quotes John Boyd-Carpenter's evidence to the Public Expenditure (General) sub-committee, where it was said that the question of raising the school leaving age and the question of continuing Concorde were not weighed against each other, but treated as wholly separate decisions.

General Eisenhower, when President, put similar choices more vividly: 'The cost of one modern heavy bomber is this: a modern brick school in more than 30 cities. It is two

electric power plants, each serving a town of 60,000
population.   It is two fine fully equipped hospitals.
It is some 50 miles of concrete highway.'
('The chance for peace', an address reprinted in the
Department of State Bulletin, 27 April 1953.   Quoted in
Hitch and McKean, 1960.) The question I wish to consider is the
extent to which these decisions could have been weighed against
each other, and what it means to say, for instance, that raising the
school leaving age, was an alternative foregone in order to
build the Concorde.   At the most basic level, it is true
that if there is a limit on public expenditure the projects
may be competing for available revenue, but it is not true
that they are competing for the same economic resources,
except in the very long run.   In the short term, the
factors of production which might be used to produce the
Concorde are quite specific - skilled labour, R & D
facilities, specialized machine tools, etc. - and these
resources could not be used to provide the class rooms,
teachers and education materials needed to raise the school
leaving age.   Therefore, in calculating opportunities
foregone it is necessary to compare only real substitutes,
where the economic resources available can be transferred
from one use to another.   It should be noted that Greenwood
is well aware of this point   See for instance his evidence
to the Public Expenditure Committee, 21 January 1975 (H C
259).

The conceptualization of Government economic management
in terms of a Keynesian income-expenditure framework has
been particularly misleading in this respect.   The use of
this one-commodity model encourages the perception of
aggregate demand as some homogenous fluid.   The problem of
demand management is then to ensure that there is the right
amount of fluid pumped into the system.   It is assumed that
the market, or some other mechanism will ensure that demand
and resources are spread appropriately between consumption,
investment, foreign trade and the state.

In fact this does not happen quickly, because the goods
produced by industry for one use and the resources used to
produce them are often quite specific and cannot immediately
be converted to other uses.

At the minimum it is necessary to use a two-sector model
of production, which distinguishes between the industries
that produce capital goods and those that produce consumption goods.   Or at a finer level of disaggregation the
input-output model can be used to quantify the industrial
structure, though this is only done at the cost of some im-
plausible assumptions. (1)

In any event economic management must take account of
the existing structure of production and the specific nature
of many inputs and outputs.   Once one has recognized that
in general it is impossible to obtain perfect substitution
between projects, then it is necessary to be clear about

264

what exactly is being given up in any particular case, i.e.
what the resource cost of a particular expenditure in fact
is. This is especially important for society as a whole,
for whereas to an individual resources tend to be in almost
perfectly elastic supply at a given price, for society they
are not; and constraints with respect to certain specific
resources can become crucial.

The relevance of these general points about choice,
allocation and the specific nature of many resources, to
decisions about military expenditure is fairly obvious; but
to emphasize the point let us consider some military examples
where the resource and expenditure costs differ. The cost
of a soldier to society, for instance, may be very much
greater or less than his wages. With a volunteer army,
recruitment often tends to be from areas or groups subject to
high unemployment rates. Then if the individual was not in
the army, he would be unlikely to have an opportunity to work
and produce useful output, instead he would have to be
supported by the social services. Apart from equipment, the
cost to society is then merely the difference between the
consumption of a soldier and of unemployed worker — which is
very much smaller than the cost in wages. Conversely, in
expenditure terms, a conscript army tends to be much cheaper
than a volunteer army, while in terms of lost output the
resource is identical — and may in fact be very high in a
situation of full employment.

2. Factors Determining Resource Cost

In trying to estimate resource cost there are a number of
factors that should be taken into account.
The most important are:

(a) The degree of utilization in the various sectors;

(b) The multiplier and linkage effects between sectors;

(c) The degree of substitution possible between commodities,
both in the pattern of production and the pattern of
expenditure;

(d) The costs involved in adjusting from one pattern of input
and output composition to another;

(e) The rates of growth of the different sectors.

a) The degree of utlization is important because if the
factors used to produce the desired output would not be
employed otherwise, then the resource cost of the project is
likely to be small. b) However, calculating this resource
cost is no easy matter since the multiplier and linkage effects
introduce considerable complications. It may be that there is

a structural imbalance in capital stock, such that plant and workers in the defence procurement industries are unemployed, while in the consumption goods industries there is full utilization of capacity. Employing the defence workers would raise their demand for consumption goods and involve shortages, inflation or other costs. Likewise if consumption goods (e.g. food) are imported, employing the defence workers may generate balance of payments difficulties. In the longer run both of these examples would require government action to correct the structural imbalance. But in the short run, and for the purpose of calculating resource cost, the appropriate measure of utilization is relative to the desired or equilibrium level. This may be less than full capacity working, and will in general depend on the structural characteristics of the economy - balance of payments constraints, capacity constraints in certain sectors, etc.

c) The distinction made in the list above between substitution in production and expenditure is important. The former emphasizes the technical specificity of the factors that produce the output - to what extent they can be used to produce different types of goods. The latter emphasizes the social and political trade-offs which determine the extent to which consumers, government and electorate are willing, at the margin, to replace one commodity by another in their total final demand. In an open economy the pattern of production and expenditure between commodities need not match, since some domestic output is exported and some domestic expenditure is met from imports.

d) Adjustment costs are important because the process of change itself causes inefficiency, loss of output, etc. In the case of military procurement, this is particularly important because of the long lead times that complexity and R & D necessitate; the need to build up and maintain highly skilled teams, and the insecurity and loss of morale that repeated changes generates. In this situation short term changes in plans and priorities are themselves disruptive and create misallocation; therefore the need for long term security in factor supplies may outweigh short term wastes. For instance weapons systems may be ordered and then cancelled prior to production just in order to ensure that design and development capacity is maintained even when there is no immediate demand. These are ways of making long-term planning consistent with short-term needs.

e) The relevance of growth is that it increases flexibility. When output is increasing rapidly it is easier to divert it to different uses, than when the diversion requires a reduction in the amount of resources available to an existing sector. Problems arise here when the rate of growth of

productivity differs between sectors, then the slow productivity growth sector needs a growing percentage of resources
to provide the same share of output.   In the case of defence,
increasing technological complexity of weapons may mean that it
requires increasing amounts of resources to maintain the same
real offensive or defensive capability relative to potential
enemies.   (See for instance Kaldor (1972).)

3. The British Case

Having examined issues raised by the measurement of resource
costs, we shall now examine the empirical evidence for the UK.
Inevitably, the quality of the data is such that many of the
finer distinctions must fall by the wayside.   But let us
begin by considering how we can apply the categories in the
preceding section to Britain.

   With respect to the first influence on resource cost,
utilization, it seems likely that there has not been
substantial under-utilization in the post war period relative
to the desired level of capacity.   This does not mean that
plant and workers have not been unemployed but just that the
balance of payments constraint was binding, and that given
the ruling exchange rate and trade policy fuller utilization
would have caused sterling crises through most of the period.
So in general we can say there were no free resources.

   With respect to input substitution, the bulk of defence
procurement is produced by industries whose main alternative
output is investment goods (e.g. aero-space, electronics,
mechanical engineering).   The 1968 Input Output tables also
seem to indicate a closer correspondence between the
commodity final demands of defence and investment, than
between defence and either consumption or other government
current spending.   Details of these calculations are given
in Appendix A.   In addition the armed forces employ a large
number of highly skilled workers in trades which have tended
to be in short supply in the investment goods industries.

   With respect to output substitution, there are very strong
pressures that tend to make the share of consumption (private
and public) in output very stable.   Politically and socially
it is difficult to increase the share of output that goes to
provide future benefits - either in terms of national
security or the development of the capital stock.   This is not
always true, since, in time of war, not being invaded is of
immediate benefit — it becomes politically feasible to
enforce large sacrifices on the population.   But in general
defence expenditure like investment is seen as provision for
the future, and as such, not highly regarded.   This would
suggest that the boundary between consumption and non-consumption
is very stable, and that the only freedom is whether certain
items of consumption - health, education, transport etc -
should be provided publicly or privately and how the non-
consumption share should be split between defence and investment.

The government is also able to plan the adjustment of shares between defence and investment since over 40 per cent of gross domestic fixed capital formation is at present done by the Public Sector. Public investment also has a large discretionary element, and has fluctuated widely as the government has postponed and advanced projects in response to current pressures. Thus if its planning is effective the government can ensure that the joint demands of investment plus defence are consistent with what is electorally acceptable and what is feasible to import or produce in the domestic capital goods industries.

Finally the growth rate of the British economy has been very low by comparison with other advanced capitalist nations so there have been very few spare resources which can be switched towards sectors which are growing relatively.

The general indication of this argument is, therefore, that <u>for most of the post war period defence and investment have been close substitutes and expenditure on one will be at the expense of the other.</u>

This argument needs to be qualified with respect to the period of demobilization and then rearmament between the end of the Second World War and the Korean War. <u>During that period there was a seller market for exports of British capital equipment and the only constraint on foreign sales was supply shortages.</u> Therefore during that period <u>it seems likely that defence expenditure and exports were also quite close substitutes.</u> In fact, the Economic Survey of 1953 noted: important sectors of the engineering industries are heavily engaged in defence work when they might otherwise be concentrating their main energies on the export trade.'

To examine these hypotheses let us consider some regressions between investment and defence expenditure. Since it is likely that macro strategic and political decisions decided the size of the military budget, this should be the independent variable. Also <u>because we have assumed that utilization was in general at its desired level, the appropriate equation will relate the shares of defence and investment in output.</u> This also has the advantage that the error term is more likely to be homoscedastic, which it would not be if we regressed investment directly on defence spending and income. The equation we shall consider therefore will be of the form:

$$I = a + BM + e$$

where I is gross domestic fixed capital formation as a percentage of GDP at factor cost.

M is current expenditure on military defence as a percentage of GDP at factor cost.

It is not obvious, <u>a priori</u>, whether the variables should be calculated from current or constant price figures. The regressions were therefore done with both sets of data.

The theoretical discussion would lead us to expect that B should equal − 1, and a should equal the share of non-consumption expenditure. That is <u>any change in defence</u>

268

expenditure is matched by an equal and opposite change in
investment and the sum of the two is almost constant.   The
results for the UK, 1952 – 73 using current price data were:

(i) Variable Coefficient   Standard Error

          a   27.93   0.9

          M   -1.180   0.1196

       R^2 =   0.8296

       Durbin-Watson statistic   0.5966

Using constant price data

(ii) a   27.16   0.7

      M   -1.012   0.0776

       R^2 =   0.8948

       Durbin-Watson statistic   0.5323

To examine the immediate post-war period it is necessary
to allow for the substitution with exports.   This was done
by including a dummy variable which took the value 1 for the
years 1947-51 and zero elsewhere.   Reliable constant prica
data were not available for the earlier period, so the
regression was run just on the current price series.   The
results for the years 1947-73 were:

(iii) a   27.14   0.91

      M   -1.073   0.12

   dummy   -4.096   0.47

       R^2 0.892

       Durbin-Watson statistic   0.9

Let us now consider what the significance of these
results is.   Firstly the constant price version does rather
better for the 1952-73 period, both in terms of degree of
explanation and the precision of the estimates.   This would
suggest that what is at issue is the division of real
resources rather than of expenditure.   However the difference
between the two is not large and the current price estimates
would provide a reasonable approximation.   Secondly the
coefficient B is very close to -1, which indicates the ex post
military expenditure and investment have been close substitutes.
Note that the regression merely provides a description of
what in fact did happen, rather than what could possibly have
happened had the government made different decisions.   Though
earlier arguments would have inhibited the government from

behaving very differently. Thirdly the dummy variable
is significant and indicates that <u>about 4 per cent of GDP
was shifted to exports, otherwise the results confirm our
hypothesis almost for the whole period 1947-73.</u>

   The only obvious problem with these results is the
high degree of positive serial correlation indicated by
the Durbin-Watson statistic. This might suggest that
<u>adjustment was not in fact instantaneous,</u> and that a
dynamic formulation would be more appropriate. But,
when the share of military expenditure, lagged one year
was included, it had a coefficient of +0.03861, and at
ratio of 0.355, thus was not significantly different from
zero. The other coefficients had values very close to
those given in (iii). The most likely explanation for
the small positive sign, is that <u>changes in military
expenditure do induce some changes in investment demand
by the defence related industries, but the effect is small.</u>
It might also be argued that the low Durbin-Watson
statistic resulted from a mis-specification of the functional
form, i.e. that the relationship between M and I was not, in
fact, linear. Experiments with other functional forms (log-
log, and reciprocal relationships) did not indicate that a
different functional specification would remove the serial
correlation. Nonetheless the serial correlation is an
area for further investigation.

   Above, it was argued that because of the low rate of
increase of productivity and the balance of payments
constraint, growth and utilization would not be important
variables in the case of the UK. This was confirmed, when
the percentage rate of growth and percentage rate of unemployment were added to regression (iii). Neither were
significant, the coefficients having t ratios of 0.5 and
0.3.

   In general these results fit very well with our hypothesis and do seem to confirm that over the post-war period
investment and military expenditure have been very close
substitutes in the UK, though one might expect this effect
to reduce when military expenditure becomes a very small
proportion of GDP.

4. International Comparisons

Our theory would also suggest that the relationship observed
in the UK should hold true between countries. That is,
nations with high military expenditure would have low
investment (both being measured relative to GDP). However,
estimating the international relationship raises some
formidable problems. Firstly much of the data is very
suspect, definitions of military expenditure, national
income, and investment differ between countries; some
nations have under-paid conscript armies, others volunteer
armies, etc. Secondly, our relationship relates to the

full utilization trade off, and for other countries, differences in utilization and growth rates may be important: for <u>if</u>
<u>there is excess capacity or spare resources being generated</u>
<u>then it may be possible to increase both investment and defence</u>
by moving onto the production frontier where there is full
utilization.   Thirdly, spin off effects and other benefits
may be important, if, for instance, production of investment
goods is subsidized by defence contracts.   Fourthly, there is
an identification problem, for even if the true coefficient
was -1, but neither the share of defence nor of investment
changed over the period, then the estimated regression coefficient would be indeterminate.   Finally, in countries
where the public sector is smaller and the government
has less influence on investment, the adjustment lags
are likely to be longer and less predictable.   All of
these arguments suggest that the relationship within some
other countries is likely to be less well determined than
the UK.   But it remains true that if our theory has any
merit, the cross section coefficient between countries
should be close to -1.   For this relation will reflect
the long run pattern when the other factors have cancelled
out.

To test the sensitivity of the results to the data two
sets of regressions were run, using different time periods,
different sources of data and different samples of countries
If data errors are not serious then the results should be
similar for both.

The first regression used pooled OECD data, for the
years 1956-70, for US, France, West Germany and Italy;
60 observations in all.   The same variables were used as
in the UK example and the data were in current prices.
The results were:

(iv)   Variable   Coefficient   Standard error

         a   25.90   0.63

         M   -0.9461   0.11

         $R^2 = 0.55$

Using NATO data on 14 countries for the years 1960-70
(154 observations) the results were:

(v)   a   27.75   0.71

         M   -1.166   0.13

         $R^2 = 0.32$

Both of these equations were estimated including a trend
term for each country, but in neither case was the trend

significant.   In both cases B is not significantly different
from -1 at any reasonable confidence level, so we can
conclude that the international evidence also confirms our
hypothesis, and that the results are not sensitive to the
source of the data, sample of countries, or time period.
As the list of problems raised by international comparisons,
which was given earlier would lead us to expect, the degree
of explanation across countries is rather lower than in the
UK alone.

A regression was also computed for the NATO countries,
1960-70, including the percentage unemployment rate (U) and
the percentage growth rate (G) for each state and year,
together with the 1964 per capita income in US dollars, at
1964 exchange rates (Y).   This last variable, which is a
constant for each country, is a proxy for the average stage
of development.   Changes in average income over the period,
and differences in the rate of development will be captured
by the growth rate.   It should also be recognized that there
are considerable problems of measurement associated with the
data on unemployment, and in particular that these figures
may be misleading for the less developed members of NATO.

The results for this regression were:

(vi)	Variable	Coefficient	Standard Error
	a	26.63	0.9695
	M	-0.9197	0.1156
	G	0.3532	0.1063
	U	-0.9857	0.1184
	Y	0.0006816	0.000299

$$R^2 = 0.5736$$

The correlation matrix was:

(vii)						
	M	1.0				
	G	0.029	1.0			
	U	0.295	0.134	1.0		
	Y	0.117	-0.311	-0.172	1.0	
	I	-0.569	0.070	-0.609	0.103	1.0
		M	G	U	Y	I

From these results, we can note that the degree of
explanation improves substantially, but that the coefficient of M remains not significantly different from -1;
that the effect of Y on I is very small (though just
significant, t = 2.28), and that between countries growth
and unemployment do have a significant effect on the share of
investment.

It is, however, necessary to interpret the coefficients of growth and employment with caution, since the relationship raises certain problems of simultaneity and direction of causality. On the demand side, rapid growth leads to full utilization and increases the desire of firms to invest (the accelerator mechanism); but investment also expands supply which makes growth possible and expands effective demand through the multiplier. Likewise, high unemployment or low utilization of capacity, releases resources which could be invested, but simultaneously depresses demand thereby inhibiting capitalist desire to expand capacity. The form that this interaction of demand and supply forces takes is an important economic question, but not immediately germane here, since our prime concern is to correct for their influence on the investment/military expenditure relationship.

The fact that the relationship holds between countries does not imply that it will hold within each country over time, and in fact it does not do so in every case. This is indicated by an F test comparing equation (v) for the pooled data with a regression where a and B differ for each country. This test rejects the hypothesis that the coefficients are the same for all NATO states. This is not unexpected for the type of reasons which were given at the beginning of this section.

When one considers the regressions for each country using the NATO data, a two tailed t test indicates that in 7 countries B is not significantly different from −1 at the 25 per cent confidence level; in 4 countries between the 25 per cent and 1 per cent level; and for 3 countries (USA, Turkey and Greece) it is significantly different at the 0.1 per cent level. It is not clear how powerful these tests are since there are only 11 observations for each country, considerable serial correlation in a number of cases, and the extent to which the share of defence explains the share of investment varies considerably between countries. A more detailed examination of the factors operating within other countries is clearly an area for further research.

Although there are large differences between countries, even a casual examination of the coefficients indicates a systematic pattern. The relationship fits best for those countries where the defence and investment goods industries are well integrated and where the state is responsible for a large share of investment. In the USA where the state undertakes relatively little fixed capital formation and the defence industries are not well integrated, the co-efficient is close to zero.

This explains the widespread conclusion in the American literature that there is no relation between defence spending and investment, e.g. see Yarmolinsky (1971, page 240). One peculiar result is the difference between Greece and Turkey, superficially very similar economies, they have coefficients of +3.5 and −3.2 respectively, and both relations are well determined (R^2 is about 0.66 in each case).

It may be that foreign military aid and foreign capital
plays an important role in these cases, and has operated
differently in each country.

In summary, it seems clear that the cross section
results do provide confirmation of our basic hypothesis.
This is all the more striking since one might expect the
problem with the data, the wide variety of social and
economic conditions in the 14 NATO countries, and the
substantial changes which took place in the period 1960-
70 to mask the underlying relationship.   It is also true
that countries differ in this respect, and that these
differences warrant further investigation. (2)

5. Benefits

None of the arguments and results presented above, which
indicated that the resource cost of military expenditure
was foregone investment, necessarily imply that military
expenditure was excessive.   To make such a judgment, some
evaluation of the benefits, political and economic, must
be undertaken.   Briefly we can list the potential benefits
that have been suggested, as arising from:

(a) domestic security considerations; based on the threat
of foreign invasion or ruling class fears of popular in-
surrection or subversion;

(b) international gains; obtained either directly from
imperialism (low commodity prices, secure investments, etc.)
or indirectly via the leverage and bargaining strength that
a powerful position in the world provide;

(c) the need to maintain full employment and adequate
profits; this is necessary when there is ideological
resistance to other, non-military, public expenditure.   The
positive correlation between the share of military
expenditure and the unemployment rate for NATO countries,
would tend to cast doubt on the importance of this factor.
For a fuller discussion of this mechanism see Purdy (1973);

(d) 'spin off', either through the adoption of innovations
or technical developments, pioneered by the military, in
the civilian economy (see Trebilcock, 1971) or through the
potential for foreign sales of military equipment.

Some military expenditure was undertaken in the light of
specific British interests, and here, although there are
considerable problems of measurement, one can in principle
weigh the benefits against the costs.   But much of the
military expenditure contributed to the general maintenance
of the international capitalist system (the defence of the
free world).   This raises a more difficult question: did

Britain bear a more than proportionate burden, and did it gain any specific benefits from its special position as junior partner to the USA in the role as world 'policeman', which did not accrue to other capitalist powers? For instance Germany and Japan, which spent little on foreign military commitments, also benefited from the low primary product prices in the period 1952-71.

Although this is not the place for a detailed discussion there is considerable evidence that Britain did incur substantial costs (in particular the disruption caused by the Korean War rearmament, and the loss to the balance of payments in the 1960s) in an attempt to maintain an untenable international position. Although the world capitalist system as a whole and the owners of particular foreign assets benefited from this attempt the specific gains to the British economy seem to have been very small in comparison with the costs.

6. Conclusions

In this paper it has been argued that account must be taken of the resource cost of military expenditure, and it may be misleading to focus on the simple expenditure or direct opportunity costs. We saw that the resource cost would depend on a variety of factors: degree of utilization; multiplier and linkage effects; input and output substitutability; adjustment costs and the growth rate. Closer consideration of these factors suggested that the primary resource cost of military expenditure would be foregone investment. This hypothesis was put into an operational form as a regression of the share of investment on the share of military expenditure in GDP. Our theory would suggest that the slope coefficient should be -1. This was confirmed both by a time series regression for the UK, and a cross section regression for 14 NATO countries. The results did not seem to be sensitive to the source of data, sample of countries within NATO, or the choice of time period. However, some qualifications had to be made with respect to exports in the immediate post war period in the UK and with respect to intra-country regressions for some other NATO members. It also appeared that there were a number of areas which deserved further investigation.

Annexe A

The argument given in the text would lead us to expect that defence-related industries would tend to specialize in production of investment goods, rather than consumption goods (i.e. public or private civilian current expenditure). To investigate this hypothesis we shall use the 1968 Input Output tables to examine the commodity composition of output for those industries in which the supply of equipment for military

use was important.   The Tables distinguish 90 industry/
commodity groups, of which 12 account for over 80 per cent
of direct military demand for industrial products.   These
12 groups are listed in table I, together with their total
output, their net output (the output that goes directly to
final demand, rather than as an input to other industries)
and their direct supplies to military expenditure.

Table II shows the allocation of direct inputs between
the various categories of final demand and supply, for the
defence related industries and all industries, each
component expressed as a percentage of net output.   Net
output, plus imports for final demand less exports, equals
domestic final demand, which is used for: private consumption, fixed investment, military expenditure, and civilian
public current expenditure.   There is also a small
residual composed primarily of stock-building and re-
exports.   (Sources Tables P and B 1968 Input-Output Tables.)

These figures do indicate that our supposition was
broadly correct.   In the defence-related industries,
consumption and civilian public expenditure comprise 14 and
4 per cent of output respectively, compared to 54 and 6
per cent for all industries; whereas investment which
absorbs 21 per cent of total net output, takes 57 per cent
of the output of the defence related industries.

Table II only takes account of the direct inputs, but
there are also indirect inputs.   Unfortunately the figures
on direct military demand for each commodity include imports
of goods which prevent an analysis of indirect inputs into
military expenditure.   However, if we aggregate military
and civil current public expenditure, we can ask (using
Table H) what percentage of each component of final demand
originates as value added in a particular industry, taking
account of both the direct and indirect requirements.   This
indicates that value added from the defence related
industries accounts for 5 per cent of consumption and 11 per
cent of current government expenditure investment.   Again
our supposition is confirmed.   It should be noted that this
calculation asks what percentage of each component of final
demand comes directly or indirectly from the 12 defence-
related industries whereas the calculation in Table II,
asks what percentage of the net output of the defence-
related industries goes directly into the various categories
of final demand.   But both approaches confirm that defence-
related industries specialize in investment goods production.

Notes

1.   An example of the use of this approach to examine the
     economic consequences of disarmament is provided by
     Leontief et al. (1965).

2.   It might be noted that these differences indicate that there
     is nothing inherent in the model which makes B = 1 by
     definition, or by virtue of the way the data were constructed.

Bibliography

D. Greenwood, 1972, <u>Budgeting for Defence</u>, RUSI.

C.S. Hitch and R.N. McKean, 1960, <u>The Economics of Defence in the Nuclear Age</u>, Harvard University Press.

M. Kaldor, 1972, <u>European Defence Industries</u>, ISO Monograph No.8.

J.R. Kurth, 1973, 'Why We Buy the Weapons We Do', <u>Foreign Policy</u>, No.11, Summer 1973.

W. Leontief <u>et al.</u>, 1965, 'The Economic Impact — Industrial and Regional — of an Arms Cut', <u>Review of Economics and Statistics</u>, August, 1965.

D. Purdy, 1973, 'The Theory of Permanent Arms Economy — A Critique and an Alternative', <u>Bulletin of the Conference of Socialist Economists</u>, Spring 1973.

C. Trebilcock, 1971, '"Spin-off" and the Armaments Industry: A Rejoinder', <u>Economic History Review</u>, 1971.

A. Yarmolinsky, 1971, <u>The Military Establishment</u>, Harper Colophon Books.

Table I. The Defence-related industries.   £m

I/O No	Name	Total output	Net output	Defence
16	Mineral oil refining lubricating oils and greases	970.0	339.1	20.4
37	Other non-electrical machinery	772.3	521.0	12.1
39	Other mechanical engineering	679.6	198.4	64.4
40	Instrument engineering	468.3	336.1	23.3
41	Electrical machinery	494.3	338.3	12.6
43	Electronics and telecommunications	1008.2	720.9	151.6
45	Other electrical goods	373.5	107.6	10.9
46	Shipbuilding & Marine engineering	509.1	418.2	164.2
48	Motor vehicles	2311.4	1552.9	37.1
49	Aerospace equipment	820.1	567.1	425.1
81	Construction	6091.4	4811.5	112.8
88	Communication	1089.0	524.1	18.3

Sources 1968 Input-Output Tables

Annexe B. Regressions for each country

	A	B	SE(B)	R^2	t(0)	t(−1)	DW
Belgium	30.35	−2.003	1.136	0.26	−1.76	−0.88	1.9
Denmark	23.25	−0.678	0.5159	0.16	−1.3	0.61	1.8
W. Germany	24.12	0.1973	0.6594	0.001	0.3	1.81	1.3
France	33.15	−1.539	0.2355	0.834	−6.5	−2.3	0.8
Greece	7.273	3.515	0.8467	0.66	4.1	5.3	1.9
U.K.	22.52	−0.7173	0.3584	0.31	−2.0	0.79	1.05
Italy	19.16	0.4846	1.842	0.2631	0.3	0.80	0.5
Canada	23.59	−0.475	0.4685	0.1025	−1.0	1.12	0.7
Luxembourg	15.76	9.3898	3.677	0.42	2.6	2.82	0.94
Netherlands	29.15	−1.002	0.5314	0.2833	−1.9	−0.0	1.9
Norway	35.81	−2.282	4.832	0.02	−0.5	−0.26	2.11
Portugal	12.99	0.6672	0.6757	0.10	0.98	2.47	1.9
Turkey	34.11	−3.296	0.7823	0.6636	−4.2	−5.4	0.9
USA	18.07	−0.1492	0.1263	−1.2	6.7	6.7	1.3

SE(B) is the Standard error of B.

t(0) is B divided by its standard error

t(−1) is B−1 dvidided by its standard erro

D W is the Durbin-Watson statistic

Table II.　The composition of final demand for the output of the defence-related industries and all industries

		Defence related %	All industries %
1	Net output	100	100
2	Military expenditure	10.09	3.38
3	Civilian public current expenditure	3.80	6.28
4	Consumption	14.72	53.70
5	Fixed invest-ment	57.94	21.22
6	Exports	19.55	21.64
7	Direct imports	-7.73	-8.11
8	Residual	1.63	1.89

8 DEFENCE COSTS AND THE DEFENCE INDUSTRY

by Mary Kaldor

It is widely agreed that the fundamental problem in Britain defence policy making is the mismatch between available resources and the range of supposed service requirements. Successive governments have attempted to meet the service requirements through the development and production of a few highly sophisticated and technically elaborate systems at the expense of many minor and possibly more essential systems. (Often this was in contradiction to declared objectives on entering office.) Thus, in the late 1970s, the procurement budget will be dominated by two projects: the Anti-submarine Warfare Cruiser and Associated Naval Systems and the Multi-Role Combat Aircraft (Tornado). This trend had been criticized from various sectors of informed opinion. A recent criticism comes from the Expenditure Committee which recommended:

> a less ambitious approach to the problem of providing new equipment. The Ministry have occasionally appeared to aim for so high a standard that either projects have had to be abandoned or sacrifices have had to be made elsewhere in the defence programme in order to limit or accommodate escalating costs. (1)

A recurrent theme in the work of the Study Group has been the need to reduce, or eliminate, expenditure on these 'high technology' projects in order to achieve substantial defence cuts. In this paper, I want to argue that this is not simply a matter of cancelling a piece of hardware. An aircraft or a ship can be treated as the embodiment of a particular technology, itself the outcome of a set of strategic requirements and a set of industrial capabilities. In the past, attempts to limit defence expenditure through the cancellation of major projects have been thwarted because governments have failed to tackle the underlying military industrial structure. Cancelled projects, as I shall try to show, have given birth to new and more sophisticated successors. If we are to achieve effective cuts, this must also involve a shift in strategic priorities and the conversion of industrial capabilities. Chapter Four of the Report deals with the former issue, although most members of the Study Group would probably agree that the Labour Party needs to formulate an alternative defence strategy in much greater detail. This paper concentrates on the second issue; the need for fundamental change in the defence industry as a pre-condition for defence cuts.

Costs and Technical Advance

The first three tables show the increased cost of aircraft and warships since 1945. The aircraft costs are based on estimates made by the Stockholm International Peace Research Institute (SIPRI) of the real resource cost in 1973 US dollars of individual aircraft; (they are used for the compilation of arms trade statistics). The results are remarkable. They show, for example, that the real cost of producing 385 Tornados will be slightly greater than the entire production costs of Spitfire before and during World War II. The warship costs are actual building costs, as provided, until last year, in the annual appropriation costs. Even if one assumes that a third of the cost increases can be attributed to inflation (which is high), one finds that the cost per ton of warships has increased by anything from a factor of 10 (the difference between an early post-war 'Bay' class frigate and the last of a long series of 'Leander' class frigates) to a factor of 15 (the difference between an early post war 'A' class submarine and a modern 'Swiftsure' class submarine or between the Vanguard Battleship and the new Anti-Submarine Warfare Cruiser).

If anything, those figures are understated because they refer only to production costs. Total life cycle costs would be even more striking since maintenance and support costs tend to increase faster than production costs owing to the unreliability of highly complex and sophisticated equipment. In particular, the increase in electronic equipment, which accounts for much of the overall cost increases involves a disproportionate increase in what are known as logistic support costs. (2) It is estimated that the building costs of a ship account for 25 per cent of the total life cycle costs. (3) A study of the A-7D American combat aircraft shows that ownership costs, i.e. operation and maintenance, base support and training, are 30 per cent higher than acquisition costs - the cost of spares being included in the estimate for acquisition costs. (4) It should be noted that ownership costs do not primarily consist of material costs. The pressure on the manpower budget is not simply due to increases in pay and allowances, it is also due to the maintenance requirements of new and costly equipment.

The increases in cost are related to increases in the sophistication and complexity of military equipment designed to meet more stringent and elaborate performance targets. Obvious recent examples are the variable geometry aircraft, i.e. swing-wing, ostensibly designed to increase the flexibility of an aircraft and multiply its possible functions; gas turbine engines for surface warships, designed to increase speed; nuclear propulsion for submarines which increase speed; and the various types of electronic equipment which can be found across the whole gamut of weapons systems and which vastly improve communications, navigation, the identification and detection of enemy targets, and the guidance of weapons and which can reduce the effectiveness of enemy equipment intended for these purposes through various devices, such as jamming. These are the major technological changes. There are also continual developments in existing technologies - increases in the thrust of a given type of jet engines, improvements in the accuracy of a particular kind of

guidance system.   Over time, marginal developments of this
kind have tended to involve a disproportionate increase in
costs.   A 15 per cent increase in the speed of an aircraft,
for example, might involve a tenfold increase in cost. (5)
Then there are all sorts of lesser additions that are made
to military equipment.   The Chieftain tank, for example,
contains facilities which would enable the crew to survive
for several days in the hatch in the event of nuclear attack.
Interesting foreign examples can also be enumerated.   An
Israeli rifle has a built-in bottle opener.   The Soviet
bayonet has a wire cutting capability while the handgrips
on its ZSU-23-4 vehicle are electrically warmed. (6)
    Such technical 'improvements' are likely to be more
expensive the more quickly they are introduced into service.
Technical advance can be defined as the amount of new
technology, however measured, undertaken within a given time
period.   A study undertaken at the Rand Corporation has
shown that cost escalation, the difference between original
cost estimates and final costs, is directly related to the
degree of technical advance sought.   By implication, this is
also true of the increase in the cost of successive generations
of military equipment since cost estimates are based on
knowledge of the previous generation.
    Technical advance was measured, crudely, through a system
of subjective ratings ranging from one to twenty.   The
ratings were known as A-factors.   It was found that programmes
with A-factors less than 12 tended to be reasonably predictable
although outcomes tended to vary above eight.

    For programmes with A-factors above 12, the predictability
    of programme outcomes lessens appreciably and the worsening
    effect is not on a linear scale.   As the A-factor value
    increases and programme duration extends past 60 months,
    the cost outcome of a given programme becomes so unpredict-
    able that initial cost estimates are very nearly worthless.(7)

    It is interesting to note that all British programmes
mentioned had A-factors above eight and nearly all the more
recent programmes - Vulcan, Victor, Kestrel and P-1127
(Harriers' predecessors).   Lightning and probably MRCA had
A-factors above 12.   Concorde had the highest rating of any
programme, British or American.   This tendency towards
increasing technical advance contrasts with the American
experience where A-factors declined from an average of 12.2
in the 1950's to 8.9 in the 1960's.   The Rand results are
shown in Chart 1.
    Technical advance can always be explained in strategic
terms.   In peacetime, as several members of the Study Group
have stressed, threat assessment is necessarily subjective.
Changes in technology, by inducing new perceptions and
counter perceptions of the threat, can propel planners into
further changes in an autistic fashion. (9)   When the
Ministry of Defence were explaining the numerous design

changes made to the Type 21 frigate, whose cost escalated by
107 per cent, they talked about 'the need to meet as quickly
as possible new threats arising from advances in weapon
technology'. (10)

The idea of defence as a race against technology originated
in the Second World War and has come to have an existence
apparently independent of changes in the political and
economic environment. Representatives of the arms sector
tend to view technology as an end in itself and regret
cancelled projects simply because they represent a loss of
technical leadership. One well established writer, for
example, explains that because the Ministry of Defence were
sceptical of the variable geometry aircraft concept 'on the
grounds of both time and price' and were unwilling to finance
continued research by Vickers during the 1950s, it 'became
imperative that some manned military application be found
for all the theories and tests'. (11) This was the origin of
the Tornado.

Yet it is questionable whether increased sophistication
really represents technical advance or improvement; whether,
in fact, it does represent the optimum approach to strategic
problems. Complexity greatly increases unreliability and
reduces manoeuvreability and flexibility. The vulnerability
of modern weapon platforms greatly increases the risk of
operating expensive aircraft and warships. One of the main
arguments against TSR-2 was the fact that its cost and lack
of manoeuvreability made it too risky to use for conventional
battlefield operations, except in very bad weather. (12) The
same will be true of Tornado. At the end of World War II,
it was widely believed that speed had become an expensive
luxury for warships with the advent of naval aircraft and
submarines and that the 'daring' Class destroyers would be
the last fast warships to enter service with the Royal Navy.
Nothing has occurred to change this perspective. And yet
gas turbine propulsion has been introduced for the Type 42
destroyers and Type 21 and Type 22 frigates at considerable
cost. Another example is the American experience in Vietnam
and the discovery that older slower aircraft were more
manoeuvreable.

The point is not to show that cheap and simple solutions
to strategic problems are necessarily more effective
militarily — although this has been done to some extent in
the papers by Dan Smith and David Greenwood. Rather, it is
to show that there is room for dispute and that the tendency
to adopt highly sophisticated and expensive solutions cannot
be explained entirely with reference to military efficiency.
It must be explained in terms of the decision-making process
and the structure of the institutions responsible for military
technology.

Industrial Capacity

The capacity to develop and produce military aircraft may be

285

defined in general terms as the amount of plant, machinery and labour available for production in a given time period. It may be measured in abstract monetary terms. In particular terms, however, it represents a particular infrastructure of skills and techniques, a set of relationships with the services and with subcontractors needed to manufacture a particular type of military equipment. Military aircraft production, for example, involves experience 'in the use of unusual materials, of specialized machining and welding techniques, of stringent quality control and reliability in standards'. (13) It involves a particular composition of skills in the workforce, a familiarity with strategic requirements, and a knowledge of available supplies.

In Britain, for the most part, the developers and producers of military equipment are private firms. With the exception of the Royal Dockyards and the Royal Ordnance Factories even those firms that are wholly or partially nationalized, like Short Brothers and Harland, operate on the same principle as private firms, namely the principle of independent viability. This means that firms must finance this capacity and cannot afford long periods of idleness.

With the increase in the sophistication of military equipment, military technology has become increasingly divorced from civilian technology and consequently industrial capacity has become increasingly specialized. Since the government is the main customer for military equipment, this means constant pressure on the government in such a way as to avoid excess capacity. As Vice-Admiral Clayton explains:

> We have to give the shipbuilders a regular rolling programme of orders. We depend on the specialist warshipbuilders. They have a very carefully balanced selection of trades — drawing office, steel workers and outfit trades — which are required especially for warshipbuilding and not for commercial shipbuilding, and of necessity we have to keep a flow of orders going to them. (14)

Similarly, Sir Michael Carey, when Head of Procurement Executive, has argued that

> if you get down to single source of supply, and if you accept my thesis that in these very high technology areas there is no civilian backing for defence technology, what you may face is a choice between preserving a capability nationally, or running a risk of seeing it disappear. (15)

Tables 4 to 8 show how this 'regular rolling' system of orders has helped to preserve the two main aircraft companies, Hawker Siddeley and BAC together with their predecessors, and three specialists warship builders, Yarrow & Co. Ltd, Vosper Thorneycroft Ltd, and Vickers Ltd.

It will be seen that the aircraft companies have been less assured of capacity utilization then the warshipbuilders and that all faced considerable difficulty in the early post-war years; these questions are discussed below.

The avoidance of surplus capacity is not simply a matter of employment, it also involves expansion. The principle of independent viability entails the principle of profit maximization and the constant striving after technical change. In a competitive situation – competition being defined in its broadest sense – firms must innovate, they must introduce new ideas, designs and products, in order to maintain their markets. This applies as much in the military as in the civilian sphere. Firms must keep up with wider developments in military technology if they are to continue to receive orders and if the armed services, in turn, are to justify those orders. In this way, of course, they contribute to the strategic developments they are trying to match.

The new technological developments must be paid for and that means an increase in orders and hence an expansion of capacity. (16) The striving for technical progress becomes the more extreme the more limited is the market.

We have seen how the cost of competition has risen in recent years in the automobile and merchant shipbuilding industry. The same is true of defence. The more the Government has reduced the numbers or types of weapons, the greater has been the effort to achieve technical progress and the greater has been the compensating cost increase. This trend is illustrated in the tables. In the last fifteen years, Hawker Siddeley has produced or developed seven military aircraft, including two that were cancelled, and BAC had developed or produced six military aircraft, including two that were cancelled. In the previous fifteen years, the companies that amalgamated to form Hawker Siddeley and BAC developed or produced 28 and 18 military aircraft respectively, as well as several research aircraft which have not been included. The number of aircraft produced of any given type has also declined substantially. Yet the increase in the cost of individual aircraft is such as to compensate for the decline in numbers. The same phenomenon is to be found in shipbuilding. Table 9 shows the output of the main warship-builders since 1945. It can be seen that in the period 1965-74, about half as much warship tonnage was launched compared with the period 1945-54. Yet real costs have increased by factors ranging from 10 to 15, resulting in a substantial overall increase in warshipbuilding capacity.

The explanation of this tendency need not be viewed simply in crude capitalistic terms. It is also a matter of redeploy-ment following numerical reductions in orders. Designers shifted from one project to another bringing with them pet ideas that can always find supporters within the services. The alternative to a rigorous elimination of unnecessary ideas and a choice between competing ideas is the compromise of doing everything. The way in which this occurred in the case of TSR-2, MRCA, and the recent shipbuilding programme is described below.

An alternative to Government orders is exports. Exports have, of course, played an important role in the British arms industry since the war but they can never be more than a supplement to Government orders. First of all, the proportion of output exported has remained relatively constant and therefore cannot provide a substitute for increased Government orders. According to Vice-Admiral Clayton,

> We plan at the moment to leave something of the order of 30 per cent of every warshipbuilders' capacity available for export orders ... This, of course, tends to vary, depending upon how the market goes in export orders. We may, in fact, find outselves wanting to support a warshipbuilder over a short period to keep him going. Similarly, there may be times when, because of the run on export orders, we may be asked to hold back a bit. (17)

Secondly, exports do not, for the most part, cover the costs of maintaining development capacity. The Ministry of Defence has had considerable difficulty in recent years in recovering the levy on the sale of military equipment developed at Government expense. Also, only about a third of the Ministry's sales of military equipment in the financial year 1975-6 achieved a 'target price of full economic cost including all overhead expenses and a return on capital'. (18) Thirdly, and perhaps most importantly, exports, for the most part, consist of the less sophisticated items of equipment. There are a few exceptions like the sale of Harriers to the United States, Lightnings to Saudi Arabia, Chieftains to Iran and submarines to Brazil and Israel. But the staple exports are aircraft like the Jet Provost and BAC 167 trainers, the Britten Norman Defender utility plane, helicopters, refurbished Canberras and Hunters, and the Shorts Skyvan transport or armoured cars like Ferret and Saladin. Apart from the sales mentioned above, none of the more recent and more expensive types have achieved much in the way of an export market. Likewise, the most successful warship exporters have been firms like Vosper Ltd, now part of Vosper Thorneycroft, which specialize in the production of small fast patrol boats for export. It can be seen from Table 9, that the share of output sold abroad is below 30 per cent in the case of Vickers and Yarrow and well above it for Vosper Thorneycroft. This unevenness in export orders only serves to increase the pressure on the Government to order the most sophisticated types of equipment.

Government Policy

The cost of providing a 'regular rolling' system of orders to the arms industry became a cause for concern from an early date. Yet attempts to stem these cuts through cancelling major programmes and to curb the growth of industrial capacity through reorganization of the industry have so far backfired.

The first such attempts concerned the aerospace industry. The first post-war Labour Government envisaged that peace would last for at least a decade and the V-bombers and Hunter were ordered for service around 1957. The Korean War speeded up the RAF re-equipment programme and Canberra and Valiant were introduced as interim combat aircraft. Aircraft design teams were kept together with work on research aircraft; the maintenance of design capacity was viewed, at the time, 'as a strategic necessity as it would enable aircraft production to be rapidly expanded should war recur'. (19) This attitude changed in the mid-1950s. With the emphasis on short cataclysmic war, the argument about reserve capacity lost much of its force. Also, the advent of missiles was expected to 'put all the airplane makers out of business'. (20) More importantly, the cost of financing design teams for 23 aircraft companies and 8 aeroengine companies was becoming prohibitive.

In retrospect, the notorious 1957 White Paper was ahead of its time. Duncan Sandys enuciated a radical change in defence policy which involved emphasis on nuclear deterrence and on missiles. The V-bombers and Lightning were expected to be the last manned aircraft to enter service with the RAF and nearly all other aircraft projects were cancelled. A possible exception was the Canberra replacement, a light tactical bomber for conventional warfare East of Suez. Simultaneously the Government enunciated a policy for industrial reorganization, which resulted in the mergers shown in Tables 10 and 11. As it turned out the aircraft companies were amalgamated but not rationalized.

In September 1957, the heads of all the airframe companies were invited to a meeting at the Ministry of Supply and were told that for future major aircraft, there would be work for three or at most four major groupings, and that the order for a Canberra replacement would only be placed with a group of firms with two or three in co-ordination. (21) In the event, a design contract was placed with a consortium of English Electric and Vickers for TSR-2, in preference to the well developed English Electric P.17 aircraft, which was widely considered the more appropriate design. This was because the consortium was considered an appropriate basis for a future air frame grouping.

From the beginning, the technical advance sought in TSR-2 escalated. First, there was the intense competition to acquire the only major aircraft project foreseeable; consequently extravagant technical ideas abounded. Secondly, the RAF extended the operational requirement to cover all the functions that might have been fulfilled by other aircraft had it not been for the 1957 White Paper. (Indeed, the requirement for a Canberra replacement by itself might have been easily met by the Buccaneer.) They were therefore open to a wide variety of suggestions. Thirdly, the attempt to amalgamate the Vickers and English Electric design teams led to a proliferation of technical 'improvements'; one might say that the historical competition between the two aircraft firms was internalized in a single aircraft project.

The description of TSR-2 that was given to the House of Commons in 1959 was already markedly different from a simple Canberra replacement. Indeed it must sound familiar to those who follow the vagaries of Tornado's history:

> What, then, must be the basic features of the TSR? It must be able to carry a useful bombload and have a radius of action of at least 1,000 miles ... This range implies flight at high level for a considerable part of the time, and to survive at high altitude the aircraft must be able to fly at high supersonic speed. At the same time, it must be capable of flying at around the speed of sound at low altitude. It must also be able to use short runways with unprepared surfaces; to carry the navigating and bombing equipment to enable it to find and attack accurately tactical targets by day and by night in all weathers; and to give Army support in a variety of situations.

> Remarkable technical advances will enable this new aircraft to take off on operational missions using as little as 600 yards of runway, and it is clear that the electronic equipment for accurate navigation, bomb-aiming and reconnaissance can be developed, although it presents many problems. This combination of characteristics will certainly be unsurpassed and possibly unequalled in any aircraft known to be in the design stage today. (22)

It turned out not to be so clear that the electronic equipment could be developed, while the STOL capability was of doubtful benefit given the sophisticated servicing requirements of such a complex aircraft. The engine, which had been selected on the basis of a strategy towards the aeroengine sector, kept blowing up and it proved extremely difficult to fit the engine into the airframe because of changes which had been made to accessories fitted round the engine. The requirements also escalated. A strategic role was added in the early 1960's and later a maritime patrol function. The result was a rapid escalation in cost from £80-90m in December 1959 to £750m in April 1965 when it was cancelled. The expected date of service entry slipped from 1966 to 1970 over the same period. According to Dennis Healey, who was Minister of Defence at the time:

The trouble with the TSR-2 was that it tried to combine
the most advanced state of every art in every field.
The aircraft firms and the RAF were trying to get the
Government on the hook and understated the cost.   But
TSR-2 cost far more than even their private estimates,
and so I have no doubt about the decision to cancel.(23)

The cost of cancellation was £125m.   About 8,000 people
were made redundant as a result of the cancellation of
TSR-2, together with the cancellation of HS 681 and P 1154.
Few attempts were made to alter the strategic requirements,
although this was clearly discussed initially.   Moreover,
certain requirements for manned military aircraft made
redundant by the 1957 White Paper were revived.   Nor was
there any attempt to restructure or convert industrial
capacity.   The reorganization following the Sandys White
Paper, which is shown in Tables 10 and 11, was evidently
considered sufficient.   Yet it has been estimated that
the cost of redeploying labour, particularly designers,
into TSR-2's successors, the Anglo-French Variable Geometry
Aircraft and Tornado, was about £250m. (24)   And this
presumably does not include the cost of bright ideas proposed
by these designers.   As is shown in Dan Smith's paper, the
history of TSR-2 is being rewritten in the history of Tornado.
   Attempts to control the cost of naval procurement are more
recent.   In the 1966-7 White Paper, the Labour Government
announced radical changes in the size and shape of the fleet
including the cancellation of an aircraft carrier and supporting systems.   The policy was accompanied by reorganization
of the shipbuilding industry, along the lines proposed by the
Geddes Report. (25)   This is shown in Table 12.   Subsequently
the Ministry of Defence decided to go ahead with a new class
of destroyers and frigates despite the fact that their design
was incomplete.   This was partly because of 'operational
need' and partly because of the 'gap in work for the specialist
shipyards caused by programme cuts'. (26)   Undoubtedly, the
second reason was the more important.   The Government
decision to phase out a carrier based air capability stemmed
from the decision to withdraw British forces East of Suez.
At that time, NATO's flexible response strategy had not been
adopted and the battle for the Eastern Atlantic which provides
the current justification for naval air power was not anticipated.
   HMS Sheffield, the first Type 42 destroyer was ordered from
Vickers Ltd for delivery in mid-1973.   HMS Amazon, the first
Type 21 frigate was ordered from Vosper Thorneycroft for delivery
in the first half of 1972.
   HMS Sheffield was finally handed over in February 1975
and excluding the increase in costs due to inflation, its
cost increased by 59 per cent 'because of delay charges, the
development of much of the equipment in parallel with the
development of the ship design and during the early stages
of construction and the complexity of the alterations which
had to be made'. (27)   HMS Amazon was delivered in July 1974

and costs were 109 per cent higher than expected; about a third of the increase was due to inflation. The increase was due to problems associated with the new propulsion system, defects in the main ship machinery, propulsion system and the vessel itself which emerged during tests. Subsequently MOD ordered 5 more Type 42 destroyers and 7 more Type 21 frigates. All the ships were subject to considerable delay, ranging from 8 months to three years and all were subject to considerable cost escalation. It is noteworthy that cost escalation on the previous generation of frigates, the Leanders, averaged 5 per cent and that the total cost of HMS Amazon, at £17.1m, represented a remarkable increase over HMS Ariadne, a Broad-beamed Leander class frigate of the same weight delivered in 1973 at a cost of £7.4m.

But the story does not end there. As the design of HMS Sheffield drew to a close, the aircraft carrier concept was revived in the form of HMS Invincible, the ASW Cruiser, ordered from Vickers in 1971. (I have described the cruiser story in an earlier paper for the Study Group.) Vickers informed the Government that it required:

> a construction contract to provide employment for the production (industrial) trades, but involving a minimum of design effort. Without such a contract some of their design capacity would have to be reserved for whatever alternative production they could obtain, which in turn might affect the timely completion of the cruiser. (28)

As a result, the Government ordered a Second Type 42, HMS Cardiff, from Vickers for delivery in May 1975 despite the fact that Vickers' quotation exceeded the lowest tender. In the meantime, costs and hence capacity absorption on HMS Sheffield and HMS Invincible had increased and an order for a Type 42 destroyer was received from Argentina. Similar pressures on industrial capacity were to be found in the submarine division with export orders from Brazil and Israel, and cost escalation on the 'Swiftsure' class submarines. In addition, skilled labour was drifting to the continent and the oil rig industry. Work slowed down on HMS Cardiff and ceased in March 1975. Swan Hunter agreed to complete the contract, although this involved more delay and increased costs. 'Vickers told MOD that unless the financial and time "gaps" in the contract could be closed' they would have to resume work on HMS Cardiff. This would cost £2m more than completion by Swan Hunter.

'In addition, labour would be withdrawn from the cruiser contract, let on a cost-plus basis, giving rise to extra costs, estimated by Vickers at about £4.75 million, and there would be consequential serious delays to the cruiser completion date, causing major operational manpower and financial penalties for MOD.' (29)

As a result, the Government renegotiated the contract.
Thus, we can see that since the cancellation of the Aircraft
Carrier, there has been a growth of industrial capacity, a
reassertion of strategic requirements and renewed pressure on
the naval procurement budget.

In the aerospace field, the next stage in the attempt to
control capacity has been international collaboration. From
the industry's point of view, this has one major advantage.
According to Sir Geoffrey Edwards:

> Our great strength now rests on the fact that the
> programmes we have got really are part and parcel
> of a major piece of international policy and are not
> any longer at the mercy of changes of national
> government policy. This means that there is a much
> more stable future than we have ever been able to
> forecast in the past. (30)

From the point of view of the public, international
collaboration means even less parliamentary control, (31)
and even greater cost escalation. The difficulties of
amalgamating the design teams of two companies and
reconciling a number of service requirements pale into insignificance when compared with the difficulty of amalgamating
design teams from several different countries and of reconciling the requirements of distinctly separate national armed
forces; hence the dramatic cost increases in Jaguar, the
Anglo-French helicopters and, above all, Tornado (not to
mention Concorde). When the Anglo-French Variable Geometry
aircraft was cancelled, General Gallois, French Director of
Military Affairs, told The Times that on a co-operative
project, 'you divided the cost by two and then had to multiply
it by three to take account of the difficulties of building it
in two countries.' (32)

Conclusions

Whether or not the Government heeds the Study Group's
recommendations on cutting defence expenditure, it is evident
that we cannot afford MRCA's successor. Already, there are
discussions about Air Staff Target 403 for a Harrier/Jaguar
replacement in 1990. Since it is expected to be a supersonic
multi-mission aircraft with STOVL capability (short take-off
and vertical landing) it is likely to involve an expensive
and complex programme of development. According to one aerospace commentator: 'The UK Government would therefore like
another international collaboration programme following on
from MRCA to spread the costs, as well as making political
cancellation more difficult.'

Such a programme will be astronomically expensive. It
will only be conceivable in the context of new forms of
military industrial organization. A few years ago, a popular
idea was the creation of European multinational defence

companies which were expected to integrate and rationalize industrial capacity; (although it was never explained why such companies would fare any better in this respect than did Hawker Siddeley or BAC). With the nationalization of aerospace, this prospect seems more remote. The only possibility for further attempts to 'control' capacity on an international scale would be the creation of a European Procurement Agency as proposed by the Commission of the European Community and, by implication, an integrated European defence policy and a European federation. Quite apart from the dangers of such a course — the creation of a European super power, the division of the Atlantic Alliance, the reproduction of expanding military industrial capacity on a Europe-wide scale — European integration is most unlikely within the MRCA time frame.

Nor is it likely that we can afford the successors to the Cruiser and the Chieftain. International collaboration had hardly affected these sectors of the armaments industry. If only to avoid entry into the international cost control charade, effective ways of reducing costs, capacity, and expenditure must be found.

The only alternative is the abandonment of certain military manufacturing capacities. This involves three simultaneous steps.

(1)　A genuine re-examination of the future need for manned military aircraft, surface warships, heavy submarines and battle tanks in conjunction, necessarily, with a re-examination of strategic and political requirements and a plan for restructuring the armed forces.

(2)　Abandonment of the principle of independent viability and hence profit maximization for armament manufacture, in order to subordinate industry to military requirements, rather than vice versa. This task is facilitated but by no means assured by the nationalization of aerospace and shipbuilding.

(3)　A reduction and/or conversion of industrial capacity through the restructuring of industry. This would include the physical disbandment of current design and management teams, and the development of alternative products and planning structures, involving the full participation of the workforce. The potential for alternative products and planning structures has already been extensively examined by the Study Group.(34)

It should be emphasized that this alternative is borne of necessity. Otherwise, Britain will be entangled in outrageously costly ventures from which there is no return and, consequently, in an inexorable drift towards a form of international integration that is suffered and not chosen.

294

Table 1 UK Combat Aircraft, 1945-75

Manufacturer	Aircraft	Type	Entered Service	Number	SIPRI Cost (1) 1973 dollars
Supermarine	SPITFIRE	Fighter	1934	21,000	300,000
Gloster	METEOR	Fighter	1943	(1000?)	600,000
English Electric	CANBERRA	Medium bomber	1951	n 1000	600,000
De Havilland	VAMPIRE	Fighter	1946	n 2250	500,000
Hawker	HUNTER	Fighter	1954	2000	800,000
Folland	GNAT	Fighter	1958	100*	700,000
A.V.Roe & Co	VULCAN	Bomber	1955	n 160	(3,000,000)
Vickers	VALIANT	Bomber	1955	108	(2,000,000)
Handley Page	VICTOR	Bomber	1957	82+	(2,500,000)
Blackburn	BUCCANEER	Bomber	1962	187	2,000,000
English Electric	LIGHTNING	Fighter	1960	n 300	2,000,000
BAC/ Dassault	JAGUAR	Fighter	1972	400+	4,000,000
Hawker Siddeley	HARRIER	VTOL fighter	1969	234	5,000,000
Panavia	TORNADO	Multi Role Combat aircraft	1977	696	(20,000,000)

* Includes purchase of GNAT trainer and sale of 25 sets of components
. to India but does not include 200 built under licence in India.

Bracketed figures are estimates

(1)   The Stockholm International Peace Research Institute has developed a method for calculating the acquistion cost of different types of weapons systems on the basis of technical criteria, such as weight, speed, etc.   These are used to construct arms trade statistics. Where SIPRI figures are not available, the costs have been calculated using the same method; these figures are bracketed.

Sources: Jane's All The World's Aircraft Annual; Stockholm International Peace Research Institute; Arms Trade Registers, 1975, informal communications; International Air Forces and Military Aircraft Directory; Aviation Advisory Services, Stapleford, Essex.

Table 2 Cost of Battleships, Cruisers and Aircraft Carriers and Submarines

Name	Type	Completed	Standard, tons Displacement	Official Estimated Building Cost £	Cost per ton £
Battleships, etc					
Vanguard	Battleship	1946	44,500	9,000,000	202.2
Sydney	Aircraft Carrier	1949	14,380	3,890,000	270.5
Eagle	" "	1951	44,100	16,335,000	370.4
Centaur	Commando Carrier	1953	22,000	10,483,000	476.5
Bulwark	" "	1954	23,300	10,394,000	446.1
Ark Royal	Aircraft Carrier	1955	43,060	21,793,000	506.1
Melbourne	" "	1955	16,000	8,309,000	519.3
Tiger	'Tiger' Class Cruiser	1959	9,500	13,113,000	1380.3
Blake	" "	1961	9,500	14,940,000	1572.6
London	'County' Class Cruiser	1963	5,440	13,935,000	2561.6
Antrim	" "	1970	5,440	16,740,000	3077.2
Bristol	Type 82 Cruiser	1973	5,650	24,582,000	4350.8
Invincible	ASW Cruiser	1978?	$\underline{n}$ 20,000	$\underline{n}$ 100,000,000	$\underline{n}$ 5000
Submarines					
Ambush	'A' Class Submarine	1947	1120	456,143	407.3
Acheron	'A' Class Submarine	1948	1120	589,298	526.2
Porpoise	'Porpoise' Class Patrol Submarine	1958	1610	2,048,000	1272.0
Ocelot	'Oberon' Class Patrol Submarine	1961	1610	2,920,000	1813.7
Onyx	'Oberon' Class Patrol Submarine	1967	1610	3,600,000	2236.0
Excalibur	'Ex' Class Submarine	1958	780	1,355,000	1737.2

Table 2 Cost of Battleships, Cruisers and Aircraft Carriers and Submarines (Continued)

Name	Type	Completed	Standard, tons Displacement	Official Estimated Building Cost £	Cost per ton £
Submarines contd.					
Dreadnought	'Dreadnought' Class Fleet Submarine	1963	3000	18,455,000	6151.7
Valiant	'Valiant' Class Fleet Submarine	1966	3500	25,200,000	7200.0
Warspite	'Valiant' Class Fleet Submarine	1967	3500	21,450,000	6128.6
Resolution	'Resolution' Class Nuclear powered ballistic missile Submarine	1967	7000	39,700,000	5671.4
Revenge	" " "	1969	7000	39,500,000	5642.9
Churchill	'Churchill' Class Fleet Submarine	1970	3500	24,661,000	7331.7
Conqueror	" "	1971	3500	30,091,000	8597.4
Swiftsure	'Swiftsure' Class Fleet Submarine	1973	3500	37,084,000	10595.4
Sovereign	"	1974	3500	31,898,000	9113.7

Sources: Annual Appropriation Accounts; Jane's Fighting Ships, Annual

Table 3 Cost of Selected Frigates and Destroyers

Name	Type	Completed	Standard Displacement, tons	Official Estimated Building Costs £	Cost per ton £
Frigates					
Morecambe Bay	'Bay' Class Frigate	1949	1600	500,000	312.5
Whitby	'Whitby' Class Frigate	1956	2150	3,155,000	1467.4
Eastbourne	'Whitby' Class Frigate	1958	2150	2,675,000	1244.2
Hardy	'Blackwood' Type 14 Frigate	1955	1180	1,469,000	1244.9
Duncan	'Blackwood' Type 14 Frigate	1958	1180	1,991,000	1687.1
Salisbury	'Salisbury' Type 61 Frigate	1957	2170	2,915,000	1343.3
Chichester	'Salisbury' Type 61 Frigate	1958	2170	3,227,000	1487.1
Lynx	'Leopard' Type 41 Frigate	1957	2800	2,885,000	1030.3
Ashanti	'Tribal' Type 81 Frigate	1961	2300	5,220,000	2269.6
Zulu	'Tribal' Type 81 Frigate	1964	2300	5,060,000	2200.0
Dido	'Leander' Class Frigate	1963	2450	4,600,000	1877.6
Juno	'Leander' Class Frigate	1967	2450	5,000,000	2040.8
Andromeda	Broad-beamed 'Leander' Class Frigate	1968	2500	6,600,000	2640.0
Ariadne	Broad-beamed 'Leander' Class Frigate	1973	2500	7,403,000	2961.2
Amazon	Type 21 Frigate	1973	2500	17,161,000	6864.4
Destroyers					
Matapan	'Battle' Class Destroyer	1945	2380	944,326	396.8
Alamein	'Battle' Class Destroyer	1948	2380	940,000	395.0
Battle Axe	'Weapon' Class Destroyer	1947	2280	1,007,961	442.1
Broadsword	'Weapon' Class Destroyer	1948	2280	991,000	434.6
Defender	'Daring' Class Destroyer	1952	2800	2,280,000	814.3
Diana	'Daring' Class Destroyer	1954	2800	2,975,000	1062.5
Sheffield	Type 42 Destroyer	1975	n 3200	26,438,000	8261.9

Sources: Annual Appropriation Accounts; Jane's Fighting Ships;

TABLE 4

HAWKER SIDDELEY
MILITARY AIRCRAFT 1945-75

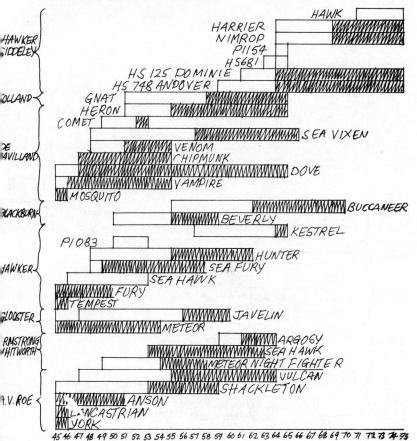

45 46 47 48 49 50 51 52 53 54 55 56 57 58 59 60 61 62 63 64 65 66 67 68 69 70 71 72 73 74 75

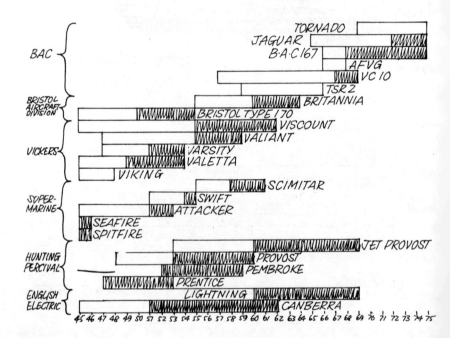

TABLE 5
BAC MILITARY AIRCRAFT 1945-75

TABLE 6

YARROW WARSHIPS 1945-75

AVENGER 21 TYPE FRIGATE
ARDENT 21 TYPE FRIGATE
BROADSWORD 22 TYPE FRIGATE
ALACRITY 21 TYPE FRIGATE
ARROW 21 TYPE FRIGATE
LYNCH 'LEANDER' CLASS FRIGATE
CONDELL 'LEANDER' CLASS FRIGATE
AMBUSCADE 21 TYPE FRIGATE
MUKUT ROJAKUWARN FRIGATE
ARIADNE BROADBEAMED 'LEANDER' CLASS FRIGATE
APOLLO BROADBEAMED 'LEANDER' CLASS FRIGATE
PROTEA SURVEY SHIP
RAH MAR 'YARROW' TYPE FRIGATE
DIOMEDES BROADBEAMED 'LEANDER' TYPE FRIGATE
CANTERBURY 'LEANDER' CLASS FRIGATE
ACHILLES BROADBEAMED 'LEANDER' CLASS FRIGATE
JUPITER BROADBEAMED 'LEANDER' CLASS FRIGATE
'YARROW' MERMAID TYPE FRIGATE

HYDRA 'HECLA' SURVEY SHIP
HECLA 'HECLA' SURVEY SHIP
HERATE 'HECLA' SURVEY SHIP
NAIAD 'LEANDER' FRIGATE
PRESIDENT 'PRETORIUS' PRESIDENT FRIGATE
DIDO 'LEANDER' CLASS FRIGATE
PRESIDENT 'KRUGER' PRESIDENT CLASS FRIGATE
ASHANTI 'TRIBAL' CLASS FRIGATE
BRIGHTON 'ROTHESAY' CLASS FRIGATE
ROTHESAY 'ROTHESAY' CLASS FRIGATE
MALCOM 'BLACKWOOD' CLASS FRIGATE
KEPPEL 'BLACKWOOD' CLASS FRIGATE
HARDY 'BLACKWOOD' CLASS FRIGATE
DIANA 'DARING' CLASS DESTROYER
DECAY 'DARING' CLASS DESTROYER
BROADSWORD 'WEAPON' CLASS DESTROYER
BATTLEAXE 'WEAPON' CLASS DESTROYER
STAVANGER CLASS DESTROYER
COMET 'C' CLASS DESTROYER

45 46 47 48 49 50 51 52 53 54 55 56 57 58 59 60 61 62 63 64 65 66 67 68 69 70 71 72 73 74 75 76

TABLE 7

VOSPER THORNEYCROFT WOOLSTON WARSHIPS 1945-75

BRECON MINE COUNTER MEASURES VESSEL
NITEROI CLASS DESTROYER
NITEROI CLASS DESTROYER
ACTIVE TYPE 21 FRIGATE
ANTELOPE TYPE 21 FRIGATE
2 PATROL BOATS
2 PATROL BOATS
AMAZON TYPE 21 FRIGATE
WILTON MINE COUNTER MEASURES VESSEL
OTOBO 'HIPPOPOTAMUS' CORVETTE
DORINA 'HIPPOPOTAMUS' CORVETTE
8 · 15 ton PATROL BOATS
FARAMOZ 'SAAM' CLASS FRIGATE
SAAM 'SAAM' CLASS FRIGATE
3 · 102 TON PATROL BOATS
4 · 30 TON PATROL BOATS
2 · 100 TON PATROL BOATS
ZELTIN LOGISTIC SUPPORT SHIP
2 · 100 TON PATROL BOATS
2 · 100 TON PATROL BOATS
MINELAYER
6 · 40 TON PATROL BOATS
JUNO 'LEANDER' CLASS FRIGATE
2 FAST PATROL BOATS
GURKHA 'TRIBAL' CLASS FRIGATE
OTAGO 'ROTHESAY' CLASS FRIGATE
'T N' CLASS MINESWEEPER
DUNCAN 'DUNCAN' CLASS FRIGATE
BLACKWOOD 'BLACKWOOD' CLASS FRIGATE
SECHURO OILER
DUCHESS 'DARING' CLASS DESTROYER
CROSSBOW WEAPON CLASS DESTROYER
CONCORD 'S' CLASS DESTROYER
COMUS 'C' CLASS DESTROYER
ACETAEON, LATER BLACK SWAN FRIGATE

45 46 47 48 49 50 51 52 53 54 55 56 57 58 59 60 61 62 63 64 65 66 67 68 69 70 71 72 73 74 75 76

TABLE 8
VICKERS WARSHIPS 1945-76

THREE SUBMARINES FOR ISRAEL

CARDIFF TYPE 42 DESTROYER
TONELERO 'OBERON' CLASS SUB.
SCEPTRE 'SWIFTSURE' CLASS SUB.
SUPERB 'SWIFTSURE' CLASS SUB.
HERCULES TYPE 42 DESTROYER
SHEFFIELD TYPE 42 DESTROYER
SOVEREIGN 'SWIFTSURE' CLASS SUB.
RIA CHUELO 'OBERON' CLASS SUB.
HUMAITA OBERON CLASS SUB.
SWIFTSURE SWIFTSURE CLASS SUB.
ROSTAM 'SAAM' CLASS FRIGATE
ZAAC 'SAAM' CLASS FRIGATE
COURAGEOUS CHURCHILL CLASS SUB.
' CHURCHILL CHURCHILL CLASS SUB.
REPULSE 'RESOLUTION' CLASS SUB.
RESOLUTION 'RESOLUTION' CLASS SUB.
WARSPITE 'VALIANT' CLASS SUB.
VALIANT 'VALIANT' CLASS SUB.
OSIRIS 'OBERON' CLASS SUB.
MOHAWK TRIBAL CLASS FRIGATE
DREADNOUGHT 'DREADNOUGHT' class SUB.
OLYMPUS 'OBERON' CLASS SUB.
ORPHEUS 'OBERON' CLASS SUB.
RIVEROS 'ALMIRANTE' CLASS DESTROYER
WILLIAMS 'ALMIRANTE' CLASS DESTROYER
HERMES AMPHIBIOUS WARFARE SHIP
EXCALIBUR " " CLASS SUB.
RORQUAL 'PORPOISE' CLASS SUB.
PORPOISE 'PORPOISE' CLASS SUB.
SPIGGEN MIDGET TYPE SUB.
ARAGUA 'ARAGUA' CLASS DESTROYER
EXPLORER EXPERIMENTAL SUB.
MELBOURNE 'MAJESTIC' CLASS AIRCRAFT CARRIER
ZULIA 'ARAGUA' CLASS DESTROYER
NUEVO ESPARTA ARAGUA CLASS DESTROYER
ANDREW 'A' CLASS SUB.
ANCHORITE 'A' CLASS SUB.
AMBUSH 'A' CLASS SUB.
ALLIANCE 'A' CLASS SUB.
AUROCHS 'A' CLASS SUB.
AURIGA SUB.
ASTUTE SUB.
TEREDO SUB.

45 46 47 48 49 50 51 52 53 54 55 56 57 58 59 60 61 62 63 64 65 66 67 68 69 70 71 72 73 74 75 76

TABLE 9 Output of main Naval shipbuilders 1945-47
appears in the hardback edition of 'Sense About
Defence' published in 1978

Name of firm	1945-1954					1955-1964				
	Ships for Royal Navy		Ships for Export		Exports as % Total Production 1945-54	Ships for Royal Navy		Ships for Export		Exports as % Total Production 1955-64
	No	Tons	No	Tons	by Tons	No	Tons	No	Tons	by Tons
John Brown & Co	6	66030	2	3810	5.4	4	13070	1	2300	15.0
Cammell Laird	13	18115	1	15892	46.7	13	66810	1	2144	3.1
Fairfield SB & Eng Co	6	14375	0	0	0.0	4	18260	0	0	0.0
Harland & Wolff	10	101588	2	10135	9.1	7	40630	4	34510	45.9
HM Dockyards: Chatham	4	5605	0	0	0.0	3	4830	0	0	0.0
Devonport	2	2590	1	14380	84.7	3	6850	0	0	0.0
Portsmouth	3	3460	0	0	0.0	3	6980	0	0	0.0
Scott Lithgow Lithgows	1	16310	0	0	0.0	0	0	0	0	0.0
Scot & Sons	0	0	0	0	0.0	0	0	0	0	0.0
Scotts SB & Eng Co	12	14375	1	1710	10.6	8	22350	0	0	0.0
Swan Hunter: Furness	4	42800	0	0	0.0	2	20000	0	0	0.0
Goole	4	1440	0	0	0.0	0	0	0	0	0.0
Hawthorn Leslie	6	25205	0	0	0.0	3	19232	0	0	0.0
Smiths DD & Co Ltd	2	3230	1	1435	30.8	0	0	0	0	0.0
Swan Hunter & Wighams Richardson	11	71432	7	49352	40.9	5	12630	0	0	0.0
Vickers & Tyne	4	18065	0	0	0.0	3	6750	2	4502	40.0
Alex Stephen: Alex Stephen	6	12690	0	0	0.0	4	7040	2	3430	32.8
W Simons	1	1200	0	0	0.0	2	1600	0	0	0.0
Lobitz & Co	10	19735	0	0	0.0	0	0	0	0	0.0
Vickers Armstrong, Barrow	12	13380	4	9010	40.2	10	39640	4	21496	35.1
Vosper Thorneycroft Thorneycroft	12	13800	0	0	0.0	9	6820	5	19984	74.5
Vosper Ltd	12	724	0	0	0.0	19	2218	23	2675	54.7
Vosper Thorneycroft	0	0	0	0	0.0	0	0	0	0	0.0
Keith Nelson	0	0	0	0	0.0	0	0	0	0	0.0
Samuel White	45	18950	1	120	0.6	6	8420	2	4760	36.1
Yarrow & Co	6	13580	2	3420	20.1	7	13050	2	4500	25.6
Totals	192	499000	22	109000	18.0	115	317000	46	100000	24.0

Name of firm	1965-1974					1945-1974				
	Ships for Royal Navy		Ships for Export		Exports as % Total Production 1965-74	Ships for Royal Navy		Ships for Export		Exports as % Total Production 1945-74
	No	Tons	No	Tons	by Tons	No	Tons	No	Tons	by Tons
John Brown & Co	1	11060	0	0	0.0	11	90160	3	6110	6.3
Cammell Laird	5	23260	0	0	0.0	31	108185	2	18036	14.3
Fairfield SB & Eng Co	2	10880	0	0	0.0	12	48515	0	0	0.0
Harland & Wolff	2	2440	1	2450	35.8	19	146618	7	47095	24.3
HM Dockyards										
Chatham	1	800	3	6600	89.2	8	11235	3	6600	37.0
Devonport	5	13540	0	0	0.0	10	22980	1	14380	38.5
Portsmouth	2	4950	0	0	0.0	8	15390	0	0	0.0
Scott Lithgow										
Lithgows	0	0	0	0	0.0	1	16310	0	0	0.0
Scott & Sons	0	0	5	650	100.0	0	0	5	650	100.0
Scotts SB & Eng Co	2	22040	7	11270	33.8	22	58765	8	12980	18.1
Swan Hunter:										
Furness	0	0	0	0	0.0	6	62800	0	0	0.0
Goole	0	0	0	0	0.0	4	14400	0	0	0.0
Hawthorn Leslie	6	34040	0	0	0.0	15	78477	0	0	0.0
Smiths DD & Co	0	0	0	0	0.0	2	3230	1	1435	30.8
Swan Hunter & Wighams Richardson	13	64210	0	0	0.0	29	148272	7	49352	25.0
Vickers & Tyne	3	10390	1	440	4.1	10	35205	3	4942	12.3
Alex Stephen:										
Alex Stephen	4	11490	0	0	0.0	14	31220	2	3430	9.9
W Simons	0	0	0	0	0.0	3	2800	0	0	0.0
Lobitz & Co	1	604	0	0	0.0	11	20339	0	0	0.0
Vickers Armstrong, Barrow	9	39500	7	6700	14.5	31	92520	15	37206	28.7
Vosper Thorneycroft										
Thorneycroft	1	2450	7	1615	39.7	22	23070	12	21599	48.3
Vosper Ltd	1	143	41	4079	96.6	32	3085	64	6754	68.6
Vosper Thorneycroft	6	3421	43	8875	72.2	6	3421	43	8875	72.2
Keith Nelson	0	0	9	156	100.0	0	0	9	156	100.0
Samuel White	1	2450	0	0	0.0	52	29820	3	4880	14.1
Yarrow & Co	10	22995	6	12810	35.8	23	49625	10	20730	29.5
Totals	75	283000	130	55600	16.4	382	1098000	198	265000	19.4

TABLE 10 Main aircraft manufacturing companies 1946-74

Company	Years 1946–74 / Notes	Group
*A.V. ROE		HAWKER SIDDELEY AVIATION
*HAWKER AIRCRAFT		
AIRSPEED	Became part of Hawker Siddeley 1960	
DE HAVILLAND AIRCRAFT		
GENERAL AIRCRAFT	Blackburn and General Aircraft 1949	
BLACKBURN AIRCRAFT	Blackburn Group 1959 — Taken over by Hawker Siddeley 1960	
FOLLAND AIRCRAFT	Became part of Hawker Siddeley 1959	
*ARMSTRONG WHITWORTH AIRCRAFT	Became Whitworth-Gloster then Avro Whitworth – Closed 1965	
*GLOSTER AIRCRAFT	Merged with Armstrong Whitworth Closed 1963	
BRISTOL AEROPLANE CO.	Aircraft Division 1951 Bristol Aircraft — Became part of BAC 1960	BRITISH AIRCRAFT CORPORATION
ENGLISH ELECTRIC	English Electric Aviation 1950 — Became part of BAC 1960 Luton Works Closed 1966	
PERCIVAL AIRCRAFT	Hunting Percival 1954 — Hunting Aircraft 1957	
VICKERS ARMSTRONGS (Weybridge and Supermarine)	Vickers Armstrongs Aircraft 1955 — Became part of BAC 1960	
WESTLAND AIRCRAFT		WESTLAND AIRCRAFT
CIERVA AUTOGIRO CO.	Helicopter interests merged as Saunders-Roe Helicopter Division 1950	
SAUNDERS-ROE	Saunders-Roe acquired by Westland 1959 (Cowes later BHC)	
BRISTOL AEROPLANE CO.	Bristol Aircraft — Bristol Helicopter interests to Westland 1960	
FAIREY AVIATION	Fairey helicopter and Aircraft interests Taken over by Westland 1960	
SCOTTISH AVIATION		SCOTTISH AVIATION
AUSTER AIRCRAFT	Beagle Auster Aircraft and Beagle Miles (1960 formation) as divisions of British Executive and General Aviation formed 1960 — Beagle Aircraft 1962 — Into Receiver's hands 1969. Closed 1970	
MILES AIRCRAFT	Liquidation 1947 — F.G. Miles founded 1951	
HANDLEY PAGE	Certain Miles Aircraft programmes and assets acquired by Handley Page 1948 — became HP Reading	Handley Page Aircraft 1969 Liquidation 1970
BOULTON PAUL AIRCRAFT	Ceased own aircraft design late 1950s — Became part of Dowty Group 1961	
		BRITTEN NORMAN (Bembridge) — Britten Norman formed 1964 — Receivership 1972 — Aircraft Interests to Fairey Group 1973

*Original member Companies of Hawker Siddeley

Source: Derek Wood. Project Cancelled. MacDonald and Jane's London 1975

TABLE 11 Main aero-engine manufacturing companies 1946-74

	1946 47 48 49 50 51 52 53 54 55 56 57	58 59 60 61 62 63 64 65 66 67 68 69 70 71 72 73 74
Alvis		Aero engine development ceased ... Rover Gas Turbine Aviation activities taken over by Alvis 1968 ... Rover aero gas turbine activities acquired by Lucas 1973
Armstrong Siddeley Motors	Armstrong Siddeley acquired Metvick aero-engine interests 1949	To BSE 1958
Metropolitan Vickers		To BSE 1958
Bristol Aeroplane Co.	Bristol Aero Engines 1956	BRISTOL SIDDELEY ENGINES — (All share capital in BSE acquired by RR 1966) — Company in Liquidation 1971 — ROLLS ROYCE (1971) LTD
Blackburn Aircraft (Cirrus)		Blackburn Engines 1959 — To BSE 1961
De Havilland Engine Co.		To BSE 1961
D. Napier and Son		Napier Aero Engines, taken over by Rolls-Royce 1961
Rolls-Royce		ROLLS-ROYCE
Powerjets (R & D)	Ceased engine design and development 1946	

*Taken over by Rover and now part of BLMC

Source: Derek Wood. Project cancelled. MacDonald and Jane's, London 1975

TABLE 12 NAVAL SHIPBUILDERS

1946   1976

```
Bartrams & Co.Ltd.
Wm.Pickersgill & Co.Ltd. ──────────────── Austin Pickersgill
John Brown & Co.Ltd ──────────────────── John Brown & Co.Ltd.
Cammell Laird & Co.Ltd ───────────────── Cammell Laird & Co.Ltd.
Wm.Denny Bros Ltd.
Fairfield Shipbuilding & Eng.Co.Ltd.──── Govan Shipbuilders Ltd.
Harland & Wolff Ltd. ─────────────────── Harland & Wolff Ltd.
HM Dockyards ─────────────────────────── HM Dockyards
Burntisland Shipbuilding Co.Ltd. ─────── Robb Caledon Shipbuilders Ltd.
Caledon Shipbuilding & Eng. Co.Ltd.
Henry Robb Ltd.
Lithgows Ltd. ────────────────────────── Scott Lithigows Ltd.
Scotts Shipbuilding & Eng. Co.Ltd.
Alex Stepbell & Sons Ltd.
Wm.Simons & Co.Ltd. ──────────────────── Alex Stephen & Sons Ltd.
Lobnitz & Co.Ltd.
Sir James Laing & Sons Ltd.
Redfen Construction Co.Ltd. ──────────── Sunderland Shipbuilders
Goole Shipbuilding & Repair Co.Ltd.
Furness Shipbuilding Co.Ltd.
R & W Hawthorne Leslie & Co.Ltd.
Smiths Dry dock & Co.Ltd. ────────────── Swan Hunter Ltd.
Swan Hunter & Wigham Richardson Ltd.
Vickers - Armstrong Ltd. Tyne
Vickers - Armstrong Ltd. Barrow - Vickers*
J. Samuel White & Co.Ltd. ────────────── J. Samuel White & Co.Ltd.
Yarrow & Co.Ltd ──────────────────────── Yarrow & Co.Ltd*
Vosper Ltd. ──────────────────────────── Vosper Thorneycroft* Ltd ┐ David Brown
John, Thorneycroft Ltd. ────────────────                           │ Holdings Ltd.
                                          Keith Nelson Ltd.*      ┘
Ailsa Shipbuilding Co.Ltd. ───────────── Ailsa Suipbuilding Co.Ltd.
Sir Wm. Awol & Co.Ltd.
Blyth Shipbuilding & Drydock Co.Ltd.
Cochrane & Sons Ltd. Selby
Cook, Welton & Gemme, Beverly
Hall, Russell & Co.Ltd.
Charles Hill & Sons Ltd.
A & J Inglis Ltd.
Philip & Sons Ltd. ───────────────────── Philip & Sons Ltd.
                                          British Hovercraft Corporation
                                          Brooke Marine Ltd.+
                                          Appledore Shipbuilders Ltd.
                                          Drypool Eng. & Drydock Co.Ltd.
                                          R. Dunstan (Thorne) Ltd.
                                          Fairey Marine+
                                          C.D. Holmes, Beverly
                                          J. Lamont & Co.Ltd.
                                          Whittingham & Mitchell +
```

Sources: Jane's Fighting Ships, annual; Who Owns Whom, annual

* Specialist Warship Builders

+ Mainly exporting patrol boats

TABLE 13

FINAL SALES OF UK, FRENCH & EED AEROSPACE INDUSTRIES IN 1974,

BROKEN DOWN BY CUSTOMER. Expressed as % of Total Sales

		CIVIL				MILITARY		
		UK	FRANCE	EEC		UK	FRANCE	EEC
DOMESTIC SALES	STATE R & D	18.5	32.3	24.2		19.3	23.0	22.9
	STATE SALES, Repairs and Maintenance	3.6	3.8	5.2		44.4	36.1	45.9
	OTHER NATIONAL END-USERS	18.4	42.4	25.3		-	-	-
TOTAL DOMESTIC SALES		40.5	78.5	54.7		63.7	59.1	68.8
EXPORTS	AEROSPACE FIRMS IN NON-EEC COUNTRIES	21.2	-	12.3		3.2	2.6	3.6
	OTHER EEC END-USERS	6.3	2.3	4.3		2.8	3.5	2.8
	NON-EEC END-USERS	32.0	19.2	28.7		30.3	34.8	24.9
TOTAL EXPORTS		59.5	21.5	45.3		36.3	40.9	31.3

Source: Commission of the European Communities, The European Aerospace Industry
Position and Figures, Brussels, May 1976.

CHART 1

TECHNOLOGICAL ADVANCE RATINGS

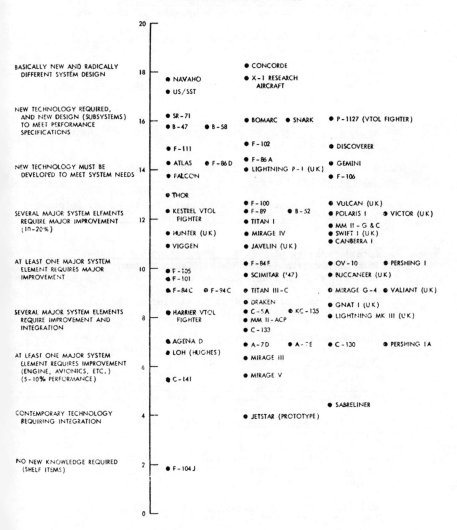

Source: ROBERT PERRY et al: SYSTEM ACQUISITION STRATEGIES, Rand Corporation, R-733 PR/ARPA, June 1971.

Notes

1. <u>Second Report from the Expenditure Committee</u>, Session (1974-5), The Defence Review Proposals, HC 259, para 28.

2. See Marco R. Fiorello, <u>Problems in Arionics Life Cycle Analysis</u>, Rand Corporation, Rand P.5136 (Santa Monica, California, Dec.1973).

3. Evidence of Vice Admiral Clayton, <u>Fifth Report from the Committee on Public Accounts</u> (Session, 1975-6), HCP 556.

4. Marco Fiorello, <u>Estimating life cycle costs</u>: <u>A Case Study of the A-7D</u>, Rand Corporation, R-1518-PR (Santa Monica, California, Feb.1975).

5. See Sir Michael Carey, 'Military Procurement', <u>RUSI Journal</u> (March 1974).

6. See Norman R. Augustine, 'One Plane, One Tank, One Ship: Trend for the Future', <u>Defence Management Journal</u> (April 1975).

7. Robert Perry <u>et al</u>., <u>System Acquisition Strategies</u>, Rand Corporation R-733-PR/ARPA (Santa Monica, California, June 1971).

8. MRCA was not included but it is similar to the F.111 which had a rating of 15-16.

9. The psychological concept of autism has been applied to deterrence theory by Dieter Senghas, in <u>Rustings und Militarismus</u> (Munich 1972).

10. <u>Report of the Controller and Auditor General</u>, Appropriation Accounts, (Session 1974-5).

11. Derek Wood, <u>Project Cancelled</u> (MacDonald's and Jane's, London, 1975).

12. See Dr Geoffrey Williams, Frank Gregory, John Simpson, <u>Crisis in Procurement</u>: <u>A Case Study of the TSR-2</u> (Royal United Service Institution, Whitehall).

13. William Walker, <u>The Multi-Role Combat Aircraft</u>, Research Policy, Vol.2, Number 4(1974).

14. Evidence to the Public Accounts Committee.

15. Sir Michael Carey, (Military Procurement).

16. It might be objected that companies could compete through process innovation, through developing cheaper solutions to military problems. The problem is that such developments might undermine the existing market causing the Government to reduce military spending and abandon the current inventory of sophisticated equipment as well as removing the raison d'être of specialized defence firms.

17. Evidence from the Public Accounts Committee.

18. Report of the Controller and Auditor General, Appropriation Accounts (Session 1975-6).

19. Dr Geoffrey Williams, et al., Crisis in Procurement, p.13.

20. Select Committee on Estimates, Session 1956-7, The Supply of Military Aircraft.

21. The Meeting is described by Derek Wood in Project Cancelled.

22. Quoted in Jane's All the World's Aircraft, 1961-2 (London, 1962).

23. Quoted in Arthur Reed, British Aircraft Industry, what went right? what went wrong? (Dent, London, 1973).

24. Williams et al., Crisis in Procurement.

25. The Shipbuilding Inquiry Committee, 1965-6, cmnd 2937 (March 1966).

26. Public Accounts Committee.

27. Public Accounts Committee.

28. Ibid.

29. Ibid.

30. Quoted in Reed, British Aircraft Industry, p.173.

31. For example, material made available to the Controller and Auditor General may cease to be available because of the nature of trans-national projects.   See John Simpson, 'The Political and Parliamentary Implications of Trans-national Defence Procurement' in Weapons Procurement, Defence Management and International Collaboration (RUSI, Whitehall, 1972).

32. Arthur Reed, British Aircraft Industry,   p.114.

34. It is worth noting that the potential of civil aircraft may be greater than suggested in Study Group papers. Compared with other EEC countries, Britain is very successful at exporting civil aircraft. Also the civil sector of the British aircraft industry has been more successful at exporting than the military sector. This is shown in Table 13. There is room for expansion in the short and medium haul market. Another possibility is inter-city transport; Britain currently has a comparative advantage in VTOL technology and quiet engines.

9   THE EMPLOYMENT AND OTHER ECONOMIC CONSEQUENCES OF REDUCED
    DEFENCE SPENDING

    (and POSTSCRIPT)

by David Greenwood

Foreword

Four years ago I wrote a monograph on <u>Budgeting for Defence</u>
which was published by the Royal United Services Institute
for Defence Studies.   In the penultimate chapter of the work
– on 'Parliament, Public and Policy Options' – I argued that
to promote more realistic debate about the priorities on
which the allocation of resources to and within defence
depends there was a pressing need for the elaboration of
'costed options'.   What concerned me in particular was the
sterility of discussion in which, on the one hand, arguments
were advanced to the effect that 'we ought to reduce defence
spending by £1,000 millions' (or 20 per cent, or whatever)
without any clear conception of what a change of this
order would really entail;   and in which, on the other hand,
officialdom sought to maintain that no judgments or assumptions
other than those underlying existing provision were even
worth considering.   There can be no worthwhile dialogue on
such a basis, only the exchange of assertions.

Given this personal commitment to the options approach to
defence analysis it was, naturally, with great interest that I
learned last year of the establishment by the Labour Party's
National Executive Committee of a Study Group expressly
charged to establish, among other things:

(a)   what reduction of the United Kingdom's defence effort
      <u>would be</u> required to meet a strict interpretation of
      the Party's manifesto commitment to bring military
      spending as a proportion of GNP into line with that
      of major European allies;

(b)   what kinds of change in force structures, force
      levels, equipment and deployment <u>would be</u> necessary
      to produ œ such a reduction; and

(c)   what the employment (and other economic) consequences
      of such change <u>might be</u>.

Furthermore, since an essential validation of academic
research on defence lies in the provision of information for
more realistic debate on policy choices, it was with a sense
of professional obligation that I accepted an invitation to
assist the Study Group by preparing discussion papers for it.

This is one of two papers written for this purpose during
the academic year 1975–6.   Both are issued now, in this
format and at this juncture   because that part of the Study
Group's work to which they relate is more or less complete.
Indeed, with the publication of the Labour Party's new
policy programme, 'the debate' is now very much in the
public domain.   The papers appear here as I originally
tabled them before the Study Group in my capacity as a
consultant to it.   Needless to say, responsibility for the
facts and opinions contained in them is mine alone.

Equally important, it should be stressed that these are research papers designed to elucidate the kind of change significant reshaping of the defence effort might require, and the likely employment consequences of such change. They are not prescriptive.

In this paper an attempt is made to gauge the likely employment consequences, and certain economic implications, of,

(a) the reductions in planned defence spending announced following the United Kingdom Government's 1974 Defence Review, and

(b) the modifications to its revised programme which were decided upon on completion of the 1975 Public Expenditure Survey 'round'.

Further, an assessment is offered — on a broader but more speculative basis — of what might follow from,

(c) the adoption of additional programme reductions designed to bring the defence budget below c. £3,700 millions (at 1975-6 Estimated 1975 Survey prices) by 1980-1.

The principal purpose of the exercise is to elucidate the last point.   The original intention, in fact, was simply to estimate the 'employment dimension' of several so-called Programme Option Modules elaborated in an earlier essay on Defence Programme Options to 1980-1.   On reflection, however, it seemed appropriate — even necessary — to try, in addition, to clarify some implications of the reductions in planned expenditures to which the Government is already committed.

Broadening the scope of the examination in this way has two advantages.   First, it lends emphasis to the fact that the manpower consequences of further diminution of the defence effort would be additional to those already 'programmed'. This is important for appraisal of the adjustment problem involved.   Secondly, it enables pertinent questions to be raised about the meaning and measurement of the employment effects of changes in planned expenditure.   This is worthwhile because confusion has arisen over the interpretation of estimates of such effects produced in amplification of the 1974 Defence Review proposals and more recent revisions.

The main conclusions of this 'composite' analysis are summarized in the final section of the paper, drawing a distinction between the likely consequences of present plans and those which might be expected to follow from further reductions.   The penultimate section notes the main differences, in terms of the incidence of employment effects, among selected options for 'further reductions'.

317

Defence Manpower: Categories

In the first instance, however, it is necessary to take a
synoptic view of the manpower directly involved in the national
defence effort. The following categories may be distinguished:

(1) Service Personnel of the Royal Navy (and Royal
     Marines), the Army and the Royal Air Force, including
     the members of the women's Services. At 1 April
     1974 these numbered c. 358,000, of whom just over
     349,000 were United Kingdom personnel, the remainder
     being 'locally enlisted' overseas.

(2) Civilian Employees of the Ministry of Defence (or
     other Departments) whose employment costs are borne
     on the defence budget. At 1 April 1974 these
     numbered approximately 310,000. Of this total
     249,000 were 'UK-based civilians' - serving at
     home or overseas; the remaining 60,000 or so were
     'locally entered civilians' at installations
     overseas.

(Categories (1) and (2) constitute the defence labour force
whose employment costs make up the personnel costs components,
other than pensions, of the defence budget.)

(3) Defence Contractors' Employees, largely in manufactur-
     ing industry. Data on employees in employment in
     manufacturing industry engaged on defence contracts
     were collected at one time, but (it appears) the
     practice has been discontinued. Projection of
     estimates from earlier years suggests that, in 1974,
     there were perhaps some 235,000 private sector workers
     whose employment stemmed from military material
     expenditures with industry. Official estimates put
     the figure nearer 200,000.

In sum, therefore, before the changes to be considered were
decided upon, there were probably 875,000-900,000 persons
directly engaged in defence-related activity; with some
750,000-800,000 of these being so engaged in the United
Kingdom itself. (That figure corresponds to around 3 per
cent of the United Kingdom's economically active population.)
     The three categories represent those who owe their jobs
more or less directly to defence expenditure. However,
for a full appreciation of defence's 'employment dimension',
account has also to be taken of secondary (or indirect)
civilian employment engendered - in both the 'home' economy
and 'host' economies overseas - because the employment and
income produced directly by defence spending prompts further
employment and income generation. (This indirect defence-
related employment occurs, for example, in the retail and
service trades in localities where there are major military

installations or 'defence' plants; and, more generally,
among defence suppliers.)   It is questionable whether any
operational relevance attaches to the concept of <u>total</u>
secondary employment of this sort, for reasons which need
not concern us here.   But it is clear that, when what is at
issue is the 'release' or 'redeployment' of manpower
resources because of <u>marginal changes</u> in defence spending,
indirect effects should enter the reckoning.   The consequences
of running down establishments or reducing equipment orders
include the indirect and the more remote as well as the direct
and immediate.

Suitable categories and nomenclature are one prerequisite
for elucidation of the employment consequences of defence programme changes.   A second is some understanding of precisely
what the 'release' of manpower resources means in this
context.   It is rather important to be clear about this.

On 'Releasing' Resources

At any one time the real resources actually engaged in
defence-related activity comprise manpower of the various
categories that have been distinguished and the stock of
material with which they work.   Current expenditures cover
the remuneration of the labour, the costs of operating and
maintaining the capital stock (and of developing and sustaining the 'human capital') together with payments for replacement of, or additions to, equipment inventories. (1) This
present provision underpins today's capabilities.

There are also plans and programmes which specify, at
least implicitly, envisaged future resources claims.
Budgetary projections define the counterpart money outlays
which will be required to consummate these claims.   The
planning, programming and budgeting systems which many
defence organizations use represent procedures and presentations whose role is to formulate and express intended
provision for capabilities tomorrow.

It follows that there are always two senses in which the
notion of 'resources committed to military purposes' can be
used.   It can refer to resources actually engaged: that is,
present provision or current commitments.   Alternatively,
it can refer to resources implicitly earmarked: that is,
intended provision which may or may not constitute firm
'commitment' - in the sense of an inescapable undertaking to
fulfil the intention.

If all this smacks of pedantry it is unfortunate.   But
the fact is that failure to distinguish clearly between these
two meanings continuously bedevils discussion of the economics
of defence.   There arise opportunities for 'double-talk',
which both politicians and commentators are quick to take
according to the dictates of expediency or presentational
advantage.   More seriously, encouragement is given to
indulgence in 'double-think', nowhere more so than in relation

to the analysis of <u>changes</u> in 'resources committed to military purposes': for, corresponding to the dual sense of that expression, there are two meanings to the related concept of 'releasing resources for other purposes'.

It is obvious what these are. In the first place there will be a 'release of resources' — in one sense — when present provision is reduced. In real resource terms cutting the scale of current commitments leads to redundancy of both labour and productive capacity. Moreover, if there is underlying growth of productivity in the economy, or if the defence effort is becoming more capital-intensive (or more technology-intensive), maintaining the level of current <u>expenditures</u> — or even allowing these to rise — may be compatible with some actual release of resources of this kind too. Over the long-term it is certainly true that British expenditure on defence (at constant prices) has remained surprisingly stable: yet thousands of men have been 'released' from uniformed service, for instance. (2)

In the second place resources are 'released' — in a quite different sense — when declarations of intended provision undergo downward revision. Deciding that in the future less substantial claims will be made on resources than had previously been planned means that manpower, material, productive capacity, technical ingenuity and organizing ability which would otherwise have been absorbed in one use will be 'available' for other purposes in due course. Scheduling more modest claims for defence 'makes room' — in aggregate terms (and using the fashionable phrase) — for exports, or investment, or consumption.

It is not merely to make a formal point that this distinction is made. There is practical significance to it. 'Release' in the first sense precipitates a more or less immediate adjustment or reabsorption problem. Or, looking at it more positively, it offers immediate opportunities for the assignment to alternative uses of such resources as are in fact 'released'. Productive potential which is being used for one purpose becomes available for others. 'Release' in the second sense is a shorthand for the emergence of new options <u>vis-à-vis</u> future claims on capacity. Productive potential <u>which would have been required</u> for one purpose will, in time, become available for different ones. If there is a 'problem' here, it is a planning problem: how to ensure that resources formerly destined for employment in one direction are indeed brought into use to satisfy other societal needs.

That having been said it should be added that, in the public expenditure planning context, it would be perverse to dwell exclusively on the problematical aspects of 'release of resources' — in whatever sense the expression is used. Adjustments to existing dispositions, or planned future ones, are not made arbitrarily, creating 'problems' to which solutions must then be sought. In principle at least, the opposite is true. Whether animated by a desire to alter the pattern of societal uses of resources (allocative motives) or to moderate public claims as part of overall restraint (stabilization motives) the aim is to effect shifts in resource-uses. Revisions to defence plans release resources - creating redundancies now or destroying job opportunities later - because that is the purpose of the exercise.

At the same time there are adjustment or transition problems which cannot be discounted. 'Resources' are not macro-economic fluid which, when 'released' in some quarter, can find its way unerringly down the very channels through which those making public choices have indicated it should flow. What is more, it is apparent that the problems that do arise in this connection are likely to be more intransigent the more specific and 'localized' to present uses relevant resources are, or have become. Thus, the motives for adjustment notwithstanding, to attempt illumination of the 'microeconomics of release' is a worthwhile undertaking - provided that 'double-talk' and 'double-think' can be avoided.

Manpower Aspects of the Defence Review

These pitfalls were not avoided in either the official presentation or subsequent public discussion of the 1974 Defence Review. (3)   And this has been a source of much difficulty.

From the outset the Government chose different bases for the presentation of expenditure 'savings' and manpower implications.

   (a)   On the expenditure side, 'savings' totalling £4,700 millions (at 1974 Survey prices) were claimed for the period 1975-84.   These were measured against the programme inherited from the Conservatives, i.e. against intended provision and over the full span of the official Long Term Costings.   (In relation to the then current provision the Review simply entailed stabilizing the defence budget, from 1977-8, near c. £3,800 millions at 1974 Survey prices.)

   (b)   Regarding manpower, first intimations were that the programme change would mean reductions of c. 38,000 in Service personnel, c. 30,000 in Civilian Employees, and a (direct) loss of industrial jobs of c. 10,000 up to 1 April 1979.   These reductions were measured against current provision, in fact in relation to manpower 'actually engaged' at 1 April 1974;   and, of course, over a shorter time-scale. Later it was estimated that the effect on industrial jobs in the period would be a 5,000 increase rather than a 10,000 reduction, because of the higher equipment share in the revised budgets.

(Measured against intended provision, however, the
industrial manpower implications were assessed at
40,000-70,000 fewer employment opportunities for
each year of the full programme period to 1984.)

Strictly speaking, it is the substantial number of 'lost
job opportunities', plus their equivalent in Service and
civilian employment, which is the manpower counterpart of
the £4,700 millions expenditure 'savings'. The anticipated
changes in labour 'actually engaged' which were emphasized
in official announcements corresponded not to the 'savings'
against intended provision but to the decision to hold the
defence budget at the c. £3,800 millions (1974 Survey
prices) level. Yet few commentators registered these
facts at first; and, even when fuller information was
made public and some official elucidation attempted,
confusion persisted. (4)

It is important not to perpetuate misrepresentation and
misunderstanding. Accordingly, the main expenditure and
manpower data relating to the Defence Review are set out in
Table 1,pp.323-4.The format accommodates all the key facts
and forecasts which have been produced, but in a way which
shows clearly what corresponds to what.

A striking feature of this tabulation is the relative
paucity of manpower data. Some information has appeared
giving a broad-brush assessment of the Government's expectations about the total job opportunities in industry
eliminated by the Review decisions; that is, about one
element of the employment counterpart of the 'savings'
which Ministers were at pains to stress. But the full
picture of the manpower implications of these 'savings'
has not been given. (This would require data about the
Service strengths and Civilian Employees' total implied by
the Conservatives' programme, as a basis for calculating
the 'loss' of 'job/years' attributable to these categories;
this figure, when added to the 'loss' of 495,000 'job/years'
among defence contractors, would yield the exact counterpart
involved.) A fairly detailed indication of the envisaged
pattern of manpower rundown from 'current' (or 1 April 1974)
levels is available. But this only goes so far as 1 April
1979; virtually no light at all has been shed on manpower
expectations over the full fiscal years 1979-80 to 1983-4.
(See the bottom right-hand quadrant of the table.)

It follows that an exhaustive and authoritative assessment
of the employment consequences of the proposals is out of the
question. Gauging the 'release' of manpower resources implied,
measured against intended provision, is possible only in the
most rough-and-ready way and on the basis of some highly
vulnerable assumptions. (5)

The 'release' that may be expected, measured against current
provision, lies within the limits of more reasonable speculation, however. For present purposes an estimate of the overall implied for the five years 1976-7 to 1980-1 is the most

322

Table 1

THE 1974 DEFENCE REVIEW PROPOSALS

A. Expenditure at 1974 Survey prices	Financial Years				Change + or − to 1 April 79 'Savings' ☐
	1975-76 £m	1976-77 £m	1977-78 £m	1978-79 £m	
1. Original Programme (Conservatives)	4000	4070	4150	4300	+300
2. 'Savings': Defence Review (74 PES)	300	270	350	500	−1420
3. (a) New Programme A	3700	3800	3800	3800	+100
(b) New Programme B	3700	3690	(3800	3800	(+100)
New Programme B (revalued at 1975 SPs) see Table 2 below	4548	4566	4725	4723	+175

B. Manpower (000s)	01Apr74	01Apr76	01Apr77	01Apr78	01Apr79	
1. Original Forecast (Cntrctrs Emplys (3))	− (235)	− 235	not known 235	− 245	− 260	− +25
2. 'Jobs Lost' (Cntrctrs Emplys (3))	− (n.a.)	− 40	not known 40	− 45	− 60	− −185
3. New Forecast (Cntrctrs Emplys (3))	− (n.a.)	− 195	not known 195	− 200	− 200	− +5
(1) Service Personnel						
RN/RM	79	76	(76)	(75)	74	−5
Army	180	176	(170)	(166)	165	−15
RAF	100	91	(89)	(85)	82	−18
	359	343	(335)	(326)	321	−38
(2) Civilian Employees						
UK-based	249	−	245	−	234	−15
LEC	60	−	53	−	45	−15
	310	(310)	298	(290)	280	−30
(3) Contractors Employees	(200-235)	195	195	200	200	+5
* TOTAL	(875-900)	(848)	(833)	(815)	(800)	(?)

Figures in parentheses are author's estimates.

Financial Years					Change + or – to 1 April 84 'Savings'	
1979–80 £m	1980–81 £m	1981–82 £m	1982–83 £m	1983–84 £m		
4400	4400	4450	4450	4500	+500	Cmnds 5879, 5976
600	600	650	700	750	−4700	Detail for 1979–84 based on data in HC 259 (1974–5)
3800	3800	3800	3800	3750	+50	
3800	3800	3800	3800	3750)	(+50)	Post-Budget:£m110 cut in 1976–77
(4725	4725	4725	4725	4670)	(+120)	Cmnd 6393

01Apr80	01Apr81	01Apr82	01Apr83	01Apr84		
–	–	not known	–	–	–	
270	270	260	260	250	+15	
–	–	not known	–	–	–	cf. boxed figure above
70	70	60	60	50	−495	for Comments see
–	..	not known	–	–	–	HC 155 (1975–76)
200	200	200	200	200	+5	
		not known				
	(310)				?	For details see Cmnd 5976 and Cmnd 6432 incldg LEP breakdown: also HC 155 (1975–76) pp.30–1
		not known				
	(275)				?	Cmnd 5976 p.20 Cmnd 6432 p.30
	(205)				?	See text p.10–11
	(790)				?	See text p.10–11

useful calculation. In the first place, the manpower
implications of the Defence Review for the next five
years are of greater immediate interest than what should
have happened in the recent past and what might be in store
towards 1984. In the second place, the main point of the
analysis is to establish the context for elucidating the
employment consequences of selected 'programme options' for
further reductions over the five years to 1980-1. (6)

On this basis the implications, derived from the presentation in Table 1, emerge as follows:

(1) <u>Service personnel</u>: a rundown of <u>c</u>. 33,000 to aggregate
Service strengths at 1 April 1981 of <u>c</u>.310,000. This is
less than the <u>c</u>. 38,000 rundown over April 1974-April 1979
'recorded' in Table 1 because, of course, almost half of
that should already have taken place by 1 April 1976 and
it is assumed that strengths will fall by only a further
11,000 or so between April 1979 and April 1981.

(2) <u>Civilian employees</u>: a reduction from current levels to,
say, 275,000. This implies 'release' of 35,000 rather
than 30,000 workers. There are no fewer employees on
the civilian payroll now than there were two years ago,
and a further reduction of 5,000 may be assumed after 1
April 1979. (7)

(3) <u>Defence Contractors' Employees</u>: although uncertainty
about the present size of this category bedevils assessment of change, there are no obvious grounds for doubting
the (eventual) conclusion of official calculations – that
the Defence Review programme change entailed a <u>rise</u> in
the labour requirement in defence-related industry.
Assuming a steady 3 per cent growth of productivity the
Ministry of Defence now think that, in the early 1980s,
200,000 industrial jobs might be attributable to the
procurement programme, some 5,000 more than at present.
Since the productivity growth estimate may be generous
the increase could be greater.

Summarizing this assessment, the Defence Review proposals
implies <u>overall</u> net manpower 'release' (<u>actual</u> release,
against existing levels) in the middle five-year period of
the order of some 55,000 jobs viz.

Employment Effect 1976-81 (overall): Actual Release (000s)			
	(1)	(2)	(3)
Total	Service Personnel	Civilian Employees	Contractors' Employees
A_0 Defence Review 55	33	35	-10 (increase)

This is the comprehensive - or 'global' - assessment.   In
categories (1) and (2), however, there is overseas labour whose
release would not pose problems for domestic manpower management.   Uniformed personnel locally enlisted overseas have been
getting fewer in recent years and now number just over 9,000.
Further contraction to 7,000 (by 1981) may be expected as
the Review proposals are implemented.   'Locally-entered'
civilians are to bear a major share of the planned rundown to
1979, as the figures show (Table 1, B.3(2) LEC).   But,
thereafter, that share must necessarily decrease.   Of the
35,000 Civilians Employees' jobs to disappear in the period
1976-81, therefore, perhaps 16,000 in all will be from among
the locally entered overseas.

Total manpower release (net) to the United Kingdom labour
market in these years as a result of the Defence Review
should thus be c. 40,000.

Employment Effect 1976-81 (UK):   Actual Release (000s)

		(1)	(2)	(3)
		Total	Service Personnel	Civilian Employees	Contractors' Employees
A_UK	Defence Review	40	31	19	-10 (increase)

This is the first element to be entered in our 'composite'
analysis. (8)

Programme Revision: the 1975 Public Expenditure 'Round'

The new programme and budgetary projections for 1975-84
which emerged from the 1974 Review were modified before the
ink was dry on the 1975 Defence White Paper.   The Chancellor
of the Exchequer announced in his Budget Statement a reduction
in the target Estimate for 1976-7 of £110 millions (at 1974
Survey prices). (9)   Of later years he said nothing.   But
one inferred that the projections for 1977-8 and beyond would
be under searching scrutiny in the 1975 Public Expenditure
Survey 'round', which was indeed to be the case.

The baseline for the 1975 Survey was, as usual, the existing programme and budget, revalued - for this 'round' - at 1975
Survey prices (SPs) (equivalent for all practical purposes to
1975-6 Estimated prices).   On this new price basis the
'flat' expenditure profile to 1983-4 lay at an annual budget
level of £4,700-£4,725 millions, corresponding to the c.
£3,800 millions level of the previous year's values. (10)

After protracted ministerial deliberations, which were
accompanied by considerable speculation and enlivened by more
than one veiled resignation threat, this programme was revised.
Cabinet decisions on the scale of the revision were not
reached until December 1975, and the contents of the required

'cuts' package were not finalized until the Christmas/New Year period. When announced, however, they amounted to reductions in intended provision of more than £500 millions over the period 1977–80, viz £178 millions in 1977–8, £193 millions in 1978–9 and £164 millions (against an original target itself revised downwards) in 1979–80. Planned spending is now programmed to level off at £4,530 millions (at 1975 SPs).

What corresponding adjustment was made to budgetary projections for the 1980's is not known. But it is inconceivable that the Ministry of Defence's reworking of the Long Term Costings should have left targets for later years unchanged. A plausible guess would be that the revised £4,530 millions 'ceiling' has been projected through the early 1980's. That the Government is adopting a 'low profile' on these medium-term intentions may reflect the somewhat different climate of opinion vis-à-vis defence in early 1976 compared with 1974–5. Because of concern over jobs expressed in debate on the original Defence Review, coupled with a growing disillusion about the nature and durability of East-West détente, ministers may have thought it inappropriate to publicize the 'Christmas cuts' of 1975 as a further round of 'savings' totalling almost £1,400 millions to 1985. Yet that is what the programme revision probably amounts to, as the presentation in Table 2 makes clear.

Table 2

THE 1975 PUBLIC EXPENDITURE SURVEY 'ROUND' PROGRAMME REVISIONS

A. Expenditure at 1975 Survey prices	Financial Years				Change + or − to 1 April 80 'Savings'
	1976–77 £m	1977–78 £m	1978–79 £m	1979–80 £m	
1. Post-Review Programme(B) (Table 1, line 3(b)revalued)	4566	4725	4723	4694	+128
2. 'Savings' Prog. Revision (75 PES)	–	177	193	164	−534
3. Revised Programme	4566	4548	4530	4530	− 36
Revised Programme (revalued at 1976 SPs) cf. Defence Estimates 1976–7	5632	5610	5588	5588	

One notable feature of the programme revision is that it involves stablization of the defence budget at a level fractionally below that of current expenditure, albeit at a current price figure of more than £5,500 millions (1976-7 Estimated Prices/1976 Survey prices). (See Table 2, p.327, final line.)

Programme Revision: Manpower Implications

What are the manpower implications of this revision, for the period 1976-81 which is of practical interest? The Government's general assertion is that the further 'savings' will be made by reduction in support, not the front-line; and that substantial reductions in civilian manpower are to be made in addition to those resulting from the Defence Review. Any more detailed assessment must be speculative in the extreme, but the following analysis is not inconsistent with such hints and indications as ministers have given:

(1) Service personnel: although no formal statement has been made of further rundown of this category of defence manpower some additional wastage or redundancy is probably inescapable as a concomitant of rigorous pruning of support arrangements. It would certainly be quite realistic to envisage a reduction of, say, 5,000 on this account, implying a revised aggregate Service strength of 305,000 for the end of financial year 1980-1.

(2) Civilian Employees: release of a further 10,000 is anticipated, about half from employment with Ministry of Defence headquarters and related establishments, about half from jobs in support of the three Services. Reduction of this category to a level of c. 265,000 by 1 April 1981 is, therefore, a relatively firm expectation.

(3) Defence Contractors' Employees: the 1976 Defence White Paper states that the programme revision measures 'are concentrated on support for the Services and will not have a significant effect on Industry; there will be a loss of about 3,000 job opportunities but no major plants will have to be closed'. (11)   Adoption of the job opportunities' formula reflects the Ministry of Defence's reluctance to give hostages to fortune in the prediction of actual 'release' of industrial manpower, given the difficulties of calculation and the confusion and embarrassment caused by its revision of ideas on the industrial implications of the Defence Review proposals (see p. 329). However, for present purposes, the inference is that the anticipated increase in industrial manpower will not be as great as

it would otherwise have been: hence a 3,000 jobs
'release' may be entered in the 'composite' analysis.

Financial Years					Change + or − to 1 April 85 'Savings'
1980–81 £m	1981–82 £m	1982–83 £m	1983–84 £m	1984–85 £m	
(4700)	(4700)	(4700)	(4675)	(4675)	(+100)
(170)	(170)	(170)	(175)	(175)	−1394
(4530)	(4530)	(4530)	(4500)	(4500)	(−66)
(5588)	(5588)	(5588)	(5588)	(5588)	

What proportion of the employment effects under headings (1) and
(2) should be attributed to the United Kingdom and what proportion to 'overseas' is a difficult question.   But support
economies would be unlikely to yield more than 1,000 posts
among locally enlisted Services personnel or more than 2,000
jobs among locally entered civilians overseas (in addition to
the release associated with the Defence Review itself).   If
this assessment is correct, the foregoing may be summarized as
follows:

Employment Effect 1976–81 (overall): Actual Release (000s)

		(1) Service Personnel	(2) Civilian Employees	(3) Contractors' Employees
	Total			
B_0 Programme Revision	18	5	10	(3)

Employment Effect 1976–81 (UK): Actual Release (000s)

		(1) Service Personnel	(2) Civilian Employees	(3) Contractors' Employees
	Total			
B UK Programme Revision	15	4	8	(3)

This is the second element to be entered in our 'composite' analysis.

Present Plans: Manpower adjustment − scale and incidence

With reference to the five year period spanning financial years
1976-7 to 1980-1, therefore, the manpower consequences of the
diminution of the defence effort already 'programmed' − expressed

as actual release, measured against estimated 1 April 1976
manning levels - appear to be,

(a) reduction in overall direct defence-related employment, Service and civilian, at home and abroad, of some
73,000;

(b) the release of around 55,000 (net) to the labour
market in the United Kingdom itself. (35,000 Service
personnel, 27,000 Civilian employees of the Ministry
of Defence and other departments, less (for this
calculation) an estimated 7,000 addition to Defence
contractors' payrolls, because of the heavier emphasis
on equipment spending in the 'new' programme and
budget.)

This is the 'release' - in the first of the senses distinguished
earlier - which creates a direct and immediate adjustment or
reabsorption problem.   As explained, the extent of 'release'
in our second sense (reduction of planned resource claims
which 'makes room' for alternative uses of resources) is substantially greater. (12)   Moreover, these figures take no
account of indirect employment effects. (13)
     What information exists on the likely incidence of rundown
within the United Kingdom?   For Service personnel the answer
is: none.   This is partly because not all the details are yet
planned.   More important, it is because those in uniformed
service have neither exclusive regional affiliations nor exact
industrial (or occupational) associations.   The employment
effects must be regarded as impinging on the labour market as
a whole and affecting a cross-section of industrial/occupational
skills. (14)
     Regarding Civilian employees some specific information is
available, because some precise decisions have been made about
closures of installations and reductions of headquarters'
staff.   But officials do not know where redundancies will
occur;   indeed, they do not expect the exact incidence of the
first stage (Defence Review) manpower 'cuts' to become apparent
until the summer of 1976. (15)   What can be said is that:

(a) the effects will be determined to some extent by the
existing geographical distribution of Civilian employees,
a feature of which is concentration in the South East and
South West regions (see Annex 'A'); and

(b) although several military installations nominated
for closure will be turned over to other defence uses
(with retention of civilian employees), the initial
regional incidence of reductions may be expected to
approximate that implied by the specific closures
announced or anticipated.   (See Annex 'B'.)

Both factors enter the reckoning because of the possibility

of formally 'non-mobile' staff affected by closure proposals moving to other locations. On balance it is likely that the main burden of adjustment will fall on the following regions: South East, South West, West Midlands, North West, East Midlands in England; and also, significantly, on Wales, Scotland, and Northern Ireland.

The net effect of the 1974 Defence Review and the 1975 Public Expenditure Survey 'round' revision on employment in defence-related industry over the period 1976-81 is estimated to be an increase of jobs of around 7,000. But this masks movements of expansion and contraction in individual firms and industries. Consolidation of fragments of evidence yields a picture of effects on <u>Defence Contractors' Employees</u> comprising:

(a) <u>Contraction</u>: (i) in the airframe and aeroengine industry at firms/plants engaged on programmes curtailed by the decisions, or simply coming to an end as planned, where there is no immediate expectation that capacity will be taken up by new domestic procurement business: examples include Hawker Siddeley Aviation (plants at Brough, Coventry, Manchester, Hamble, Kingston), Hawker Siddeley Dynamics (Hatfield), Rolls Royce (plants at Coventry, Derby, Leavesden, Hillington and East Kilbride), and Westlands (Yeovil and Cowes), (16)

(ii) generally throughout industry as a result of 'cuts in the works, fuel and clothing programmes', 'certain reductions ... in the equipment programme' and 'economy in extra-mural research', (17)

(b) <u>Expansion</u>: in the electronics and general equipment sectors of the aerospace industry and, more broadly, throughout 'the systems areas: aircraft systems, ship systems and army systems'; perhaps in shipbuilding and marine engineering. There is a general expansionary effect attributable to the higher expenditure on equipment budgeted for the later 1970s compared with the recent past. (18)

It does perhaps bear repeating that, even following the 1975 'Christmas cuts' and the decision to stabilize the defence budget at more or less its current level, no 'shake out' of manpower from defence-related industry is anticipated over the next five years. Such reductions in numbers actually engaged in the defence effort as are envisaged are confined to Service personnel and civil servants. Labour resources in manufacturing industry will be 'released' to help promote economic recovery only in the sense that many of the claims which would have arisen under the Conservative pre-Defence Review programme will not now materialize.

Further Reductions: Options

The employment (and other economic) consequences of further
reductions in planned defence spending over 1976-81 would be
additional to the already 'programmed' effects discussed
above.   In this part of the paper a speculative analysis is
attempted of the manpower implications of adoption of three
specific 'programme options' for the period.   The characteristics of these 'options' were delineated in an earlier paper.
Briefly, they represent alternative 'packages' for attainment
of a level of defence spending of c. £3,700 millions at
1975/6 Estimated/1975 Survey prices, which corresponds
(approximately) to a level of c. £4,525 millions at 1976-7
Estimated/1976 Survey prices.   That is to say, they are
options for diminution of the defence effort to a budget
level some £1,100 millions below that set for the current
financial year (1976-7).   This is the scale of further
reduction that would be necessary to bring defence's proportion of the United Kingdom's forecast GNP for 1980/1
down to rough equality with the defence/GNP proportions of
France and West Germany which may be expected to hold at
that time.   The expenditure profiles associated with these
options for the financial years 1976-7 to 1980-1 are set out
in Table 3 on alternative prices bases and against two bench-
marks, (1) the 'Defence Review programme and budget' (cf.
Table 1 A.3(b) as updated in Table 2 A.1) and (2) the revised
projections following the 1975 Public Expenditure Survey
'round' (cf. Table 2 A.3) (19)

   Compared with the first (Defence Review) benchmark, the
essential features of the three Programme Options may be
expressed as follows:

   (a) each is predicated on phasing-out all provision
   for the Polaris force and eliminating budgetary provision
   for any Army garrisons outside Europe;

   (b) the options are differentiated according to the
   emphasis in Service major mission programmes which they
   reflect.   Option I presupposes adoption of a 'Smaller
   ship Fleet' philosophy and reductions in European Theatre
   Ground Forces but the retention of a 'full spectrum'
   Royal Air Force.   Option II is based on adoption of a
   Smaller ship Fleet and a lower cost Air Force with the
   preservation of existing European Theatre Ground Force
   levels. Option III maintains the existing naval programme
   with its Balanced Fleet philosophy (including 'ships of high
   quality') but embodies the Lower cost Air Force and
   substantial cuts in ground forces in both Germany and
   the United Kingdom. (20)

Table 3

PROGRAMME OPTIONS TO 1980-81
£m on alternative prices bases
(all figures rounded to nearest £m25)

Programme	1976-77	1977-78	1978-79	1979-80	1980-81
Part A					
1975-76 Estimated/ 1975 Survey prices Benchmark (1)*	4575	4725	4725	4700	(4700)
Option 1	4565	4475	4200 ·	3950	3700
'Savings'	10	250	525	750	1000
Option II	4560	4525	4125	3800	3700
'Savings'	15	200	600	900	1000
Option III	4570	4575	4200	3800	3700
'Savings'	5	180	525	900	1000
Benchmark (2)	4575	4550	4525	4525	(4525)
(1) less (2)	-	175	200	175	(175)
Part B					
1976-77 Estimates/ 1976 Survey prices Benchmark (2)	5625	5600	5600	5600	5600
Option I	5575	5450	5125	4850	4525
Option II	5575	5525	5025	4650	4525
Option III	5575	5575	5125	4650	4525

* See text, note 19

It should be added that these programme options were
developed to illustrate the kind of change in the defence
effort that 'equalizing' defence GNP/proportions over the
next five years would require.   Needless to say they do
not represent the only, nor necessarily the most advantageous,
ways of effecting future defence expenditure reductions to
this end.   Nor is it implied that such reductions should be
made.   Similarly the assessment of manpower implications in
what follows has been undertaken for the purpose — and for
the sole purpose — of elucidating the likely employment
consequences of reductions of this magnitude.   What weight
these should have in thinking and choice about 'alternative
futures' is beyond the paper's remit.

Further Reductions: Manpower Aspects

What would adoption of each of the 'packages' imply for the
three categories of defence manpower which have been
distinguished and used in assessment of the effects of
changes already 'programmed'?   What actual release, measured
against manpower currently engaged, might be expected over
the period to 1 April 1981?

   Option I would mean a much reduced naval manning requirement for the surface fleet and would make possible related
manpower economies to training and support.   Bearing in mind
the rundown of the Polaris force incorporated in this 'package'
(as in the others) a reduction in the strength of the Royal
Navy/Royal Marines by as much as 17,500 is implied.   This
option also entails a massive reduction in size of the army,
to say, 120,000-125,000.   Associated with contractions on
this scale would be a reduced demand for civilian support
effort (including employment in HM Dockyards and at Army
facilities in the United Kingdom, Belgium and Germany).
Civilian employees released to the home labour market might
number 40,000.   If this seems low compared with the Service
strength reductions at issue it should be remembered first,
that much of the effect of a 40 per cent reduction in
ground troops in Germany would be borne by the West German
labour market; and, secondly, that measures already — or
soon to be — in train mean that support manpower levels
should be dropping anyway over the next few years.   How
employment on defence contracts in industry would be affected
is harder to gauge.   Implementing this option would,
however, end any anticipation that jobs associated with new
ship construction might increase over the next few years.
Rather one would foresee reductions affecting, in particular,
the firms involved in the ASW cruiser programme (Vickers
(Barrow), Swans (Wallsend)), in the SHEFFIELD class
construction effort (Swans again and, possibly Cammell Lairds),
and in Fleet submarine building (Vickers (Barrow) again);

the associated propulsion machinery manufacturers (e.g. Rolls
Royce, David Brown and the Vickers Engineering Group at
Barrow); and several guided weapons and other 'systems'
manufacturers. In addition a smaller ship fleet would have
no place for the maritime version of the Harrier aircraft:
cancellation of this project would affect mainly Hawkers
(airframe), Rolls Royce (engine) and Ferranti (radar and
avionics). So far as Army procurement is concerned the
impact would probably fall on the workload of the Royal
Ordnance Factories and other makers of armoured fighting
vehicles; on the guided weapons manufacturers (reduced
requirement for BAC's Rapier, Short's Blowpipe and the
Milan anti-tank missile, for instance); and on firms
developing improved communications systems of various
kinds. To try to quantify these effects is hazardous in
the extreme. At a guess, some 30,000-50,000 jobs might
be involved. Thus, summarizing the position in (very)
round figures one might have,

Estimated Employment Effect 1976-81 (UK) Actual Release 000s(21)

	(1) Total	(2) Service Personnel	Civilian Employees	(3) Contractors' Employees
c_I Option I say	140	55-65	40	30-50

The basis of <u>Option II</u> is adoption of both that smaller ship fleet
philosophy which is a characteristic of Option I and the concept of a
lower cost Air Force. The latter means a force equipped to cover
less than the full spectrum of tactical air missions, without
the MRCA but with more aircraft for 'battlefield' roles. It
follows that the employment consequences would include,

(a) all those associated with the reduced naval provision,
     which have been outlined in the previous paragraph;
     these require no further elaboration;

(b) those that would accompany the transition to a more
     modest air force, i.e. one operating more types of
     aircraft in production (and perhaps, in time, aircraft
     purchased from abroad as part of European or Atlantic
     collaborative arrangements) but not acquiring the MRCA
     in either its 'common' or UK-only interceptor variants.

An indication of the scale of the industrial impact of the
MRCA cancellation implied here can be gleaned from some basic
facts. First, <u>in early 1976</u> some 5,500 BAC workers (mainly
in Lancashire) were employed on the airframe and about 4,000
Rolls Royce employees (mainly in Bristol) were engaged on
engine work; in addition development of the aircraft's
avionics accounted for several thousand jobs among a number of
prime contractors, including Ferranti, Marconi-Elliott and

Smiths. (22)   Secondly, and perhaps more to the point, it has
been estimated that by 1981 'at least 25,000 people could be
employed on this project in the United Kingdom'. (23)
However, this does not mean that assessment of manpower
release on this account is a straightforward matter.   The
labour market consequences of cancellation would depend on
the circumstances, notably on whether a unilateral withdrawal by the United Kingdom caused abandonment of the entire
undertaking.   For the purpose of speculation about the
employment effects of this second option, there is the added
complication that prolongation of production, of say, Jaguar
and Harrier - together with any new developments would
obviously mitigate the impact.   Thus, any quantification of
the net effect on jobs in the aerospace and related industries
is subject to even wider margins of error than surround some
of the other assessments here.   For the Programme Option
overall an industrial manpower effect of 55,000-65,000 jobs
(roughly half attributable to the naval 'savings' and half
to the air side) has been assumed for the purpose of the
composite analysis.   As for the service personnel and
civilian employees categories, the manning requirements of
a Lower cost Air Force might be met with a total service
strength of 65,000-70,000 and some 15,000-20,000 fewer
civilians would be required in support functions.   Taking
these numbers together with the appropriate naval elements
in the Option I calculations, the overall assessment for the
employment effects for Option II comes out as:

Estimated Employment Effect 1976-81 (UK):   Actual Release 000s

	Total	Service Personnel (1)	Civilian Employees (2)	Contractors' Employees (3)
C_{II} Option II say,	125	30-40	30	55-65

Finally, it is a simple task to set out the manpower
consequences of Option III.   These would be:

(a)   those associated with the provision for reduced
European Theatre Ground Forces - as in the Option I
assessment;

(b)   those associated with the lower cost Air Force
concept of Option II.

In the standard format adopted for the presentation of employment effects the figures are:

Estimated Employment Effect 1976-81 (UK):   Actual Release 000s

	Total	Service Personnel (1)	Civilian Employees (2)	Contractors' Employees (3)
Option III say,	145	55-60	40	45-50

336

The three sets of data for the selected Programme Options are the third and final input to the 'composite' analysis in the closing section of this paper.

Further Reductions: Employment Impact

Before proceeding to that denouement, however, some summary observations are in order on the likely incidence of the employment effects of further reductions achieved along these lines.   In the first place there is the matter of the differing distributions among defence manpower categories of the 'actual release' which might be expected from adoption of the alternative 'packages' analysed.   The estimates reflect that time-honoured adage: the Royal Navy and Royal Air Force man equipment, the Army equips men.   That is why,

(a)   the options which incorporate a smaller army (I and III) could be expected to produce a relatively large release of service personnel: up to 40 per cent of the labour 'shake out' they would entail would be from the uniformed ranks;

(b)   option (II), being based on a much diminished navy and air force, i.e. the capital-intensive arms, would yield the greatest release in defence-related industry, perhaps fully half the total employment effect.

Secondly, some tentative ideas may be advanced concerning the geographical distribution of the employment impact:

(a) in the case of service personnel the scale of man-power rundown implied for each of the three services is substantial in relation to their present strengths. Thus, although no one can predict exactly where the men and women might go, the regional pattern of recent recruitment to the forces and the pattern of current home deployments may give indirect clues.   This is because there is a tendency for persons leaving the services to concentrate their search for civil employment near either the family home or the last (perhaps a recent) duty station.   Relevant data suggests that some regional concentration of impact might be expected in Scotland, Yorkshire and Humberside, the South East, South West and North West.

(b) Defence's civilian employees are fairly tightly concentrated within the United Kingdom, as the figures in Annex 'A' indicate.   More specifically, the Navy Department's civilians, who include dockyard workers, are almost exclusively in the old home ports dockyard towns; the Army's civilian support is partly spread

337

around (as army units are), partly concentrated in the
Midlands and South (where the main support establishments
are); most of the Royal Air Force's civilians are either
at air stations lying in a 'boomerang-shaped' zone
stretching from (roughly) Wiltshire through Oxfordshire
into East Anglia/Lincolnshire and thence to Yorkshire and
the North East or at maintenance units and other facilities, some within, some 'behind' this zone. (24)

(c) Official reticence about predicting the geographical
incidence of the impact on Defence Contractor's Employees
notwithstanding, there seems little doubt that

    (i) options I and II would bear particularly heavily
    on the North and North West, because of the effects
    on warship building and related activities,

    (ii) options II and III would bear particularly
    heavily on the North West and South West, because
    of the effects on MRCA-related aerospace activities,

    (iii) since many of the electronics and 'systems'
    manufacturers' plants are in the South East this
    region cannot escape an important share of the burden
    of adjustment to any cuts in defence procurement.

Perhaps the most significant point about the spatial aspect
of employment effects, however, is that the regional
perspective may not be particularly illuminating. Not only
is defence spending (and the jobs it generates) unevenly
spread among regions, it is unlikely ever to be spread
evenly within any given region. Thus, its impact is local
and specific rather than general. Hence, as Timothy Stone
has observed, 'a minor change as viewed from the regional
level may be a momentous one for a particular town. It may
entail economic disaster (in the short-run anyway) or it may
offer a most welcome economic opportunity.'

and, following on from this, he concludes,

    analyses which assume the regional perspective cannot
    reflect such phenomena. This ... is an argument for
    micro studies to complement the macro picture — at
    least where there is a strong presumption, or other
    evidence, or intra-regional concentration of defence
    activity. In such circumstances, indeed ... only
    painstaking micro studies can 'tell it like it really
    is' within the region. (25)

Needless to say this is very much to the point in the light
of the direct knowledge we have of how defence installations
and defence contractors' plants are located in the United
Kingdom and the importance they assume in their local
economies. (26)

Attention to the _local_ perspective is important because of a prevailing tendency to discuss the impacts of defence demand changes in terms of national aggregates. It is appropriate to do this, not least because the _purpose_ of reshaping a national defence effort is to make possible the use of resources for worthwhile civil purposes, on a national scale. Moreover, preoccupation with local adjustment problems can lead to the absurd situation where local labour market tails wag national economic dogs; or even to industrial or provincial carts being put before national interest horses. At the same time factors of production are _not_ perfectly mobile in the modern world. Macroeconomic measures to maintain or increase national product and income are a necessary but _not_ a sufficient basis for adjustment by people and places to the temporary disruptions of defence changes. Impacts on localities can create severe and long-lasting problems. (27)

A third observation concerns the _timing of employment effects_. Projections of the incidence of 'savings' associated with the Programme Options under discussion were set out in Table 3 p. 333. In the first one or two years of the period there would be only limited scope for financial 'savings'. In part this is because an orderly process of contraction is envisaged. But more important, it is because the short-term costs of creating redundancies among service personnel and defence civilians are fairly high — perhaps high enough to offset 'early' savings on wages and salaries; and because the abandonment of procurement programmes involves cancellation charges and nugatory expenditures. Thus, the fact that estimated 'savings' appear to accrue in a fairly even 'wedge-like' fashion does _not_ connote an equally leisurely manpower rundown. On the contrary the 'savings' pattern mapped out, based on assessment of the budgetary margin for manoeuvre, presupposes early decisions, early redundancies and early project cancellations. The manpower rundown which is consistent with the budgetary information in Table 3 is one concentrated in the early years, tapering off towards April 1981. (28)

Finally, a most important matter: are the employment effects that have been estimated commensurate with the expenditure 'savings' of Table 3? The employment effect figures for the three Programme Options were built up by the simple technique of looking at the manpower attributable to the implied deletions from the defence programme and aggregating them. A check on whether this procedure yields numbers of a plausible order of magnitude may be made by an elementary proportionality calculation. The expenditure 'savings' amount to a 22 per cent reduction in the budget level over 1976–81. The manpower 'release' totals for the Options lie in the range 125,000–145,000 (or 140,000–160,000 if one counts in the effects attributable to the 1975–5 programme revision, as perhaps one should for this purpose). This represents a 15–17 per cent (or 17–19 per cent) rundown measured against 1 April 1976 'total defence manpower' figure of $\underline{c}$. 850,000. The estimates may, therefore, err on the low side.

Conclusion: The Composite Analysis
(Key conclusions numbered thus: 1 - 7)

In conclusion it remains only to consolidate the estimates of
employment effects to show how further reductions in defence
spending over 1976-81 would add to the defence manpower
absorption 'problem' implied by changes already 'programmed'.
The data are set out in Table 4, from which one notes that:

(1)   The effects of the 1974 Defence Review decisions
will be 'working through the system' during 1976-81 and
should produce an <u>actual release</u> (net) of manpower
resources for other uses, to the United Kingdom's labour
market, of 40,000.

(2)   The programme revisions decided upon following the
1975 Public Expenditure Survey 'round' may be expected to
add a further 15,000 to this figure.

(3)   Further expenditure reductions, were they to be made
on a scale necessary to 'equalize' defence/GNP proportions
among France, West Germany and the United Kingdom by 1981,
would entail actual release of an additional 125,000 -
145,000 workers (service and civilian) - if sought by one
or other of the three Programme Options examined.

(4)   A decision for 'further reductions' on this scale
would, therefore, imply a redeployment 'problem' during
the five years to 1 April 1981 involving 180,000-200,000
persons now directly engaged in defence work of one sort
or another.   Among the three principal categories of
defence-related manpower, the numbers involved would be
as follows:

        70,000-95,000 service personnel
        <u>c</u>. 55,000-70,000 civilian employees
        up to 55,000 defence contractors' employees.

The overall size of the 'problem' and the incidence among
the different manpower categories would depend on the
nature of the reduction option chosen.   So too would the
incidence within categories - individual services and
specializations for the first two, individual industries
and plants for the third (see pp. 331-2 above).

That the figures relate to <u>direct</u> employment connected with
the defence effort and to <u>actual release</u>, measured against
current manpower levels (1 April 1976), must be stressed.
No specific analysis has been done to gauge the total,
direct <u>and</u> indirect, employment effects of the changes
considered.   However, case studies of defence installations
in the United Kingdom have indicated that a local employment
multiplier of 1.25 (one indirect job for every four direct

340

Table 4

SUMMARY OF ESTIMATED EMPLOYMENT EFFECTS, 1976-81
(United Kingdom)   (000s)

Direct Employment Only/Actual Release

	Total	(1) Service Personnel	(2) Civilian Employees	(3) Contractors' Employees
Already 'Programmed' A. Defence Review (1974)	40	31	19	−10 (increase)
B. Programme Revision (1975 PES)	15	4	8	3
sub-total	55	35	27	−7 (increase)
Further Reductions C_I Option I	140	60	40	40
C_{II} Option II	125	35	30	60
C_{III} Option III	145	57.5	40	47.5
Total with C_I	195	95	67	33
Total with C_{II}	180	70	57	53
Total with C_{III}	200	92.5	67	40.5

defence jobs) may be appropriate for estimating aggregate installations-related defence employment. (29) Official calculations, based on inter-industry relations data for 1970, suggest that the relevant multiplier for jobs on defence contracts in industry is as high as 2.3 (1.3 indirect jobs for every direct one). But this is a higher value than most other impact studies have arrived at and a multiplier of 2.0 (every direct job generates one more indirect job) may be a sounder basis for estimating aggregate industry-related defence employment. (30) On this basis,

(5) Applying appropriate multiplier factors to the separate manpower categories in Table 4 yields a rough-and-ready estimate of the scale of the overall redeployment 'problem' implied by the changes considered affecting 235,000-280,000 workers directly or indirectly involved with the defence effort.

And, what follows from the above,

(6) Over the five-year period this would represent 47,000-56,000 job-changes per year, although the phasing assumed in the calculations for expenditure 'savings' presupposes manpower release earlier rather than later so that a peak absorption of c. 60,000 in two or three years might be necessary.

For certain purposes the concept of 'lost job opportunities' has been adopted in presenting the employment consequences of changes in intended provision for defence. This is the logical counterpart to the 'savings' concept which it is customary to use in presenting the financial aspect of such changes. Unfortunately absence of full information on the benchmark 'defence manpower budget' for the next few years makes it impracticable to produce a comprehensive estimate of this alternative measure of resources released. (But see note 5, p. 346).

The regional incidence to the employment effects reviewed in the paper cannot be predicted with any great confidence. However, because of the existing geographical distribution of defence installations in the United Kingdom, which determines the distribution of service personnel and civilian workers, one would foresee most of the main adjustment 'problems' for these categories — in quantitative terms — arising in the more prosperous regions of the United Kingdom. The location of defence-related industries obviously determines the distribution of contractors' employees: for certain options there would be potentially troublesome concentrations of redeployment needs for this category. This leads to the most important general conclusion on the 'spatial aspect':

(7)   Both defence <u>installations</u> and <u>contractors' plants</u>
are frequently dominant in their local economies:
therefore, the most illuminating perspective from which
to evaluate absorption or redeployment 'problems' is
the local one.   It is at the local level that they
occur, must be confronted and (in due course) solved.

There is, of course, an alternative to 'solutions': avoid
the problem in the first place by retaining the 'resource'
in its existing use — in this context its defence use.   One
must presume, however, that no-one would seriously suggest
that manpower and productive capacity should be committed,
or remain committed, to military purposes solely because
redeployment would be troublesome.

Yet to arrive at this point is, in fact, to close the
circle on the argument with which this essay is concerned.
A defence effort confers security benefits but imposes
opportunity costs.   If Governments plan reduced expenditures,
or pressure groups demand them, it is 'to release resources'
with fulfilment of other worthwhile purposes in mind.
Unfortunately the transfer of resources is not a costless
process and other aspirations — preserving employment,
regional balance or industrial expertise — may be, incidentally, put at risk.

Thus, there is no escape from the exercise of choice
according to value judgments among societal objectives.
Exploration of some consequences of planned and advocated
change should have illuminated one facet of this central
calculation.   That is all.

April 1976

SELECTED REGIONAL STATISTICS ON DEFENCE MANPOWER
Percentage Distributions: Various years 1972-73

Region (a)	Defence Manpower Categories			Other Statistics	
	UK Armed Forces in the UK (c) %	Civilian Employees (d) %	Contractors' Employees (e) %	Service Recruitment (f) %	Economically Active Population (g) %
Northern	4.0	3.6	2.9	8.4	5.7
Yorkshire & Humberside	4.9	3.0	3.4	10.1	8.9
E. Midland	3.9	3.7	10.3	7.4	62.
E. Anglia	6.7	1.5	1.6	2.9	2.8
S. East	42.9	47.3	40.9	20.7	33.9
S. West	22.0	19.3	13.6	7.9	5.9
W. Midland	4.7	5.1	6.2	8.6	10.0
N. West	0.7	4.6	12.2	14.3	12.8
Total Eng.	89.8	88.0	9.26	80.3	86.2
Wales	2.6	3.6	1.6	4.9	4.3
Scotland	7.6	8.4	5.3	12.2	9.5
Total G.B.	100.0	100.0	100.0	97.4	100.0
N. Ireland (b)	6.1	1.8	0.5	2.6	2.3
ACTUAL TOTAL UK	270.4	263.5	245.0	38.2	23,135

000s

Notes

(a) Regions as defined in 1971 Census.

(b) Northern Ireland figures as % of UK total for the first three columns and the last.

(c) 1972 data.

(d) 1972-73 (estimated).

(e) 1973 (estimated)

(f) 1972-73 data.

(g) 1972 data.

Source: Aberdeen RIDE project worksheets

CLOSURE PROPOSALS 1975-76
(Sources: Cmnds.5976 and 6432)

1. Naval Installations
   HMS GANGES
   Antrim and Bandeath Armament Depots
   Lyness Oil Fuel sites
   Llanion and Llanreath Oil Fuel sites
   Aircraft Yard, Wroughton

2. Army Installations
   JTRs Rhyl and Troon
   Oswestry
   'a number of storeholding depots'
   'a main repair facility'
   'a number of smaller logistic establishments'

3. Royal Air Force Stations
   Biggin Hill
   Church Fenton
   Cottesmore
   Manby
   Upwood
   (To be vacated but taken over by the Navy: Thorney Island)
   (To be vacated but taken over by the Army: Bicester, Chessington, Colerne,
   Driffield, Hullavington, Leconfield, Little Rissington, Medmenham, Ternhill)

4. Other Royal Air Force Installations
   Aldergrove and Sydenham Maintenance Units (Northern Ireland)
   Hartlebury Maintenance Unit

5. R & D Establishments
   (a) Naval testing facilities: Cobham, Holton Heath, Teddington,
        West Drayton.
   (b) RARDE outstation, Woolwich
   (c) ERDE sites: Waltham Abbey (North), Potton Island

Notes

1. In United Kingdom practice they also happen to include
    what may be regarded as payments for productive
    services rendered in the past, i.e. pensions.

2. See the argument and data in D. Greenwood, 'Constraints
    and Choices in Transformation of Britain's Defence Effort
    since 1945', British Journal of International Studies, 2
    (1976), 1-22, especially pp.4-5 and Tables 1 and 2.

3. However, see the author's Memorandum in Second Report
    from the Expenditure Committee, Session 1974-5, The
    Defence Review Proposals, HC 259 (1974-5 especially
    p.2 (para 4) and p.6 (para 19).

4. Data in this paragraph are from Statement on the Defence
    Estimates 1975 Cmnd 5976, esp. pp.20-2 and Second Report
    from the Expenditure Committee, Session 1975-6, Defence,
    HC 155 (1975-6) esp. pp.185-8. A graphic illustration
    of the 'confusion' is to be found in the article in The
    Times dealing with the Expenditure Committee Report.
    The paper's Defence Correspondent wrote: 'The White Paper
    had estimated that the Government's decision to cut £4700
    millions from defence spending over the next nine years
    would mean the loss of 10,000 jobs in defence industries.'
    (There are at least two major - if understandable -
    misrepresentations there.) See also HC 155, Evidence,
    Q's 691-704, 734-48.

5. If it is assumed that the Conservatives' programme
    envisaged total Service strengths of c. 350,000 through
    to the mid-1980s and a total for civilian employees of
    c. 300,000, a 'release' of some 700,000 'job/years'
    would be implied in these categories, making a total
    (direct) defence-related employment 'release' - adding
    in 495,000 for the contractors' employees element -
    amounting to 1,200,000 'job/years'. Over the nine
    years to which the figures relate that would be an
    annual average 'loss' of 135,000 jobs, 55,000 of them
    accounted for by direct employment in defence-related
    industry.

6. See the earlier paper Defence Programme Options to
    1980-1.

7. Note that the reductions under discussion here are those
    implied by the 'original' Defence Review proposals as
    formally set out in the Statement on the Defence Estimates
    1975, Cmnd 5976. The civilian manpower reduction exer-
    cise which figures in the revisions decided upon in the
    1975 Public Expenditure Survey 'round' is considered in
    the next section of the paper.

8. The consideration of secondary (or indirect) employment
   effects — see p.318 above — is deferred to a later stage.
   It may be noted in passing, however, that the rule of
   thumb 'loss of 4 direct defence jobs results in the loss
   of 1 indirect job' (or the assumption that the value of
   the relevant employment multiplier is 1.25) yields an
   aggregate employment effect estimate of 50,000 (jobs).
   See HC 259 (1974–5) p.21. But note the official
   calculation, for industrial jobs, based on an employment
   multiplier of 2.3 (in HC 155 (1975–6), pp.263–7).

9. See Table 1 above, cf. lines A.3(a) and A.3(b).

10. See Public Expenditure to 1979–80, Cmnd 6393, Table 2.1,
    p.16–17 and Statement on the Defence Estimates 1967,
    Cmnd 6432, p.31. Note that this means that some of my
    own earlier estimates of the forward programme on this
    prices basis were up to £50 millions — say 1.1 per cent —
    a year too low (see Defence Programme Options to 1980–81,
    p.8 and the Benchmark Programme line in Table 4). Some,
    not all; for there have been official downward revisions
    of the profile, producing among other things a lower target
    Estimate for 1979–80. (See Cmnd 6432, p.31 and Table 2
    below.)

11. Cmnd 6432, pp.29–30.

12. See note 5 above.

13. See note 8 above.

14. There is some precise information about the RAF rundown.
    It will entail fewer than 4,000 actual redundancies, well
    over half of them among engineering tradesmen. The re-
    absorption of the several hundred redundancies among air-
    crew with specialist experience of the transport force may
    be the most problematical aspect of the rundown. (See
    HC 155 (1975–6) pp. 30–5 and Cmnd 6432, p.30.)

15. See HC (1975–6). Evidence Q's 88–97.

16. See HC 155 (1975–6) pp. 185–8 and Cmnd 6432, p.27.

17. Cmnd 6432, p.19.

18. HC 155 (1975–6) and Evidence Q.704. Note that these
    observations refer to the impact of equipment expenditure
    from the United Kingdom's defence budget. Needless to
    say the overall industrial position is dependent on export
    business too. (Presumably contraction at Hawkers, for
    example, would be moderated if there were to be substantial
    overseas orders for UK-produced Harriers and Hawks.)

19. See <u>Defence Programme Options to 1980-1</u>, especially Table 4, p.41. The Expenditure profiles for the Options at 1975/76 Estimated prices as set out in Table 3 of the present paper are higher than those in this Table because of the updating of the Benchmark (1) programme. The 'savings' levels are unaffected. This adjustment was necessary because, of course, the Options were developed by estimating 'savings' and deducting these from the (estimated) Benchmark (1) programme.

20. Ibid, pp.35-7.

21. This summary, like those for the other Programme Options, includes elements based on rundown of the Polaris force viz. 2,500 Service personnel and 3,500 Civilian employees (almost exclusively at the Clyde Submarine Base and the Rosyth Dockyard). 'Maintaining the effectiveness' of the force doubtless calls for some industrial effort too (which cannot be assessed).

22. For details on the prime contractors, and their European associates, see <u>Flight International</u> (28 March 1974), pp. 395-404.

23. <u>House of Commons Debates (Hansard)</u> vol. 891, no.120 (7 May 1975), col. 1446. (The Under-Secretary of State for the Royal Air Force, Mr Brynmore John.) Later in his speech Mr John said: 'In cost terms ... the MRCA currently accounts for 10 per cent of the total United Kingdom aerospace industry workload. <u>That figure will rise substantially when the programme is established.</u>' (Emphasis added.)

24. See Cmnd 6432, Figure 11 (p.74).

25. T. Stone. <u>Analysing the Regional Aspect of Defence Spending: A Survey</u>, ASIDES, No. 3 (December 1973), pp. 21-2.

26. See also D.B. Suits, <u>Survey of Economic Models for Analysis of Disarmament Impacts</u>, University of Michigan for United States Arms Control and Disarmament Agency (USACDA), who writes, 'The most direct regional analysis is probably the best. For example common sense consideration of the geographical distribution of industries affected by a change in defence expenditure... is both cheaper and probably more accurate than what could be had even from an accurate regional model.' (p.44.)

27. See G.G. Gordon, _A Study to Measure Direct and Indirect Impacts of Defence Expenditures on an Economy_, University of Washington for USACDA, especially pp.17-18.

28. On the redundancy costs associated with the Defence Review manpower economies see HC 155 (1975-6) especially pp.27-9 and 32-5. On cancellation charges etc. see Second Report from the Expenditure Committee, Session 1974-5, _The Defence Review Proposals_ HC 259 (1974-5) pp. 58-9.

29. See for example, D. Greenwood and J. Short, _Military Installations and Local Economies_: A Case Study - the Moray Air Stations, ASIDES No.4, and J. Short _et al._, _Military Installations and Local Economies: A Case Study - the Clyde Submarine Base_, ASIDES No.5.

30. For the official calculations see HC 155 (1975-6) pp.263-7. For discussion (and full citations) of other impact studies see T. Stone, _Analysing the Regional Aspect of Defence Spending._

POSTSCRIPT

My original paper was an attempt, first, to gauge the employment
consequences of both the 1974 Defence Review and the revisions
to the post-Review programme made in the 1975 public expenditure
survey round; and, secondly, to assess what further actual
release of labour in the United Kingdom might be anticipated
from additional spending 'cuts' along the lines set out in an
earlier analysis of <u>Defence Programme Options to 1980-81.</u>
At the same time I took the opportunity to differentiate
among the several categories of manpower involved in defence
and tried to elucidate that distinction between 'actual
release' of manpower and 'jobs lost' about which a lack of
clarity in official statements had led to confusion in
Parliamentary (and popular) appreciations of the employment
consequences of reduced spending.

In the year that has passed since the paper was completed
there has been another scrutiny of public spending plans –
the 1976 survey round – in which defence expenditure projections were again amended. The question arises: with what
implications for defence-related employment? In addition,
because of the lapse of time itself, my assessment of the
budgetary consequences of further programme changes no longer
holds, as regards both <u>scale</u> and <u>timing</u>. Hence revision is
necessary to my calculations of the employment effects of
selected options for change. (My differentiation among
defence manpower categories and the distinction between 'actual
release' and 'jobs lost' remain valid, however, indeed they
define the terms in which any revised assessment of employment
consequences should be presented.)

Programme Revision

It is convenient to record the 1976 public expenditure survey's
effects on the defence programme by an analysis similar to that
used in the original paper to portray the 1975 budgetary
revisions. (Table 2, which in turn is comparable with the
upper part of Table 1, covering the Defence Review.) This
done in Table P.1 here. The presentation shows that
projections for the immediately forthcoming financial years
were revised downwards in the 1976 survey, but it was
envisaged that in the early 1980s increased defence spending
would be possible. That is to say, Ministers regarded the
changes as short-run inter-temporal adjustment. (1)

According to policy statements this adjustment is being
achieved, at least for 1977-78, by 'miscellaneous reductions'
on several support programmes (including reduced outlays in
research and development), by postponement of a number of
works schemes in both the United Kingdom and Germany, and by
'savings' on the equipment programme. (2)

It seems unlikely that there will be far-reaching additional

employment consequences in the United Kingdom, for the
following reasons,

(a) there is no indication that further reductions in
Service personnel strengths are envisaged beyond those
foreshadowed by the 1974 Review and 1975 programme
revisions.   Indeed if this were the case it would
contradict official assertions that, because of the
short-term nature of the adjustment, emphasis is being
placed on 'cuts' which can be restored later.

(b) it has not been suggested that the rundown in numbers
of Civilian Employees (which is already in train) should
be accelerated or taken further.   And here too the fact
that the Government seeks a reduced level of spending in
the immediate future only makes it unlikely that labour
'shake-out' is contemplated.

(c) it is on Defence Contractors' Employees that deferments
of equipment and works projects will impinge most.   Moreover,
since temporary cuts in R & D spending tend to be sought by
withholding extra-mural contracts rather than reducing
intra-mural effort, the main employment effects of these too
should occur in industry.   But this, of course, is the
category of defence-related manpower in which numbers were
expected to rise during the later 1970s.   The conclusion
is: the 1976 revisions probably mean that the anticipated
increase in industrial manpower will be less than it would
otherwise have been;   but they emphatically do not amount
to enough to produce an actual (net) release of labour
from defence-related industry.

In any event when looking for the overall employment
effect over a five or six year period the implications of
marginally reduced expenditure in the immediate future have
to be set against the likelihood of additional manpower
demands when, as is now envisaged, defence spending (in
real terms) begins to rise again in the early 1980s.
In the parent paper estimates of the net manpower release
to the United Kingdom labour market attributable to particular
policy changes are presented as individual elements which are
brought together at the end of the piece in a 'composite'
analysis.   Expressing the rough-and-ready assessment of the
preceding paragraph in this way yields,

Employment Effect 1978-1983 (UK):   Actual Release (000s)

	Total Service Personnel (1)	Civilian Employees (2)	Contractors' Employees (3)
B* Programme Revision (76)	2	-	-	2

Note that the effect is assigned to 1978-83 and not to 1976-81 as in the
presentations in the original text.

Further Reductions: Options

The original paper explored the manpower implications of
selected options for change in the defence programme and
budget as delineated in an earlier analysis of <u>Defence
Programme Options to 1980-81.</u>   In a Postscript to that
exercise I have pointed out that the expenditure reduction
(in real terms) which would follow from adoption of the
principal options considered must now be assessed at
slightly less than when they were first elaborated.
Further, I have shown that it would be unrealistic to
assume that any 'savings' against current plans could be
achieved by major programme change before 1979-80.
However, allowance for higher cancellation charges and other
nugatory expenditure accounts for an important part of this
revision of earlier computations.   It does not follow
therefore, that estimates of the impact on employment
should be recast in a precisely corresponding fashion.
Rather the manpower release associated with programme
decisions of the kind identified might be almost as great
as originally calculated and it would be likely to occur
earlier in time than the main 'savings' effects.
Accordingly in <u>up-dating</u> my 'composite' analysis of employment effects I have incorporated only marginal reductions
to the previous assessments of the actual release of labour
which the main options for change would entail.   And in
<u>re-dating</u> I have thought it appropriate to assign the
effects to 1978-83, even though the first financial impact
might not be evident until 1979-80.   The new 'composite'
analysis is given in the accompanying Table P.2, which
corresponds to Table 4 in the parent paper.
     It should be stressed that the figures in Table P.2,
like those in the original Table 4, relate to direct employment connected with the defence effort and to <u>actual release.</u>
I can offer no more refined analysis of the possible total –
direct and indirect – employment effects of the changes
considered than I did before, citing plausible multiplier
values for installations-related and industry-related
defence jobs.   And, as before, the absence of a reliable
defence manpower budget precludes calculation of a satisfactory
measure of 'lost jobs opportunities'.

<u>Notes</u>

1.   See the argument in my Postscript to the paper <u>Defence
      Programme Options to 1980-81.</u>

2.   See <u>Statement on the Defence Estimates 1977</u>, Cmnd 6735,
      para 105, p.l.   The equipment programme 'savings' will
      arise partly from planned re-phasing – i.e. further
      'stretching' – of procurement plans, partly from un-
      intended rephasing attributable to more realistic assess-
      ment of the likely progress of particular developments.
      There will also be 'selective cancellations' according to
      the White Paper.

352

Table P.1

THE 1976 PUBLIC EXPENDITURE SURVEY 'ROUND' PROGRAMME REVISIONS

A. Expenditure at 1976 Survey Prices	Financial Years				Average 1980-81/ 1983-84 £m
	1976-77 £m	1977-78 £m	1978-79 £m	1979-80 £m	
1. Revised Programme (Table 2, line 3) (revised)	–	5647	5633	5636	(5875)
2. Savings Programme Revision (76 PES)	–	200	230	(?)	(?)
3. Revised Programme	–	5447	5403	(5636)	(5875)
Revised Programme (revalued at 1977 SPs) cf. Defence Estimated 1977-78	–	6329	6275	6550	6800

Table P.2

SUMMARY OF ESTIMATED EMPLOYMENT EFFECTS, 1978-83
(United Kingdom)   (000s)

Direct Employment Only/Actual Release

	Total	(1) Service Personnel	(2) Civilian Employees	(3) Contractors Employees
Already 'Programmed'				
A. Defence Review (1974)	40	31	19	-10 (increase)
B. Programme Revision (1975 PES)	15	4	8	3
B*. Programme Revision (1976 PES)	2	-	--	2
sub-total	57	35	27	-5 (increase)
Further Reductions				
C_I Option I	130	55	37.5	37.5
C_{II} Option II	115	32.3	27.5	55
C_{III} Option III	135	52.5	37.5	45
Total with C_I	187	90	64.5	32.5
Total with C_{II}	172	67.5	54.5	50
Total with C_{III}	192	87.5	64.5	40.5

10 NOTE ON THE EMPLOYMENT CONSEQUENCES OF A £1,000 MILLION CUT (AT 1974 PRICES) IN MILITARY EXPENDITURE OVER FIVE YEARS

by Frank Blackaby

Summary

This note is simply concerned to present certain <u>aggregate</u>
quantifications of the employment and unemployment consequences
of a cut in military expenditure of the size agreed.   It
is not concerned with the specific problem of unemployment
in particular plants, or regions, or among those with particular types of skill;   these questions will be dealt with by
other authors in other papers.
   The general conclusion, which is set out in more detail in
the following paragraphs, is clear.   In the continuing ebb
and flow of industrial change which is going on all the time,
a reduction in military expenditure of this order would be
only a ripple.   In this country, there are about 4 million
'voluntary quits' each year - that is, people who decide to
change jobs, or who retire, or who leave a job and return to
keeping house.   The number of job changes needed to
accommodate the reduction in military expenditure on the
scale envisaged is only 60,000 a year.   This is insignificant
against the total amount of job changing which is going on
anyway   Further, we have the lesson of past experience.
Between 1953 and 1957, over 4 years, there was a cut in
military expenditure bigger than the one which is now
envisaged over 5 years.   In 1957, the national employment
percentage was lower than in 1953.

General points

Two general points.   First, economic resources are limited;
wants are not.   There are immense unsatisfied demands for
higher public sector expenditure in the social services:
slums to be cleared, antiquated hospitals and schools to be
replaced - the list is endless.   It is an obvious gain if
resources can be shifted from 'regrettable necessities' like
military expenditure to the production of goods or services
which increase human satisfaction.   There is no economic
problem in increasing Government expenditure on the social
side to compensate fully for the reductions in Government
expenditure on the military side.
   Secondly, the main thing the Government must do to ease
the problems of transition is to keep the general demand for
labour high.   Given this condition, the problems of transition
tend to disappear of their own accord;   without it, they
tend to appear insoluble.   A policy of high labour demand
certainly has problems of its own - it requires a tough
incomes policy for one thing - but there is no doubt that it
helps very substantially with problems of industrial transition.   We have learnt this lesson with regional policy.
The best way to get down the level of unemployment in the
high unemployment areas is to produce a shortage of labour
in the low unemployment areas.   Firms are then forced to go
to the areas where there are reserves of labour, or to sub-

contract to firms which are already there.   With a general
high level of demand for labour, special measures for
particular groups affected by industrial change are much
less needed.   Conversely, _without_ a high level of demand
for labour, special measures to help any one group of unemployed merely serve to shift unemployment from one group
to another.

Specific Quantifications

Against this general background, what is the size of the
disturbance which would be caused by a cut of £1,000m (at
1974 prices) in military expenditure over the 5 years
1974-9?   It would require roughly 300,000 people to leave
their present jobs, either in the armed forces or in the
industries which supply them - that is, some 60,000 a year.
    This amount of job-change is negligible, set against the
total amount of job-changing which goes on in any case.   It
is not generally realized how many people do change jobs
every year. $2\frac{1}{2}$-3 million people change jobs from one
employer to another every year.   This number does not include
those who retire, or those who joined the labour force for
the first time, or the women who leave the labour force to
keep house; and it counts as one only those who change jobs
more than once a year.   If one adds in to the figure those
who retire and women who return to keeping house, this
brings the figure to around 4 million, or 15 per cent of the
labour force.   Against these figures of the job changes
which go on anyway, the 60,000 job changes which would result
from the military expenditure cuts are insignificant.
    The same point can be made another way.   The military
sector in this country employs in all, directly or indirectly,
some 1.2 million people.   Normally each year at least
180,000 of these can be expected to change their jobs.   That
is, the annual 'natural wastage' in the military sector is
about three times the size of the job changes which would be
required by the reduction in military expenditure which is
envisaged.
    There is another way of demonstrating the same point.   We
have already had - incidentally, under a Conservative
administration - a cut of military expenditure of £1,000m
at 1974 prices; further, this was over 4 years, not 5.
The table on p.358 gives the figures for military expenditure
in the 1950s; they will appear unfamiliar, because they have
been adjusted to 1974 prices, so that the magnitudes are
appropriate for the point which is being discussed here.
From 1953 to 1957, military expenditure was cut by £1,150m
at 1974 prices.   In 1953, the unemployment percentage (1)
was 1.5 per cent.   In 1957, it was 1.4 per cent. (2)   It
is quite clear, therefore, that cuts in military expenditure
of this order can be made without any consequent increase in
the general level of unemployment.   It can be done, because
it has been done.

Table 1.   Public authorities' current expenditure on
                military defence, revalued to 1974 prices.

1952	5,717	1958	4,336
1953	5,912	1959	4,314
1954	5,709	1960	4,374
1955	5,285	1961	4,509
1956	5,085	1962	4,712
1957	4,757	1963	4,691

Sources: National Income and Expenditure, 1963–1973, and
           Preliminary Estimates of National Income and
           Balance of Payments, 1969 to 1974.

   The argument in this paper is about the general size of
the employment and unemployment consequences of a cut in
military expenditure of £1,000m at 1974 prices, over 5
years.   It would only be a small addition to the industrial
change which is going on all the time;   past experience shows
that it need not entail any increase in the general level of
unemployment.   This is in no way contrary to the view that
specific measures should be taken to help those put out of
work in specific plants.   This of course is a general problem,
and not a problem which results exclusively from reductions
in military expenditure.

Annexe

1. To calculate the employment consequences of a cut in
military expenditure of £1,000m at 1974 prices, the 1974 gross
national product at market prices was taken (£81,371m) from
Cmnd 6019.   This was divided by an estimate of the total
employed labour force – 24,900m   (Economic Trends, July
1975, table 36) – to give gross national product per employed
person of £3,268 in the economy as a whole.   GNP per head,
in money terms, is known to be rather higher in the military
than in the civil sector;   so the figure of £3,300 was taken
for the military sector, to derive the employment counterpart
of about 300,000 for military expenditure of £1,000m.   The
same method was used to derive the employment counterpart of
total military expenditure in 1974.

2. The figures for labour turnover were derived from an
article 'Labour turnover – new estimates' in the Department of
Employment Gazette for January 1975.

3. The estimates of military expenditure.   1952–63, at 1974
prices, are based initially on the series at 1970 prices in
National Income and Expenditure, 1963–73, table 14.   They
are raised to 1974 prices by multiplying them by the price
index for public authorities' current expenditure as a whole.
This went up 60 per cent from 1970 to 1974.   The price index

358

for expenditure on military defence alone, judging from past
experience, probably went up more than this, so the figures
in table 1 if anything understate the fall from 1953 to 1957.
However, the price index for military expenditure alone is
not yet available for 1974.

Notes

1. Wholly unemployed, excluding school leavers, as percentage
   of total employees, Great Britain.

2. In 1958, unemployment rose; but this rise had nothing to
   do with the cut in military expenditure.

C.   INDUSTRIAL CONVERSION AND ALTERNATIVE EMPLOYMENT

(Study Group Papers nos 16 and 17, on the Lucas Aerospace
Corporate Plan and the Vickers Proposals, are not strictly
speaking Study Group Papers as they were not actually
presented to the Group.   However, the Study Group was
able to read the Lucas Plan, and was kept informed of
developments during the research for the Vickers Paper.
Thus the ideas and conclusions of both papers were familiar
to the Study Group and played a part in its deliberations.)

11 MILITARY EXPENDITURE CUTS : NOTE ON THE TRANSFER OF RESOURCES

by Frank Blackaby

General

This note considers some policies for easing the problems
of the transfer of resources resulting from military
expenditure cuts — and in particular the problems of the
resulting unemployment.

First, two general points. Of course, this type of
problem does not simply arise as a consequence of military
cuts; in civil industry, plants are frequently closed, and
workers displaced. However, the government has two advantages in dealing with the consequences of defence cuts.
First, it can decide on the timing itself; it is not at
the mercy of arbitrary decisions on timing by private
industry. Secondly — and more importantly — since it is
government expenditure which is being cut, the government
can afford to be very generous in the transitional finance.
Indeed, the government could, if it wished, pay the workers
concerned their previous wages indefinitely — and the
government balance of revenue and expenditure would still be
no worse than it was before the cuts were made.

Another general point is that the problems of transition
will be immensely easier if the government is pursuing a
full employment policy at the time. The basic demand
management requirement is quite simple. If we assume
that, at the time of the cuts, the economy is being run at
the 'target' level of full employment, then any reductions
in demand resulting from the expenditure cuts (after allowing
for the demand consequences of any transitional arrangements)
should be compensated for by increases in demand elsewhere —
either by cuts in taxes, or increases in government civil
expenditure, or some mixture of the two.

Objectives

The government should have two objectives in mind: first, to
ensure that the resources are transferred into the production
of wanted products — that is, either products which consumers
(here or abroad) wish to buy, or products which the government
needs for its non-defence objectives Secondly, the government should try to ensure that the workers displaced find
jobs which use their skills, and provide equivalent work
satisfaction.

The big problems of transfer are provided by the large
plants specifically devoted to military production. There
would be smaller problems with sub-contractors supplying
general purpose components; given a full employment policy,
sub-contractors should themselves have a reasonable chance
of finding alternative markets. But sub-contractors could
also be helped by the policies suggested below.

362

Subsidizing the Search for New Products

It is wrong to think that the Department of Trade and Industry could, overnight, pick out from a list some suitable civil products to replace the defence products which are being phased out.   It is not that easy.

The search for alternative civil products is, of course, much more difficult now than it was in 1945 — though, fortunately, the size of the transfer required is very small indeed compared with the transfer of resources then.   In 1945, a plant could sell virtually any civil product it chose to produce;   that is not the position now.

The search for an alternative product for a defence plant is a long, complex business.   It involves a massive search of the technical literature, to discover products which have the right technological profile.   That probably leads on to an examination of the patent position, and possibly negotiations for licences.   The next stage would be an examination and projection of the world market;   an estimate of the production and costs of competing producers;   estimates of the costs of establishing a distribution chain;   and so on.   It is the general experience that preparations for producing and launching a new product take a long time, and a lot of work.

This search should be decentralized to the firms concerned. Central planning agencies — even if we had one, which we do not — cannot cope with the immense complexity of assessing the possibilities of the myriad products which exist, or could exist. It is noticeable that in the Socialist countries — and certainly in the more efficient ones — decisions on product choice, apart from major products such as steel, cars, trucks, and so on, are increasingly being devolved away from the centre down to the firm, or (in the case of East Germany) to the industrial combine.

The policy proposal, therefore, is this.   The government should require firms which have appreciable defence output to set up joint management/worker committees to plan alternative civil production;   and the government should indicate that it is ready to pay the verified costs of product search. Further, this should be done soon, because it can take a long time to find a suitable alternative civil product.

The odds are that plants which are now producing military items would be more suitable for producing goods for the private market than producing goods for government consumption — given that government civil expenditure is heavily weighted towards the purchase of services and buildings.   However, the government could certainly provide general guidance about the direction in which it proposed to expand its civil expenditure when military expenditure was cut back, so that the firms concerned could consider whether there were products in the government shopping list which might be suitable for them.

If this preparatory work has been done, then when particular defence cuts come on the agenda, the firm concerned should have ready a set of proposals for alternative civil production; these could form the basis of negotiations between the government and the firm on the finance of the costs of transfer to civil production.

The main burden of this short paper, therefore, is as follows. In an advanced economy like our own (and these points would still be valid if the means of production in this country were publicly owned) it is not practicable for a central decision to be taken about the alternative civil products which could be substituted for the present output of defence plants. The main function of the government, therefore, is to subsidize the process of product search, and to ensure that this product search is conducted in good time, so that a range of alternatives is ready when the time comes.

Other Transitional Arrangements

It has already been pointed out that the government can afford to be generous about the transitional arrangements. The problem here is not so much a problem of finance; it is rather a problem of equity. Would it be right to have very generous transitional arrangements for those displaced from defence production, if the same arrangements were not available for those displaced by shifts in civil production — for example, the Chrysler workers displaced at Linwood or Ryton?

I think there is a case for special provision. As it is, the provision in this country for employees who lose their jobs is very uneven; some private firms are much more generous than others; some nationalized industries — for instance, the National Coal Board — have been more generous than the private sector as a whole. So the government could set a standard of best practice in the range of transitional arrangements offered to defence workers; for of course, some will be displaced. No matter how good the process of searching for alternative products, it is too much to hope that the fit will be exact. There could be heavy subsidies for alternative training (including university courses); subsidies for moving house; and maintenance of the previous wage or salary for at least a year until alternative work is found.

The point needs to be stressed heavily — and this would be part of the process of persuasion that defence cuts would not mean economic catastrophe for individuals — that even if all the people engaged in producing military items for the armed forces in this country were paid their present wages and salaries indefinitely for doing nothing, instead of producing war material, the government would still be no worse off.

In fact, of course, displaced workers would find other jobs in time; very substantial savings would soon emerge. So it should be possible to offer the same transitional arrangements to workers displaced by the loss of export markets for weapons. It must be accepted that there would be cases where,

364

if production of an item for British military forces were discontinued, it might no longer be economic to produce it for export. This is apart from the arguments, presented elsewhere, for adopting more restrictive policies in the sales of arms abroad.

12   ASPECTS OF CONVERSION OF ARMS INDUSTRIES

by Dan Smith

Introduction

Consideration of the question of conversion of arms industries to other uses is important for a number of reasons.

Without conversion, workers and servicemen affected by arms cuts will be thrown out of a job, and left to find their way to other work. In the course of time many, particularly those with skills, and those lucky enough not to live in depressed regions, might be expected to find other work. In the meantime, and for those who found no work, there would be a disastrous effect both on individuals and on communities which we should not be prepared to countenance. A programme of industrial conversion, which removed the element of chance and provided other work without needing the worker to move to another area, would be both more humane and more sensible.

A second reason was identified by the Study Group at an early stage: 'The defence cuts of the 1960s had proved transient. Although the programme had been cut back, the capacity to produce the major projects had remained to comprise a permanent pressure for more spending.' (1) To relieve at least one of the pressures for increased spending, the arms industries will have to be given something else to do. Eliminating industrial capacity for certain types of arms production could not, of course, guarantee that arms cuts would remain permanent, but it would make it more likely.

A third reason has been very clearly stated by Seymour Melman of Columbia University: 'Competence for industrial conversion to civilian work is a precondition for ability to consider peace or disarmament proposals on their own merits.' (2) Some might wish to qualify such a bald statement, but it does at least locate the importance of the conversion question within the context of efforts to end the arms race.

Fourthly, industrial conversion of arms industries has an important part to play in a socialist industrial strategy. One reason why this Study Group exists at all is the concern of the Labour Party members at the squandering of valuable resources - workforce, plant, machinery, and raw materials - on military production, when there are so clearly other areas to which increased resources should be allocated. Few people would disagree with the need to devote more resources to building, education, health, and numerous other public programmes. Equally, the problems of British industry are, in a general way, clear to many, as are its needs for more investment, a major degree of re-equipping, new ideas, and imaginative planning. The conversion of arms industries could help solve both problems, although it is important not to imply that it would be the cure for all our ills, particularly as we are talking of a relatively limited cut amounting to £1,000 million per year (at 1974 prices) (3) in order to fulfil the Labour Manifesto.

However, a conversion programme could be a part of an

industrial strategy aimed at reshaping British industry, both
in its structure and its products.   The National Enterprise
Board and planning agreements would have a crucial part to
play in assuring public accountability and participation of
all levels of the workforce in framing policy. (4) Public
ownership, beginning with the already planned nationalization
of the aerospace and shipbuilding industries would also be
crucial in facilitating the changeover from one kind of
production of another, and ensuring both detailed planning
and effective control throughout.

In short, arms cuts and industrial conversion are best
conceived of within the context shaped by the policies of
the 1974 October election manifesto.   They should be
conceived of as a programme, and as part of a general
socialist programme for British Industry.

This section has stated the context in which the
problem of conversion is considered in this paper.   It is
a relatively modest paper, aiming not to provide a blueprint
for industrial conversion, but to discuss the problems
likely to be encountered, and suggest possible solutions.
Just as the Study Group has decided to present options for arms
cuts, (5) it is probably wise to present a range of options
for industrial conversion;   the initial outline of such a
range for aerospace is present here, together with more
general observations and an attempt to identify some of the
basic requirements of a conversion programme.

Reducing Military Spending

In the course of debates over arms cuts inside and outside
the Labour Party, those defending a level of spending at or
near the present level have argued that reductions of £1,000
million per year will necessarily cause large-scale redundancies;
(6) on the other hand it has also been argued that redundancies
need not be caused if the matter is gone about in the right
way. (7)

A distinction between two types of arms cut — what might be
called piecemeal and wholesale — is generally recognized;   for
the purpose of this paper it is necessary to make a further
distinction.   Put simply, it is a distinction between arms cuts
which do cause unemployment, and arms cuts which do not;
numerous examples of both kinds abound.

If the Government decides that a reduction in public
spending is desirable, and if arms cuts are carried out as a
part of that policy, then unemployment is almost certain to
result.   The exception would be when such cuts were made in
a period of expanding industrial enterprise, when displaced
workers could immediately find work;   that is not the case
today.   Unemployment would be the lot — both of arms
workers and of servicemen, depending upon how the cuts were
carried out — i.e. what spending was reduced or removed.

368

On the other hand, the government may also cut arms spending in order to reallocate resources, and this re-allocation would need to be carried out by government; in that case we are not talking about reducing public spending, but about re-ordering the scale of priorities for public spending. It is the second kind of arms cut which is the subject of the Study Group's considerations.

That arms cuts need not cause unemployment is attested to by the past experience (in the 40s and 50s), as well as by some authoritative studies. (8) On the other hand, experience of contract cancellation, especially in the USA where some detailed studies have been carried out, (9) has shown that arms cuts can have direct unemployment consequences. The cuts envisaged by the Study Group would include important contract cancellations, as well as general cutbacks; they are, however, nowhere near the scale of those made post-1945.

Because the economic situation at the end of World War II was very different from that of today, many have argued that it is not good enough to draw a parallel between the general ease of conversion then and the problems we would encounter now. Seymour Melman has described the difference between the two periods as a distinction between 'conversion' and 'reconversion' of arms industries, arguing that the firms were able to go back to what they had been doing before the war. (10) True — but that is a comment more applicable to problem of demand (see below) than to the technical question of conversion. Were it possible to maintain demand (and the conclusion of this paper is that it would be possible), the conversion task of the 1970s would be similar in nature but smaller in scope compared to 1945 conversion. After World War II British industry had been on a war footing for 6 years, US industry for 4; in the USA a lot of industry did not convert, but where it did, as in Britain, old skills had to be relearned, and assembly lines etc needed as much conversion as they would now.

If we are to match the conversion efforts of the 1940s, we shall do so because the conversion programme has been planned so that the economic problems we shall encounter will be no greater than those of 30 years ago. In successfully carry-ing out the second of the kinds of arms cuts contrasted above, the key is planning, which begins from the point of identifying the problems and needs of a conversion programme.

Identifying Problems

(a) Employment Effects

The aggregate number of jobs likely to be affected by a cut of £1,000 million per year has been calculated at 300,000; (11) government ministers have previously used figures as high as 350,000. (12) The government has also presented figures showing that the Ministry of Defence is responsible for the

employment of 349,000 servicemen, 250,000 workers in arms
industries, about 44,000 in ROFs and Royal Dockyards, and
?10,000 other civilians in the UK (13) — a total of under
b00,000. However, we may take it that the figure of
250,000 workers in arms industries is a rough calculation,
rather than a detailed quantification. It is, in any case, a
figure which is bound to fluctuate wildly depending upon what
sub-contracts and sub-sub-contracts are being fulfilled by
what firms at any given time.

A paper from David Greenwood (14) and another from Albert
Booth, (15) both forthcoming, should help in the identification
of the numbers of people whose work will be affected, the
location of their work, what they work on, and what skills are
available to them. Forthcoming papers from Mary Kaldor and
myself will help further by a more detailed identification of
the employment effects of cancellation of the ASW Cruiser and
MRCA. (16)

A crucial part of any conversion programme is the production
of this kind of information as a basis upon which the programme
can be planned. Compiling such information should already have
begun; if it has not it should start immediately; it is only
on the basis of such data that the government could produce an
adequate conversion programme.

(b) Base Conversion

One problem which this paper does not examine is the problem
of conversion of occupations at and around military bases.
This is, however, a problem which should be gone into. Some
detailed studies have been done, notably by the Aberdeen
Defence Studies Unit and Scottish CND which would be useful to
the Study Group. (17)

(c) Industrial Structures

A problem encountered in the USA, where there have been
various efforts at conversion, has been the inability of a
firm, or of the government department with which it deals, to
adapt to the demands of non-military production. Efforts
at producing environmental systems, pre-fabricated housing,
artificial hearts, and a rapid transit system for the Bay
Area of San Francisco, have all failed to one degree or
another because of this problem. (18) Arms corporations
are used to a method of production and business which depends
upon continuing and ever-advancing technical sophistication;
it is an imperative that is part of the structure of arms
production. (19) Civil production may not demand such
sophistication; while use may be made of sophisticated
techniques, or experience with those techniques, in many cases
it may be a basic simplicity of design and production which is
the keynote.

Administration and management of converted industries is not,
therefore, something that should be left to chance. A strong

370

case can be argued for accountability of converted industries to the NEB, and for planning agreements, purely on this basis. The limited success in the USA is a warning of what can happen if due account is not taken of this problem.   If we conceive of industrial conversion as more than a method of keeping firms in business or people in work, if, indeed, we conceive of it as a way of providing resources to fulfil some of Britain's pressing social needs, then public involvement throughout the NEB, and worker involvement through planning agreements, at the very least, must be a central part of a conversion programme.   In other words, we must think in terms of changing industry; we are intending to change what is produced, and therefore must face up to the question of changing the industrial structures which at present control what is produced.

(d) Maintaining Demand

One problem which immediately presents itself to those considering conversion is the question of maintaining demand for the products of converted industries.   Will the government have to stimulate demand artificially, and if it does, how will it maintain demand?
    A general answer was provided by the UN Department of Economic and Social Affairs in its 1962 report upon the consequences of disarmament:

> Monetary and fiscal policy could be used to offset the effect of a shortfall in total demand that might result from a decline in military expenditure to the extent that it were not offset by a rise in civil government expenditure.   Bearing in mind that a substantial part of military expenditure would probably be replaced by other government expenditure in most countries, it may be concluded that the maintenance of effective demand in the face of disarmament should not prove difficult. (20)

Arising from this we should note two points.   First, that the Study Group is not considering total disarmament so the problem of a surfeit is less likely to present itself; second, that the crucial element identified in the paragraph is the role of government spending.   At no stage in the conversion programme can the government afford not to be involved.
    However, it is not enough to leave it there.   A central feature of the provision of information on which to base the necessary planning should be a 'market survey'.   Such a survey need not be restricted to examining the profitability of certain types of production; instead it could ask the question, what do we not have that we need?   The result could then be matched against the production ability of the converted industries.   A study of this nature was carried out by McDonagh and Zimmerman of Columbia University (21)

371

with reference to the American airframe industry and
identified conversion options into the following areas:
rapid transit, construction, electric vehicles, hydrofoil
boats, and hovercraft. A detailed study in Britain with the
aim of identifying various areas for conversion is a necessary
step in moving towards a successful conversion programme.
Ideally the criteria of such a survey would emphasize
social need. It is undeniable that we need more hospitals,
better schools, more new houses; we also need better
transport systems, cheap methods of providing energy, methods
of protecting and cleaning up the environment; we could also
manufacture products needed for the Third World, in the fields
of transport, power etc. In other words the survey should
seek to indicate where we can most valuably allocate more
resources, how much of our resources could be allocated to
given areas to meet their needs, and the resources we have
available for that task. In this way a conversion programme
could be a chance to make sure that industry serves the people
and their needs.
Another comment from the conclusions of the UN reports on
disarmament is useful here:

> There are so many competing claims for usefully employing
> the resources released by disarmament that the real
> problem is to establish a scale of priorities. The most
> urgent of these claims would undoubtedly already have been
> largely satisfied were it not for the armaments race. (22)

There should, in fact, be no problem in maintaining demand
following arms cuts of the scale this Study Group is envisaging,
the real problem is whether or not £1,000 million per year
is enough of a reduction to release the resources we need to
fulfil numerous urgent demands.

(e) Non-industrial Employment

It will have been noted from the figures taken from the 1975
Defence White Paper in section 3(a), that industrial workers
form a relatively small proportion of the total workforce
whose employment is at present dependent upon the Ministry
of Defence. Also affected by arms cuts will be civil
servants and servicemen. So any programme which aims to
employ all those whose jobs are affected by cuts will have
to deal with, and harness the abilities of, many from outside
industry.
Occupational conversion is looked at in more detail in the
next section. It is worth saying here that in order to employ
those who are now involved in industrial production or development, we shall need, among other things, to think in terms of
expanding industry.
Arms industries are peculiarly capital intensive, and
becoming increasingly so; despite an increase in the procurement section of the military budget for 1975-6, the government
was expecting about 10,000 jobs to be lost from the arms

industries. (23)   In other words, it takes a disproportionate
weight of physical resources, compared to the average in non-
military industry, to provide an equivalent number of jobs,
and this at a time when certain other sections of industry,
particularly chemicals, are themselves becoming more capital
intensive.   Under adequate planning it should be possible for
the resources released by arms cuts to provide more jobs than
are provided in industry at present;   this could both aid
displaced servicemen and civil servants in securing new work,
and begin to cut into the unemployment figures.   If the
converted industries begin to expand, then there should be
further scope for reducing unemployment.

   Thus, although redeployment will involve non-industrial
personnel, industrial conversion is at the heart of the
matter.

Occupational Conversion

(a) Production Workers

'It is an oversimplification to suggest that you can tell
workers in the arms industry to go out and dig a sewer.'(24)
Not only would arms workers refuse to go out and dig sewers,
but also we would not want them to do so.   Nor is there any
need for them to dig sewers.   It should be possible for arms
workers to be re-employed into different production, in the
same area as they were working before, using their skill and
experience to produce needed items.

   The available literature generally accepts that there is
not a major problem for production workers in changing from
military to civil manufacture.   In the 60s in California,
where there is a high concentration of military aerospace
production, a detailed examination was carried out by the
State Department of Employment of 127 production occupations.
(25)   28 of these were described as of basic craft type
(electrician, plumber, carpenter, etc) and were readily matched
to occupations in other industries.   93 of the remaining 99
could be immediately matched to one or more non-military
production occupations.   Thus only 6 occupations required retraining (for a period of around 6 months) if direct occupational
conversion were not possible.   Similar conclusions have been
drawn from work done by the US Federal Department of Labour.
(26)

   People from industry have informed me that the relative
flexibility of most plant machinery will go a long way to make
occupational conversion easier.   In fact, as one product
follows another, all industry experiences conversion of a kind
without major problems, and there appears to be no reason to
believe arms industries incapable of a similar task.   Job
skills in the engineering industries have wide applicability,
as do component manufacturing skills.   Most assembly lines
are pretty flexible.   Workers may find themselves handling

373

different materials, and in some cases working to different standards but a major amount of retraining should not be necessary. Where retraining would be necessary would probably be in the jobs which are special to the arms industries; these are also the highly skilled production tasks, and the workers concerned will almost certainly have a good general grounding and expertise which will again make retraining less of a problem than might at first seem the case.

In its report on alternative employment at the Clyde Submarine Base (27) Scottish CND, apart from proposing general categories also put forward a specific suggestion for Submerged Production Systems, and commented:

> For the less skilled Faslane workers, some of the work can probably be broken down in semi-skilled component and assembly work. (Redpath Dorman Long have been forced on Teeside to develop a friction-bolted system for producing platform modules, because of a shortage of craftsmen-like welders.) At a more precise level of engineering, something like this could be done to accommodate semi-skilled workers here.

In fact, in a conversion programme an attempt could be made to approach the problem from the other angle — instead of making the work simpler for the workers, training the workers in the necessary skills. However, the report's thoughts on this matter, and the industrial experience it refers to, indicate the possibility of an imaginative approach to a solution of occupational problems in new production.

It is important, both from the point of view of the workers and to achieve the aim of using released resources to fulfil social needs as fully as possible, that available skills are utilized to the greatest possible degree. On the face of it there does not seem any reason why any production worker of any grade should be out of a job following arms cuts, and as long as the work is available a major amount of industrial conversion could be accomplished in a matter of days or weeks, rather than months or years. Within the context of the 5-year programme of reduction of arms spending envisaged by this Study Group, occupational conversion for these workers does not appear to be a major technical planning problem.

(b) Research and Development Employees

It is normally thought that the problem of specialization would be a major obstacle in the way of occupational conversion for the scientists and technicians employed on military research and development. In fact, the problem seems to have been over-stated.

A study by Lloyd Dumas of Columbia University concluded that in the event of general and complete disarmament in the

USA only 13 per cent of military research and development staffs would be able to find employment in their converted former employment,by moving for example from military to civil aerospace. (28)   Problems similar in nature, but smaller numerically would be raised by arms cuts in Britain. Although the Dumas study may not be entirely and in detail applicable to Britain, it is worth noting its conclusions about redeployment of these people;   it notes that of the remaining 87 per cent, redeployment to the following areas was possible given periods of retraining or study:

> 23.5 per cent would be employable in High School teaching maths and science subjects given 9 months study; (29)

> 12.3 per cent could work in urban redevelopment with little or no retraining; (30)

> 5.5 per cent could work in environment protection programmes given 12-17 months retraining;

> 33.5 per cent could work in transport and other public utilities given 3-9 months study;

> 16.5 per cent could find employment in food and related industries after periods of study ranging from nil to 9 months;

> 4.1 per cent were immediately employable on basic research by the Federal government.

This left a shortfall of 4.6 per cent or 3.8 per cent of the total military research and development personnel.   Dumas based these conclusions on available qualifications, necessary qualifications, places available, and on the assumption that the rest of American industry would wish to utilize the talents and qualifications available.   He did not posit planning to any significant degree over and above the provision of places at university and college for study and retraining.

The value of this study for our work is twofold.   Firstly, it indicates that the problem of specialization is distinctly soluble.   The maximum period of study or retraining necessary is placed at 17 months, and it is likely that this is for a very small minority;   Herbert Striner of the Upjohn Institute has calculated that the military scientists and engineers in the USA could get the qualifications necessary for re-employment after only 3 or 4 months work. (31)   Secondly, it indicates that the people concerned here would be able to make a significant contribution to many areas of production and society.   There is no reason why we should think of limiting the occupational conversion opportunities to those sectors specified by Dumas, and we are certainly not bound by the percentages which he has identified – they can be changed by planning, according to our needs and priorities.

An outer estimate of the time necessary for study and re-training in Britain might be, in order to be on the very conservative side, not more than two years for any, and much less for most. The necessary teacher training, for example, is a one year course. This kind of timescale gives plenty of room for manoeuvre and scope for a phased transfer out of military work and into civil work as part of a five year programme.

(c) Administrative Personnel and Officers

A more serious problem is posed in considering transfer of employment for administrative personnel in the Civil Service and industry and for officers. Many observers of all political shades have commented that the British economy is over-administered; it would hardly be a gain if arms cuts resulted in administrative top-heaviness in converted industries and the appropriate government department, particularly since this would be likely to have a stultifying effect on the development of new ideas in those industries, which is precisely what we should be seeking to avoid. So, perhaps, unfortunately for the individuals concerned, there may be a case for doing away with some administrative jobs altogether and not converting them. When this was argued in a previous paper on the subject, (32) it not surprisingly drew down some howls of anguish. It may be, however, that we shall have to face up to the prospect that many people, possibly numbering some thousands, will need a significant degree of retraining after arms cuts, because there is no place for them otherwise.

To be sure, many administrators would find posts in public programmes expanded after arms cuts; management sectors of converted industries would be expected to absorb many (although the problem of industrial structures referred to in section 3 c) would have to be borne in mind. As with occupational conversion for others, this should not be left to chance, and where different management techniques are appropriate retraining should be made available, possibly for periods of around 6 months.

One problem which presents itself has been raised in relation to nationalized industries, notably and recently in The Times where a feature article based around interviews with one senior civil servant, and an ex-civil servant, questioned whether Civil Service experience provided the expertise and ability necessary for administration of industry. (33) In order to operate a socialist industrial strategy, the involvement of the NEB will be crucial; public ownership is also an integral part of conversion, when it is conceived of as a programme to provide what the country needs. We might therefore consider retraining also for government administrators, and, if it is indeed the case that management

376

of other nationalized industries is outside the scope of
civil servants' expertise, then two sets of retraining could be
dove-tailed together.

Those who do not find a place in the administration of
the conversion programme or converted industries should at
no time be asked to pay a price that is not demanded of
production and development staffs.   A clear commitment to
providing the retraining, over a period of time, necessary
to gain qualification for other jobs, should be forthcoming.
It is conceivable that a number of aspiring teachers would
be left unplaced in industry or government;   the provision
of a two year course (one year to regain familiarity with
the subject, and one for teacher-training) especially for
these people could be considered;   for those with degrees
in, particularly non-science subjects, this might suffice.
Pension and superannuation schemes should be maintained
during the period of retraining and later employment;   the
Civil Service unions would need to be involved in the
negotiating process for such agreements.   Early retirement
on full pension might also be considered for a proportion of
these people.

Former officers may also find themselves in a difficult
position if the qualifications and abilities they have
developed in the armed forces are not immediately applicable
to civilian life.   Again, opportunities for study and retraining should be made available.

It is unfortunately the case that the problem has not
been examined in enough detail, either in Britain or the USA,
to make any more concrete proposals than the above.   What is
needed as a basis for planning occupational conversion for
administrative personnel and officers is a genuine care for
their welfare and a commitment to providing education
opportunities (which may include university or college
courses of three or even four years duration) for them to
gain the necessary qualifications for new jobs.

(d)   Servicemen - Other Ranks

Numerous skills are available in the lower ranks of the armed
forces, in civil and mechanical engineering, communications,
electronics, etc.   Some of these skills are directly
applicable to civilian employment, while others would provide
a sound basis for a short period of retraining to build upon;
given the likelihood, referred to above, that arms cuts
should provide more jobs than are at present provided it
should be possible to utilize these skills in converted
industry to everybody's advantage.   Many other servicemen,
however, will not have these skills, and will need to start
almost from scratch in retraining.   Adequate provision for
this should be guaranteed, and could most valuably include
opportunities for these servicemen to enter apprenticeships.

In the examination of this question, it is clear that

the numbers of servicemen, and their service background
(and thus the skills available to them), released by arms
cuts will depend entirely upon the shape of these cuts.
For this Study Group, David Greenwood has set out a number of
options for arms cuts amounting to a budgetary saving in the
region of £1,000 million per year by 1980-1. (34) Although
his 'Module' 'C' - reductions in European Theatre Ground
Forces and proportionate Home Force reductions - provides
savings of the same magnitude as his 'Module' 'E' -
reductions in the RAF including cancellation of MRCA - the
manpower implications would be very difficult. 'Module'
'C' would reduce ETGF by some 30,000 troops with further
reductions in the Home Force; 'Module' 'E', since it
includes provision to extend the life of Phantom and expand
the Harrier and Jaguar programmes, would, despite suggested
base closures, have a far smaller effect on manpower.
However, since this latter option includes cancellation of
a major procurement programme, MRCA, industry would be
affected to a significant degree - to the tune of 7,500
jobs at present and an estimated 25,000 by 1981. (35)

It can therefore be concluded that we are not likely,
in the course of a programme of cuts amounting to £1,000
million per year after 5 years, to have to face major
occupational conversion tasks in all of the areas enumerated
in this section. We might be faced with large numbers of
production workers, and research and development staffs,
released for other work, but relatively small numbers of
servicemen and civil servants. Or the mix could be the
other way around, depending upon the mix of 'Modules'.
This consideration may also suggest to the Study Group
that certain types of arms cuts would be more valuable than
others, in the necessary process of reallocating resources;
in that case although the Group could finally present its
conclusions in a range of options, it might also be valuable
to indicate a preference in cuts because of the type of
resources it would release.

Finally, with regard to occupational conversion for
servicemen, over a period of five years, a sensible handling
of recruitment would make the task considerably easier.
Reductions in recruitment levels, taking into account normal
quitting of the services as the period of employment comes
to an end, could provide, or very nearly provide, a reduction
in the armed services' manpower equivalent to that they would
be expected to face from the cuts. Rather than facing a
task of occupational conversion, we would then have the task
of providing work for those (particularly young working-class
males from depressed areas) who would normally have joined
the armed services. This would depend upon an adequate
regional industrial strategy, which could overlap with a
conversion programme.

378

Alternative Production

(a) Two Approaches

In general there have been three approaches to the job of
identifying options for alternative production after arms
cuts or disarmament.   Two are briefly discussed here, and
the third in the following section under the name of 'target
industries'.
   One approach has been to consider specific products,
applicable to the skills and industrial processes of arms
industries, as in the second half of this section.   A
second, wider approach, has been merely to propose general
fields of production.   This was the approach taken for
example by John Ullman in an important article about conversion for the electronics industry in 1970, (36) when he
described the more specific approach as a matter of producing
a 'wishing list'.   Following this approach, a recent CND
publication said:

> engineers <u>do</u> make what the country must have if the
> crisis is to be solved: investment goods, exports,
> and import substitutes
>
> The technical possibility of solving the country's
> balance of payments problem is there.   Once that is
> done, improvements in the standard of living, in social
> services and in cultural facilities will again be
> possible. (37)

Thus, increased and more efficient production will result in
a general increase of prosperity allowing increased public
spending programmes where we need them.
   While this approach is correct in stressing the general
contribution to economic strength that conversion could make,
it does tend to duck the hard question — what will be made
and who will use it?   In a crisis situation that is an
important question.   A conversion programme should not
leave to chance the new products of converted industries. It
is also the case that the more specific approach can identify
products which could make an immediate contribution to, for
example, the NHS, and thus increase the options available.
It therefore seems right that, while a fairly general approach
has its own validity, it should be supplemented by the
specific identification of new products.

(b) New Products

Products listed in this section relate to aerospace;   a
forthcoming paper from Mary Kaldor may be able to provide
similar possibilities for a converted military shipbuilding
industry.

379

The best ideas for new products have not come from academics or researchers, management or civil servants. They come from people with experience on the shop floor and in design and development, people who are aware of the wider applications of their skills and expertise. The only service which is performed here is to pass on, with comments where relevant, a very limited number of suggestions which have emerged. The Lucas Aerospace Combine Committee's corporate plan, containing 150 products (a few of which are included in the list below) has shown the potential for identifying alternative products that can be found in industry. Other sources for alternative products include the American literature, and discussion with trade unionists.

Transport

*Short Take-Off and Landing passenger and freight aircraft. Giving STOL a civil as well as a military application might well, for example, open up a Third World market, where remote areas with difficult terrain might be more easily accessible to STOL aircraft than to road or rail transport.

*Jet propulsion of ships. Gas turbine jets, if adapted for oil pumping and electricity generation, could be used as maritime engines. In Japan there has been considerable interest in adapting the RB211 engine for this purpose. It would create a problem with the marine engineering industry (although work is being done on this at Anstey), but it is a problem the industry would face if jets are successfully adapted abroad. At present, there is limited potential for this because of the state of world trade.

*Retarder brake system for trains and coaches. This is an aerospace braking system; there is no reason why it cannot be used for trains and large road vehicles. While its effect might be rather brutal, it would be worth it if it could save lives, as it could have done in previous disasters.

*Speed/distance related warning systems. For road traffic a device could be placed upon the front and rear of each vehicle to give an automatic warning to the driver when vehicles were too close; the distance at which the alarm is given would be related to the speed of travel. This might have special application for driving in fog. On trains a device could be fitted to measure the time taken to pass two points a fixed distance apart. If the speed were too great for that part of the track a warning would be given to the driver; this could be backed up by an automatic braking system if the driver failed to respond (and could thus have prevented the Moorgate crash). There would be little problem in grafting such a system onto the London Tube system.

*Battery cars. Present versions can only run for about 40 miles in cities (100 miles cruising on the flat) before they need recharging; a 4 hp diesel driven generator running constantly at maximum revs would keep the battery

fully charged at all times, thus also providing the torque
necessary for acceleration, and making electric cars a real
possibility.   Near-silent engines would be possible; fuel
savings should amount to 50 per cent and an 80 per cent
reduction in toxic fumes could be achieved.

*Helium airships for airfreight.   Helium is inert so the
disasters which brought airship development to an end could be
avoided.   Ballast could be on and off-loaded to maintain a
constant weight.   2 engines of 200 hp could move 400 tons
at 100 mph between custom-made loading and unloading points.

*New rolling stock.   Present stock utilizes 100 year old
design concepts, and the heavy rigid structures have exacerbated the effect of rail crashes.   Use of aerospace
experience in light-weight high-stress materials and design,
together with more traditional coach-design expertise, and
a willingness to break through old concepts, could provide
a faster and safer rolling stock.   Equipped with pneumatic
tyres it would raise the possibility of drive-on and drive-
off integrated transport systems.   Aerospace technology
and experience in hydraulics could also be utilized for the
development of appropriate buffer systems.

*Monorail development.   This has always been regarded as
a stock option for aerospace conversion.   Despite its
problems (switching and swaying), it is still worth pursuing
because its potential for high speed travel is so great.

There is also the possibility for further work on air
safety, air traffic control, and other transport safety
developments than the ones suggested above.

Health and Safety

*Personalized machinery for the disabled.   At its best this
could be labour intensive work, with teams of three working
together with disabled people to produce machinery suitable
for their particular needs.   One aspect of the work could be
use of vastly simplified aerospace technology, especially in
the controls. (38)   Similarly, experience in the development
of 'blind landing' techniques could be utilized in the
production of aids for the blind.

*Telechiric machines.   These are robots which 'mimic'
instructions given them by the human operator; they can be
operated from great distances.   They could be an ideal
development for use in dangerous environments such as pit
disasters, natural disasters, firefighting, North Sea oil rig
underwater work etc.   The human element is not removed, but
the operator is safe and does not need to jeopardize his own
life.

*Industrial sound proofing.   The use of fibre discs on
gears in industrial machinery could bring noise down to
tolerable limits and reduce the danger of deafness.

*Further production of pacemakers and kidney dialysis
machines.   At present it is reckoned that there is one

dialysis machine for four potential users.   These machines are
made by Vickers and by Lucas and industrial conversion would
provide the ideal opportunity to expand the medical division
of those firms.

There would also be the possibility of further development
of medical equipment, including hospital communications
systems, use of computers in hospital (as happened in Sweden),
and electronics for Intensive Care Units.

Energy, Heating, Power

*Submerged Production Systems.   These do away with the need
for oil rigs, and would be ideal for use in deep water;   they
lie on the sea-bed and are remotely operated.   The prototype
was installed in the Gulf of Mexico over a year ago.   This was
proposed in the SCND report on the Clyde Submarine Base, (39)
but could be manufactured elsewhere as well.

*Nuclear material disposal.   If we are committed, as we
appear to be, to nuclear energy, more effort could be devoted
to safe disposal of waste products, using airframe experience
in metallurgy and high-stress.

*Integrated energy systems.   These need not be, although
they could be, a replacement for nuclear energy;   they could
be a supplement.   Sources of energy could include heat-
pumps, taking the heat out of rivers and storing the energy
in accumulators;   solar panels could also be used and wind
generators.   Aerospace technological experience could be
utilized to ensure a constant power level from solar panels
and wind generators.   Such a system might provide 80-90 per
cent of the energy necessary for a large housing estate,
and would be very useful in areas where electricity power
supply is subject to the vagaries of the elements.   At the
least, a pilot project is worth the effort.

It would also be worth devoting some effort to development
work in fuel cells utilizing oxygen and hydrogen.

Others

*Prefabricated parts for building.   This has been regarded
as a major conversion option for airframe manufacturing.
There can be no running away from the problems this would
raise with construction unions if they saw traditional crafts
threatened.   But they may be problems that should be taken on,
together with the unions, given the urgency of expanding
building programmes for housing, hospitals, schools.   It is,
after all, unlikely that in the wake of arms cuts the amount
of conversion necessary in the airframe industry (even if
all conversion was to prefab building) would be enough to
drive more traditional building methods out of existence,
and the two might be able to exist side by side.

382

*Marine mineral exploitation and agriculture.   Both air-
frame and submarine technological experience could be used in
producing work nodules (either manned or remote-controlled)
for, at the outset, a pilot project and research work in this
field.   There is no doubting the potential for mineral
exploitation on the sea bed where there are nodules (like
pebbles 20-40 mm in diameter) containing manganese, iron,
copper, nickel and cobalt.

*Wider manufacture and use of ball-screws.   These are
high efficiency precision components, now used in aerospace,
but capable of replacing nuts and bolts in just about everything.   It has been estimated that there is an untapped EEC
machine-tool market which alone is worth over £60 million
annually.

In addition, the whole field of communication would bear
examination as would areas such as office machinery,
electronic libraries, self-teaching devices, etc.

The status of the products listed above varies: some
are good ideas which could do with further development and
study, others are immediately needed or desired goods which
could be produced almost immediately; some are mainly for
export, others concentrate on filling domestic need but
could also be exported.   Industrial conversion would
provide the opportunity to examine seriously the more far-
reaching projects, as well as to make more of products such
as dialysis machines and ball-screws.

Target Industries

(a) The Concept

In the course of a conversion programme it would be possible
for the government to select a particular industry as the
target for a major effort.   It would not need to be an
industry which has an especial affinity to any present
military industry, nor would it need to be in the same
location or employing the same workers as a section of a
former arms industry.   This would provide the chance for
a conversion programme to make a contribution outside what
many might regard as its immediate frame of reference.   It
should not, I would suggest, replace the use of resources
released through arms cuts, or prevent the re-employment of
affected workers and servicemen; but the excess of physical
resources over the level of job provision in redeploying arms
workers and servicemen might make possible a wider contribution.
In particular, the selection of one or more industries for
such an effort could be part of a regional policy particularly
directed at depressed areas, either by revitalizing an industry
in such an area, or by taking a completely new industry to the
area.   In the course of this, the NEB and/or full public
ownership, together with planning agreements, would have a
crucial role to play.

The general concept is not a new one.   McDonagh and
Zimmerman (40) identified target industries in their work on
conversion options for the American airframe industry,
although they concentrated on options where the skills and
industrial processes were close to those of airframe manufacturing, a limitation which need not be imposed.   The more
generalized approach towards alternative production outlined
in the preceding section contains elements of the idea, and
Ullman noted that subsidiary aspects of air transport had
been identified by many as a target followed conversion.(41)
     Candidates for pride of place as the target industries
are numerous, as might be expected.   Any of the following
could be considered: machine tools, construction, communications, transport, energy, medical and educational equipment,
and so on.   To a lesser or greater extent the skills and
processes in those industries could draw on those of present
arms industries.   Below a brief look is taken at what
might seem a less obvious candidate in order to demonstrate
the possibilities.

(b) The Motorcycle Industry

The almost complete demise of the once flourishing British
motorcycle industry was caused by the appearance of better
machines, made in Japan, and by the identification by
Japanese motorcycle firms of a market which the British
industry had failed to recognize.   The British industry was
not alone in suffering — American and Italian industries
were also hard hit.
     Essentially, the Japanese firms recognized that the
ruggedness of British machines, which was a central part of
the appeal to their market, did not appeal to many who would
be attracted by a cheap quick form of transport, but deterred
by the difficulties of riding and maintaining British bikes.
Apart from mopeds and motor scooters, that market had been
left to itself by the British industry.   The Japanese firms
introduced a range of machines which spanned the gap between
mopeds and the 175 cc size.   Their machines from 175 cc
upwards were better engineered (although the bodywork was
sometimes less sound) than British competitors; they were
faster, with better acceleration, equally economic and
generally easier to maintain.   Ultimately it was only at
the very top end of the range (500 cc and up) of engine
capacity that British bikes could compete with the Japanese
products and British firms foundered.
     However, the one great problem with Japanese machines has
always been availability of spare parts.   According to first
reports the Meriden Co-op's products (at the top end of the
capacity range) are competitive against the larger Japanese
machines.   There is no reason to suppose that bikes could
not be produced in Britain towards the lower end of the range
(say, 250 cc and down) that could be competitive in terms of
performance, and outdo the competition in availability of

spares.

The important market for motorcycles is towards the lower end of the capacity range. A detailed market survey is not needed to know that the number of bikes on the road is increasing; their advantages over cars are faster travel in towns and lower petrol consumption. Many people are either replacing cars with motorcycles, or buying motorcycles as a supplement to their cars, for travel in towns. Assuming urban traffic congestion does not ease and petrol continues to be so expensive, that trend will not reverse. While the market for new cars is badly hit by the economic crisis, the new motorcycle market does not appear to be; indeed, it may be that it has expanded during the crisis. With such a large and clearly identified market on our very door-step, it would seem daft not to take advantage of it, re-penetrating the British domestic market, with the possibility of exports as an extra.

Motorcycle engineering is not especially complicated (though, ideally, it should be rather more sophisticated than the engineering on some of Britain's long-lasting post-war models) so for workers with any experience in engineering, little if any retraining should be necessary. A good deal of component manufacturing is involved, which would allow the beneficial employment effects of revitalizing motorcycle manufacturing to be spread through a number of localities.

Neither the concept of selecting target industries, nor the specific example of the motorcycle industry, is presented as a concrete and detailed plan. It is merely put forward as an example of the kind of opportunity which is open follow-ing arms cuts. It is the kind of direction which a conversion programme might be able to take, as long as the direction is identified and the necessary planning measures are taken.

Planning and Politics

(a) The Wider Implication of Industrial Conversion

It has been stressed throughout this paper that detailed planning and effective control of the implementation of planning are crucial for successful conversion of arms industries. Industrial conversion should be conceived of as a programme within the context of socialist policies; it should have a clear sense of direction, of the tasks it needs to fulfil, which are not just re-employment of workers and servicemen, but also sensible and useful allocation of resources.

A 'rule of thumb' guide to the need for planning can be seen by glancing at the government departments which would be likely to be involved.

Obviously involved would be the Ministry of Defence, the Departments of Industry and of Employment, and the Treasury. A programme which included re-education opportunities and/or the provision of teachers would involve the Department of Education. A programme involving increased resource allocation to building programmes would bring in the Department of the Environment, and, if schools and hospitals were to be built, or medical or educational equipment manufactured, also the DHSS and the Department of Education (again). If new products were attempting to open up new markets, the Department of Trade would expect to be involved, and if some products went to the Third World in the form of aid the Ministry of Overseas Aid at the Foreign Office would be hard at work. If, from the available alternative products, anything to do with transport were taken up, the relevant ministry would join the throng, as would the Department of Energy if the relevant projects were to be taken up. The Department of Prices and Consumer Protection might have a valuable role to play in the process. It is conceivable that the Home Office and most of the Foreign Office would not need to be involved.

The prospect of industrial conversion is an exciting one, but the list of government departments which would be affected by and drawn into the planning of a conversion programme should give sufficient indication of the need for careful and detailed planning.

(b) Levels of Planning

Time and again in the literature about conversion it has been emphasized that planning from central government is not enough. (42) At least two other levels of planning are necessary: at a local or regional level, and on an industry basis.

The main reason for local planning is twofold: firstly, to involve production and development workers who have more and better ideas for conversion than anyone else; secondly, because certain communities will be deeply affected by the arms cuts and conversion, care for the well-being and the shape of those communities must be a part of a conversion programme. Although the regional distribution of arms contracts is not generally significantly disproportionate to regional population distribution, (43) particular localities depend heavily upon arms work; this is the case, for example, in Preston, Barrow, and Birkenhead, despite the fact that the North West as a whole, in the year to March 1975, received only 6.2 per cent of contracts placed compared to the region's 11.8 per cent of the country's population. There is no way that a conversion programme will be acceptable in those localities unless the communities, including the parts not directly involved in arms work, are involved in planning.

Equally, planning on an industrial basis has a two fold reason. Again, it will be a mechanism for the involvement of people presently involved in the arms industries. Secondly, not all of a particular industry affected by cuts will be able to convert one single new type of production;

386

the airframe industry, for example, could be involved in STOL civil aircraft; prefabricated housing, undersea work modules, and production of new rolling stock.   The distribution of work through converted industries will need to involve industrial personnel of all grades in the planning.

Participation in planning from communities and industries should be more than token.   It should not be a matter of being handed down ideas from the government and working out the best way to cope; ideas should also be handed upwards, and taken seriously.   This of course will be a major job, but it is only on this basis that conversion will be acceptable and it will almost certainly be more successful as a result. It should also be pointed out that among those affected by industrial conversion there will be bodies presently not involved, directly or indirectly, in arms production; an example is the construction unions.   They too should be able to participate in the planning.

The question remains of the best structure at government level for planning and co-ordination.   Various methods have been proposed, including Richard Barnett's suggestion for the USA of 'a National Conversion Commission with broad powers', (44) and John Ullman's offer of a 'National Technology Foundation'; (45) both suggestions owed much to Senator McGovern's attempt in 1964 to have a National Economic Conversion Commission established in the USA.   I do not believe it should be necessary or would be desirable to resort to such measures in Britain.   It would be more apt, and more efficient, if the government measures necessary for industrial conversion were carried out as a part of, and not apart from, the wider planning of British industrial and economic life. The suggestion of a body such as a Cabinet Committee for Economic Planning is therefore more attractive. (46)   A structure is not everything, but it is important; if a special commission, agency, or department, were grafted on top of the kind of economic and industrial planning to which the Labour Party has given its support, the result would be an unwieldy and unwholesome mess.   The general direction identified by the Cabinet Committee, on the basis of data coming up both from government departments and from regional and industrial planning bodies, could be planned in detail by department or inter-departmental committees, again using data from regions and industries, and implemented by the NEB or nationalized industry.   At the same time, a watchdog study group set up by the Cabinet Committee to report back on the progress of conversion at regular intervals might not be a bad thing.

There is no doubt that a major planning exercise will have to be taken on.   In my view, ensuring flexibility, imagination, and common humanity in the planning is more of a problem than occupational conversion, new production, demand maintenance, or full employment provision.   It is, however, a soluble problem once the principle of wide participation is accepted.   For a Party committed to planning the general direction of the economy, and to planning in detail and controlling certain important sectors of it, there should be nothing new or especially daunting in the task.

(c) Politics

The problems in the way of arms cuts and a successful conversion programme can be categorized as technical, economic and political. The technical problems (conversion of occupation and machinery, and identification of new products) appear to be soluble; economic problems (resource allocation, demand maintenance, full employment) are equally soluble. Political obstacles in the way of arms cuts are outside the scope of this paper, but once arms cuts have been agreed and the conversion programme is in process, that is not an end to political obstacles. One example is the problem that might be encountered with construction unions, previously referred to. Another is the problem of converting arms sections of multinationals or other major diversified corporations. A decision to halve further development of the nuclear energy programme pending examination of safe disposal methods and/or community energy systems would also cause problems; one can already see the lobbyists looking down the list of MPs. A decision to re-penetrate the domestic motorcycle market under government direction might cause something of a fracas with the Japanese government. Were Lucas or Vickers instructed to expand their medical divisions, or face the consequences of government action to achieve more production of medical equipment, without a guarantee of the same profits as made on arms contracts (which might be neither possible nor desirable), something more than a storm in a tea-cup should be expected.

There appear to be two major solutions to this problem. The first is for the government to stand firm. The second is to ensure the widest involvement in planning and implementation of a conversion programme, to ensure political support for the government's stand and policies. Crucially this means involvement of trade unions and communities.

Next Steps

Before a conversion programme can go ahead, a number of detailed studies are necessary, including the following:

1) The nature of each industry and company involved in arms production; its structure and workforce; the resources it utilizes.

2) The communities in which arms production is carried out; other job prospects in the locality; present unemployment levels; skills and resources available in the communities.

3) Occupational conversion prospects, based on present
skills and qualifications, and on likely demand, for everyone
whose employment depends upon military work.

4) Specific and detailed case studies of the effects of
past cancellations; similar studies of hypothetical future
cancellations.

5) Products available for converted arms industries;
adaptability of plant and machinery; levels of retraining
necessary; need or demand for these products at home and
abroad.

6) Examination of specific industries which could be
given a boost as part of a conversion programme; market
prospects.

Each of these is a major project. No government should
be considering arms cuts unless it is amassing this
information. Trade unions could go some way towards
providing data necessary, especially for numbers 1), 2) and
4); unions at both national and local level might also
valuably turn their attention to numbers 5) and 6); whether
or not the studies are utilized for a conversion programme
they could still be extremely useful; it was not in order to
help work on conversion of arms industries that the shop
stewards at Lucas Aerospace drew up the corporate plan, but
as part of a strategy to combat redundancies.

Summary

This paper has considered industrial conversion within the
context of overall socialist policies and as a central part
of: redeploying workers and servicemen affected by arms cuts;
reshaping British industry; reallocating resources to produce
the things we really need.
It has throughout insisted on the need for planning, and
has considered ways in which this could happen, emphasizing
especially the importance of meaningful local and industrial
participation in the planning process, in order to carry out
arms cuts and conversion.
It has drawn on the experience of past arms cuts and on
studies in the USA, to conclude that the problems normally
identified as obstacles to conversion are less than they
might seem, although they should be taken seriously, and are
distinctly soluble.
It has identified possibilities for a conversion programme's final products, as a way of demonstrating the potential
benefits of conversion.
Finally, it has proposed a programme of further detailed
studies to prepare the way for a successful conversion programme.

References

1. Minutes of the Study Group on Defence Expenditure, the Arms Trade and Alternative Employment, 22 July 1975.

2. 'Characteristics of the Industrial Conversion Problem', Seymour Melman, in The Defence Economy, ed. Melman (USA 1970).

3. See Frank Blackaby, 'Note on Military Expenditure and National Product: UK and certain other countries', July 1975, and Study Group Minutes, 22 July 1975.

4. As the paper goes on to suggest, such participation would also increase the efficiency of a conversion programme.

5. See Study Group Minutes, 22 July and 15 October 1975.

6. E.g. Roy Mason, Secretary of State for Defence, 3 February 1975, and Robert Brown, Under-Secretary of State for the Army, 11 May 1975.

7. E.g. Arms Jobs and the Crisis, CND, September 1975; The Defence Review: an Anti-White Paper, Fabian Society Research Series, No.323 (November 1975); defence debate amendment and the arguments used, Hansard, 7 & 8 May 1975.

8. Economic and Social Consequences of Disarmament (UN Department of Economic and Social Affairs, 1962); Economic effects of Disarmament (Economist Intelligence Unit, 1963).

9. A variety of studies on this subject are available from the US Arms Control and Disarmament Agency, Washington DC; in particular, though dealing with cancellation of a space contract; the useful The Dyna-Soar Contract Cancellation (US ACDA, July 1965).

10. Seymour Melman, The Defence Economy.

11. Frank Blackaby, 'Note on the Employment Consequences of a £1,000 million cut (at 1974 prices) in Military Expenditure over 5 years', September 1975.

12. Roy Mason, 3 February 1975, and Robert Brown, 11 May 1975.

13. Statement on the Defence Estimates 1975, Cmnd 5976.

14. 15. and 16 See 'Progress Report on the Work of the Study Group', December 1975.

17. See: Asides 'Analysing the Regional Aspects of Defence Spending: a Survey' (December 1973); 'Military Installation

and Local Economies: a case study – the Moray Air
Stations' (December 1973); 'Military Installations
and Local Economies: a case study – the Clyde Submarine
Base' (August 1974) – from Aberdeen University, Defence
Studies Unit.   Also: Replacing Employment at the
Nuclear Bases (Scottish CND, February 1975).

18.   These comments are based on unpublished material,
      based on interviews, compiled by Mary Kaldor.

19.   For a full discussion of this see Mary Kaldor, European
      Defence Industries – National and International
      Implications, ISIO Monograph No.8 (1972).

20.   Economic and Social Consequences of Disarmament, para 176.

21.   James J. McDonagh and Steven M. Zimmerman, Mobilization
      for Peace: A program for civilian diversification of
      the Airframe Industry (1951).

22.   Economic and Social Consequences of Disarmament, para 169.

23.   See Cmnd 5976.

24.   James Milne of the Scottish TUC speaking at Scottish CND
      Conference in Glasgow, February 1975, quoted in Arms Jobs
      and the Crisis.

25.   The Potential Transfer of Industrial Skills from Defence
      to Non-Defence Industries (California Department of
      Employment, April 1968).

26.   J.R. Cambern and D. Newton, 'Skill Transfers: Can
      Defence Workers Adapt to Civilian Occupations?'
      Monthly Labour Review (US Department of Labor, June 1969).

27.   Scottish CND.

28.   Lloyd J. Dumas, 'Re-education and Re-employment of
      Engineering and Scientific Personnel', in Melman ed.   The
      Defence Economy.

29.   In fact, 67 per cent have similar qualifications; the
      figure of 23.5 per cent was reached on the basis of
      demand for teachers.

30.   30 per cent are civil or mechanical engineers but job
      opportunities in this field were again restricted.

31.   Referred to in Richard J. Barnett, The Economy of Death
      (USA, 1969).

32.   Dave Griffiths and Dan Smith, Arms Cuts and Industrial

Conversion (CND, November 1975).

33. 'In these days of State takeovers do we need a new breed of civil servant', The Times, 6 January 1976.

34. David Greenwood, Defence Programme Options to 1980-81 (September 1975).

35. Speech by Brynmor John, Under-Secretary for the Air Force, in debate on White Paper, Hansard, 7 May 1975.

36. John E. Ullman, 'Conversion and the Import Problem' in IEEE Spectrum (USA, April 1970).

37. Arms Jobs and the Crisis.

38. See, for example, the description of the 'hob cart' designed by workers at Lucas Aerospace in New Scientist, 3 July 1975.

39. Scottish CND.

40. McDonagh and Zimmerman, Mobilization for Peace.

41. John E. Ullman, 'Conversion and the Import Problem'.

42. E.g. Arms Jobs and the Crisis; The Defence Economy; The Economic Consequences of Reduced Military Spending, ed Bernard Udis (USA, 1973); Seymour Melman, The Permanent War Economy (USA, 1974).

43. Answers to Parliamentary Questions by William Rodgers, Minister of State for Defence, Hansard 22 May 1975 and 9 June 1975, reproduced in a chart in Arms Jobs and the Crisis.

44. Richard J. Barnett, The Economy of Death (1969).

45. John E. Ullman, 'Conversion and the Import Problem'.

46. The role of such a committee is dealt with in more depth in chapter 8 of Stuart Holland, The Socialist Challenge (1975).

13 ALTERNATIVE EMPLOYMENT FOR NAVAL SHIPBUILDING WORKERS: A CASE STUDY OF THE RESOURCES DEVOTED TO THE PRODUCTION OF THE ASW CRUISER

by Mary Kaldor and Albert Booth

Part 1: Background to the Project

The Anti-Submarine Warfare (ASW) or through deck cruiser is a euphemism for 'small aircraft carrier'. The euphemism is required because the last Labour Government took the decision to phase out the British carrier force and to allocate all fixed-wing aircraft to the RAF. Lest the RAF be offended or thinking people be critical, the latest carrier is disguised as a cruiser.

The functions of the cruiser are twofold: (a) command, control and co-ordination of British and NATO maritime forces and (b) the deployment of ASW aircraft - the Sea King helicopter and the Harrier VTOL aircraft. The Ministry of Defence claims that to combine these functions in a single large hull is the most 'cost-effective' solution. There are, however, two fundamental weaknesses in their case. First, because all surface ships are vulnerable to attack from the air or from below the surface, a single 'cost-effective' solution in peacetime may prove an expensive disaster in war. The Ministry argue that 'the vulnerability of the cruiser will be no greater than that of any other ship in the force'. But to justify its expense the cruiser must be considerably less vulnerable. Secondly, it is not clear in what circumstances these functions will be carried out. The new naval scenario that has emerged since withdrawal from the Far East is the Battle of the Eastern Atlantic. So critical is this battle considered that only a few tugs could be spared for the defence of North Sea oil. The exact nature of the battle has not been defined. There is vague talk about providing a 'mix of naval forces with a capability across the whole spectrum of possible naval operations'. And the cruiser is described as a unit 'with a greater capability for operations at the higher level'. When a member of the Expenditure Committee asked if the planners were 'contemplating a war with an eastern bloc power which will go on long enough for us to be involved in convoy protection work and in the protection of amphibious forces crossing the channel', the Ministry of Defence witness produced a useful catch-all formula: 'We are contemplating a deterrent concept of operations at sea - a situation in which the Soviet fleet could be deployed to bring either military or political pressure to bear. NATO's concept of operations at sea is exactly parallel with its concept of operations on land.' (1)

This formula has emerged since the 1966 White Paper which envisaged that the tasks for carrierborne aircraft in the late 1970s could be 'more cheaply' performed in other ways. (2) In particular, anti-submarine protection would be given by helicopters operating from ships other than carriers, while early-warning aircraft would eventually operate from land bases. The cost of a carrier force was estimated at £1,400 million over ten years, hardly more than the estimate

394

of £1,200 million for the planned cruiser force (see below).

Denis Healey, then Minister of Defence, explained the thinking behind this decision in a lecture to the Royal United Services Institute on 2 October 1969:

It was obviously necessary to see whether it was really essential to spend these enormous sums on so limited a capability. It emerged rapidly that the role of the carrier in support of land operations could in most places which concerned us, be carried out more cheaply and effectively by land-based aircraft; and that if we renounced the strategic option of landing or withdrawing troops against sophisticated opposition outside the range of friendly, land-based aircraft, this would have little important effect on our commitments. So the case for maintaining the carrier force depended critically on its role in maritime operations − a requirement which had been regarded up to then as simply a convenient by-product of the carriers' main role. This turned out to be a difficult nut to crack if one envisaged high-intensity maritime operations against a sophisticated enemy in the Indian Ocean in the next decade. On the other hand the value of a single carrier on station in such operations was open to doubt. While it was a difficult judgment to decide against a carrier force for maritime operations East of Suez, once we had decided to withdraw from major military responsibilities in that area in the middle seventies I do not believe that the decision was easy to contest.

The decision to abandon carriers was not simply a decision to abandon carriers. It was a decision to abandon a sizeable chunk of the navy. Carriers are at the apex of the British naval structure. They justify the existence of the Fleet Air Arm, of a number of frigates and destroyers and hunter-killer submarines needed for protection, and of supply ships needed for replenishment. The navy would inevitably protest.

Almost as soon as the decision was taken, the case for small carriers was aired. (The idea had been mooted as far back as 1960.) 'Only if the Fleet Air Arm is deployed from a large number of carriers can it provide world-wide air cover; only if those carriers are small and simple can there be any hope of approaching the number needed ... Nelson never had enough frigates either, but at least he did not have to contend with some economic genius intent on concentrating all his escort tonnage into three or four super frigates on the ground of cost-effectiveness.' (3) Healey was adamant in opposing the small carrier, but the notion was taken up almost as soon as the Conservatives took power in 1970. The new small carriers turned out to be three super frigates or cruisers, justified on grounds of cost-effectiveness. Conceptions of maritime operations were shifted from the Indian Ocean to the Eastern Atlantic, along with the all-embracing doctrine of 'flexible response'.

The change of heart was not simply due to the change of government, to the naval fantasies of Lord Carrington. There was considerable concern in 1971 and 1972 about excess naval capacity in the shipbuilding industry. In particular, the bankruptcy and work-in at Upper Clyde was followed by an accelerated naval building programme in Northern shipyards. In announcing the programme, the Minister of State for Defence Procurement made it clear that 'this was a special exercise to help not only the navy but employment'. (4) The lead items for the cruiser were ordered shortly afterwards. It was not only the shipbuilding industry that stood to gain; substantial orders for electronics, missiles and aircraft were also involved. For example, the cruiser is designed to operate the maritime Harrier. A number of prospective overseas customers informed the British Government that orders for the maritime Harrier were dependent on a British order.

The industrial aspects will be discussed in more detail in Part 2, (p.399), but they may perhaps explain why the Labour Government has not found it possible to reverse the decision. For the strategic objections still remain. They were summarized by Vice-Admiral Sir Ian McGeogh:

> Professional opinion, and especially that of experienced naval aviators, remains extremely sceptical of the wisdom of the TDC (Through Deck Cruiser) concept. It is pointed out that to combine in one ship the functions of Force flagship, area defence, anti-submarine helicoptership and fixed-wing V/STOL carrier is to ensure that none of these functions will be effective. In addition, the cost ... will ensure that not more than three of them will be built. Furthermore, despite her size and armament, most of which, in any case, is defensive, such a ship would have no armoured protection and be just as vulnerable as any other to torpedo attack or mining. In action, she would inevitably be the main target for all kinds of attack and once damaged, let alone sunk, she would cease to be an asset to the Force Commander and become his biggest liability. Unlike the Cruiser, in its original environment, which could steam anywhere in safety, being fast enough to evade the only superior forces it might encounter, the TDC could not be allowed out of harbour in time of international tension or hostilities, without a screen of anti-submarine frigates at least. Her own helicopters would be quite inadequate, unassisted, to give the TDC reasonable anti-submarine protection. (5)

Cost

The table below is an estimate of the life-time cost of the cruiser programme at current prices, including the cost of associated equipment such as aircraft and protective vessels.

(The cost of missiles, Sea Dart and Sea Wolf, is included in the basic cost of the cruiser and the frigates.)

	£m
Basic cost of 3 cruisers	390[6]
Basic cost of 25 Harriers	85[7]
Basic cost of 37 Sea Kings	83[8]
Basic cost of Support Ships (including Type 21 or Type 22 frigates, supply ships, a hunter killer submarine)	250-300[9]
Running cost of 3 cruisers (over 20-year lifetime)	320[10]
Running cost of support ships (over 20-year lifetime)	200[11]
Cost of associated shore personnel (over 20-year lifetime)	390[12]
Cost of aircraft spares (over 15-year lifetime)	168[13]
Cost of new aircraft (after 15 years)	336[14]
Cost of aircraft spares (over 5 years)	112

Total functional cost of cruisers over 20 years   2,360

Thus the cruisers will cost approximately £120 million a year. This is based on current estimates and does not take into account cost escalation, which inevitably occurs and has already occurred to some extent. The final cost may be as much as three times this figure, i e. around £360 million a year.

Notes and References

1. HCP 99-V111, Session 1973-4.

2. Statement on the Defence Estimates, 1966, Cmnd 2901 and 2902.

3. Lieutenant-Commander F.P.U. Croker, RN, 'David or Goliath? An Essay in Cost Effectiveness', RUSI Journal, May 1966.

4. House of Commons Report, 11 November 1971, col.1228.

5. Ian McGeogh, "Command of the Sea in the Seventies", The Waverly Papers, University of Edinburgh, Occasional Paper 1: Series 4.

6. A figure of £330 million for a classified number of cruisers was given in evidence to the Expenditure Committee, HCP99-Vlll. It is widely assumed that the number is three since this is the minimum necessary to ensure that at least one cruiser is continuously in operation. £390 million allows for inflation.

7. Eight Harriers are needed for each ship, plus eight in reserve and one trainer. See Flight, 19 June 1975.

8. Nine Sea Kings are required for each ship. It is assumed that a further nine are kept in reserve, plus one for training purposes. The unit cost of Sea King is £2.25 million, according to the Daily Telegraph, 10 March 1975. (They may cost more if a fee for development work is included.)

9. Estimate, based on similar figures calculated for aircraft carriers in the 1960s (see Neville Brown, New Scientist, 27 January 1966), and on cost per ton of different kinds of ships calculated from Jane's Fighting Ships and other sources.

10. This is based on figures provided to the Expenditure Committee. They seem rather low. For example, personnel costs work out at around £1,500 per man. Yet the average cost in pay and allowances of the armed forces in general works out at £3,200 per man.

11. Estimate arrived at by extrapolating the running cost of the cruiser; it could, therefore, be low.

12. Estimate based on assumption of one man ashore for every man on board ship.

13. Over the lifetime of an aircraft, spares are generally reckoned to cost roughly the same as the original acquisition cost.

14. Estimate based on conservative assumption that the next generation of aircraft will cost twice as much as the present generation. No sum is included to cover conversion of the cruiser to take new types of aircraft.

(The author is grateful to Paul Cockles and Major Elliott of the IISS for help in obtaining information for this paper.)

Part 2: Alternative Employment for Naval Shipbuilding
         Workers (1)

The ASW Cruiser is being built by the Vickers Shipbuilding
Group at Barrow-in-Furness. The group employs around
13,000 workers and has an annual output of £64 million. The
cruiser probably takes up about a fifth of Barrow's shipbuilding capacity and, over its lifetime, will involve around
7,000 to 8,000 man years of work. A further 28,000 to 32,000
man years will be taken up in the supplying industries -
steel, marine equipment, etc.
   The conversion problem, for all these people, is not so
much technical as political. Any manner of alternative
products could be made with the particular skills and
talents, plant and facilities necessary for the production of
the ASW Cruiser. The problem is the choice of products - a
choice which depends on the local organization of the shipyard
and its co-ordination with national policies towards industry,
energy, transport, health, etc. This paper looks at both
aspects of the problem. The first section deals with the
people and resources employed at Barrow and the alternative
production lines that have been suggested. The second section
describes the contractor, Vickers Ltd, and why its current
status and relationship with the Government must be changed if
a successful conversion programme is to be carried out. The
third section makes specific recommendations which are of
relevance to the problem of defence conversion in general.
   The conversion potential of the resources engaged in the
production of the ASW Cruiser is not just of particular
interest. The Barrow shipyard is fairly typical of naval
shipyards; the different requirements for submarines and smaller
warships mainly concern the size of berths and the quantity
of technical equipment. Equally, the problem of warship
building conversion is not so different from the problem of
finding work for surplus capacity in shipbuilding generally.
Vickers Ltd is more than typical. If one includes its 40
per cent holding in BAC, it is the largest armaments company
in Britain, producing the whole range of armaments from
small-arms to ASW Cruisers and the Multi-Role Combat Aircraft.
It is also an important multi-national company. In so far
as its military and overseas divisions have expanded rapidly
in recent years, while civilian British output has stagnated,
it can be said to be typical not just of the armaments
industry but of the British economy as a whole.

The Technical Problem of the Conversion: Alternative Products
for Barrow Shipyard

The shipbuilding process, and the skills and facilities
available at Barrow, are essentially commensurate with any
kind of large-scale relatively labour-intensive construction
and assembly activities which involve heavy metal fabrication
and materials handling and complex logistical problems of

supply, storage and scheduling. Preferably, alternative
activities should be sea-based, partly because of the marine
experience of the workers at Barrow, and partly because of
the poor road and rail access to Barrow. Possible types of
conversion can be broadly divided into three: merchant
shipbuilding, alternative land-based manufacturing activities,
and new sea-based technologies.

Merchant shipbuilding: The most obvious alternative activity,
and the one preferred by boilermakers, is merchant shipbuilding.
The Booz-Allen and Hamilton Report on British Shipbuilding
1972, (2) concluded that between £210 million and £250 million
in new capital investment would be required to make British
shipbuilding competitive. Clearly, much more would be
required with today's recession in shipbuilding and today's
price-level, but, even so, it is not a large sum when compared
with the cost of three ASW Cruisers, i.e. £330 million.
Furthermore, the UK has one advantage in the general shipbuilding
gloom. The main collapse in shipping has occurred in the
tanker market. Tankers are a relatively low proportion of
total British output, and only Harland & Wolff have built the
expensive capital-intensive facilities needed for tanker production. The age of British plant and facilities could
conceivably prove advantageous in the specialized markets of
product carriers, container ships, etc., which are expected
to dominate future orders.

Nevertheless, there is a strong case for arguing that
money could be better spent in other ways, especially as
regards the use of naval shipyards. First of all, there is
tremendous over-capacity in shipbuilding. Industry spokesmen
have estimated that, over the next ten years, world shipbuilding
capacity will be twice world demand. This means that, in the
excessive competition that can be expected, excessive amounts
of money in terms of the social or economic return will be
necessary to improve the UK competitive position. It also
means that success in Britain will be at the expense of workers elsewhere. Secondly, the Barrow shipyard is not the most
appropriate for initiating such a competitive thrust. There
are very expensive overheads to be borne and relatively large
numbers of skilled labourers to be employed. Thus a naval
shipyard typically employs twice as many salaried staff as a
merchant shipyard, and a ship like the ASW Cruiser involves
twice as many electricians and 50 per cent more boilermakers
than a passenger liner. Furthermore, test equipment is
over-sophisticated, while craneage and steel handling facilities
are inadequate because of lower weights of structural steel
in warships.

It has been suggested that naval technology might have
useful application to merchant shipbuilding in the future.
Such applications include nuclear propulsion, gas-turbine
propulsion, or the use of lighter steel in merchant ships.
It is likely that these ideas will prove to be too complex
and expensive for commercial success. To take one example,

the main advantage of gas-turbine propulsion is the ability to leave port in twenty minutes and to reach high speeds at sea rapidly.   Few shipowners are likely to want these advantages in exchange for high cost, high fuel consumption and lack of tested reliability.   Such problems are inherent to military technology.   It is a mistake to confuse 'high' technology, which is costly and complex, like Concorde, with technology which is capable of serving socially useful ends and is likely, therefore, to be relatively cheap and simple.

Alternative land-based manufacturing activities: There is a large number of alternative technologies for which the skills and facilities available at Barrow are suitable. Indeed, both the shipbuilding and marine engineering group have manufactured a wide range of products in the past. These include:

Cement kilns (using submarine technology)
Pumping plant and pipeline system for the Sadovia Corabia
 irrigation scheme in Romania
Sulzer diesel engines for British Rail during the change-
 over from steam to diesel engines
Sugar-beet crushers
Commercial boilers for power stations
Cable laying
Machinery for North Sea oil projects

Elsewhere, shipyards have been engaged in various kinds of construction activities, including watergates, locks, bridges, dams and even large buildings, structural steel-work, industrial machinery, including construction and mining equipment, metal-working machines and material handling equipment, and various kinds of transportation equipment.   In some cases, as indeed for Vickers itself, these represented the post-war conversion routes.   In others, they were merely means of filling excess capacity in the interim between naval or merchant-ship orders.
    A study undertaken for the US Government identified fifty-five industries (2) suitable for conversion in the event of a one third cut in defence spending.   The industries were chosen on the basis of two criteria: that they utilized similar skills and facilities, and that they would be relatively unaffected by defence cuts.   It was found that, over a three-year period, one third of total naval building capacity could be converted by capturing 10 per cent of the annual growth of the market for these industries.   Unfortunately, it has not yet been possible to calculate the growth of these industries in Britain.   Although it is clear that, in the mid 1970s in Britain, the growth potential will be much smaller than in the United States in the mid 1960s, the American result indicates the scope of the conversion potential in this area.

In addition to existing industries, there is a whole range
of new technologies being discussed that could prove suitable
for Barrow. _Ad hoc_ suggestions include:

Heat pumps: these are refrigerators in reverse, which draw in
   heat at low temperatures from water, soil or air and deliver
   it at a usefully high temperature. Particularly if it was
   combined with a solar collector, this could represent a
   relatively cheap and reliable method of home heating
Solar panels
Containers
Prefabricated houses
Tanks for fish farming
Decompression chambers for hospitals
Pre-design bridges for disaster relief, etc.
Skips for cement
Recycling technologies, e.g. crushers, domestic refuse
   collection such as large drums for the production of high-
   quality compost
Fluidized bed boilers using pulverized coal for industry.
   These are small, portable and pollution-free.
Heat exchangers: these are assemblies of pipes placed by
   boilers to collect waste heat
New kinds of energy-saving capital equipment, such as
   continuous casting for steel
Hover trucks.

   The main difficulty with these schemes is the existence of
other factories or plants more appropriate for their development Given the current unemployment in engineering
generally, and the need to find conversion opportunities for
the vehicles, ordnance and small-arms sectors, it would seem
less appropriate to invest in Barrow where access is poor and
marine experience would be wasted.

New sea-based technologies: it is widely considered that
future developments in such fields as agriculture, mining,
energy and transportation will be based on exploitation of
the sea and the sea bed. As an island with considerable
experience in sea-based technologies, Britain is in an
ideal position to participate in these developments. Indeed,
the greatest asset of our heavy naval building programme is
the preponderance of naval designers. There are 3,200
people employed by the Ministry of Defence on R & D in
warship construction. They cost £54 million, ten times
as much as is spent on R & D into merchant shipbuilding.
Although the technology developed by them in the past may
have little immediate application, they represent a powerful
creative force for new technologies in the future.
   Possible ideas that might be developed at the Barrow shipyard are as follows:

Wave power: a wave-power generator has been designed by

Stephen Salter of Edinburgh University. Known as the
'Nodding Duck', the generators are designed on exactly
opposite principles from naval architecture, in order to
rock as much as possible with the waves. Each unit weighs
50,000 tons and is comprised of fifty smaller units,
largely made from concrete, with hydraulics and electrics
inside. Anchored across 300 miles of sea around the Hebrides,
100 units could provide sufficient power for the entire
United Kingdom. Each unit would cost £10 million, would
be labour-intensive to produce, employing very similar skills
to shipbuilding. The test and design programme is ahead of
schedule, and Dr Salter estimates that the generators will be
ready to enter production in two years' time. The main
problem is that of transmission to the main energy-using
centres of Britain. Dr Salter and his colleagues have
developed a scheme which, according to their estimates, will
bring the cost of wave power to 1 to 1.2p per kilowatt
hours, currently in the middle range of energy costs. Wave
power is based on energy income rather than energy stocks
and is therefore indefinite. It is also pollution free.
     The main obstacle to wave power, as with the other energy
projects listed below, is current energy requirements and
policy. There is considerable over-capacity in electricity
generation and this is estimated to reach 38 per cent by
1980. Furthermore, current pressures to preserve a coal
industry, as well as from the oil and nuclear energy lobbies,
put wave power, together with tidal power, at the bottom of
the list of current energy priorities.
     Tidal power: the main proposal for tidal power currently
under consideration is the Severn Barrage. It is estimated
that the barrage could provide one fifth of the energy
consumed in the UK. This site has been chosen largely
because of the existence of a local pressure group and an
enthusiastic local MP. In fact, because of its effects on
Bristol port and on holiday beaches, the Severn Estuary
might not prove to be the most suitable site. Another
suggestion is the Morecambe Barrage, which would, of course,
be ideal for Barrow. It would involve very suitable large-
scale construction technology.
     Ocean thermal gradients: this would involve large
condensers and evaporators designed to tap the temperature
difference between deep and surface waters in the tropical
seas. There are enormous transmission problems and the
scheme would not be suitable for Britain. However, it
might be possible for the Barrow shipyard to participate in
schemes currently being proposed in the United States.
     Submersibles for firefighting on oil rigs, nodule
collection on the sea bed, deep-sea mining and marine
agriculture. Clearly, these are ideal for Barrow, with its
long experience in submarine manufacture. Vickers Offshore
Engineering, which has recently been detached from the shipbuilding group, is at the forefront of this technology with
its fleet of five submersibles.

Ocean-going tub-barge system: barges that could be used at sea as well as on Britain's neglected canal system could be of particular use to Barrow in improving its communications with the rest of the UK. More generally, this kind of system - a sort of water-based lorry - could greatly ease cargo handling facilities and road and rail traffic. A related suggestion is a container barge.

River power: low-head, low-pressure turbines could be used to tap stream flows and deliver just a few kilowatts. Such a scheme, which would be suitable for development by the marine engineering group, would have enormous scope for villages in the Third World.

Other suggestions for new sea-based technologies include various types of deep-sea mining and farming equipment, modules covering the superstructure of drilling rigs, catamaran container ships, sea-skimmers for dealing with oil pollution, etc.

These projects will probably yield at least as much foreign exchange as our current expenditure on armaments. Currently, we export around a quarter of our total arms production, which is rather low compared with manufacturing generally, where the share of exports is one third. It seems unlikely that these new technologies, which have in any case an immediate social benefit, should generate directly or indirectly, through the general effect on the British economy, fewer exports than the average manufacturing activity.

The Political Problem of Conversion: Vickers Ltd

In a sense, the history of Vickers is a history of attempts at conversion. Vickers, originally a steel company, became a manufacturer of armaments and armour plate in the 1880s, and from there diversified into shipbuilding, marine engineering, steel and other component supplies. After each war, Vickers Ltd has attempted to adjust to peacetime conditions by expanding into such diverse areas as power presses, medical engineering, optical instruments, lithographic plates and supplies, bottling machinery, etc. It is only since the mid 1960s, when the steel business was nationalized and the company bought printing machinery and office equipment, that these efforts at diversification were successful.

The period coincided with increased overseas expansion by the company. Over the last ten years, overseas sales have increased much faster than exports. In 1965, overseas sales were roughly twice the size of exports. In 1974, they exceeded exports by a factor of three. In particular, the company acquired a number of subsidiaries in Europe for the manufacture of office equipment (the Roneo Vickers group), and chemical engineering (the Howson-Algraphy group). In 1974, British acquisitions by the company amounted to £790,000, while foreign acquisitions in Australia and Sweden amounted to £2,310,000.

This policy will be continued after the nationalization of

the shipbuilding group. The armaments and naval shipbuilding division, together with the overseas subsidiaries, are the only sections of Vickers to show continued growth and profitability The prospects for British civilian establishments are considered poor. Leeds Water Lane factory, which manufactures newspaper printing machinery, is threatened with closure. The engineering factories at Otley, Scotswood, Elswick and South Marston are all on short time of one kind or another. Other factories, in engineering and lithographic plate, are threatened with short time.

The Chairman of Vickers, Lord Robens, has already implied his intention to spend the compensation monies abroad. In his Annual Report he wrote:

> A large part of Vickers' activities in the United Kingdom will ... continue without interruption, and in addition to the Engineering Group, Howson-Algraphy Group, Offshore Engineering and Vickers Instruments, we have extensive and successful operations in Australia and Canada. It will be the Board's first priority in use of compensation monies to add to the strength of these activities, whether by internal investment or by acquisition.

One British division which might benefit is the Offshore Engineering Group which, presumably to avoid nationalization, has recently been detached from the Shipbuilding Group, where it resided, according to the Chairman, 'for reasons of administrative convenience, though not engaged in shipbuilding'. (3)

Diversification through overseas expansion of Vickers makes sense on the criterion of private profit. It is, of course, a self-reinforcing policy since it contributes to the low level of investment in Britain and further limits the prospects for domestic diversification. In addition, the direction of such investment as does occur in Britain is dependent on the overall structure of British industry and reinforces current industrial trends. Thus the success of the Offshore Engineering Group reflects the fact that the oil industry is one of the few growth areas in Britain and, at the same time, creates a vested interest in further growth.

Nearly all the alternative projects described above are dependent on a high level of investment in Britain and a reversal of current industrial trends. The energy projects depend on an increased demand for electricity and a reversal of the current emphasis on oil and nuclear power. The transportation projects depend on the level of external and internal trade and a reversal of the current emphasis on road transportation, itself a reflection of the power of motor-car manufacturers and road constructors. The projects involving mining, marine agriculture and various kinds of capital equipment depend directly on the levels of investment and income and may also involve less emphasis on oil technology.

405

Recognition that the Government must enter industry as an investor is implicit in the nationalization of shipbuilding. But, by itself, it is insufficient to ensure a successful programme of conversion. We have seen in the past how nationalization has been used as a tool for the orderly contraction of declining industries. This is made possible by the preservation of existing organizational structures where local employees have little opportunity to make their views felt and where the direction of central government policy remains largely unchanged. Especially where the growing sectors of the industry, e.g. offshore engineering, are excluded from the nationalization, this kind of approach can be justified by the lack of alternative investment opportunities.

Conclusions

The following recommendations emerge from this survey of conversion potential of Barrow shipyard.

1. While alternative technologies could be enumerated, the most attractive opportunities lie in the development of new sea-based technologies, of which the best prospects are wave-power, submersibles and other equipment for nodule collection, mining and agriculture on the sea bed, and ocean tug-barge systems. There is no reason to suppose that the export potential of these projects would be less than that of armaments.

2. If these opportunities are to be exploited, the Government must:

(a) Enter the industry as a direct investor. Development contracts for new technologies are an essential and immediate alternative to defence spending or unemployment.
(b) Coordinate investment in these areas with other related policies such as energy, transport, health, agriculture, etc.

3. The Government can only act as an investor and reverse the direction of existing policies if current organizational structures are changed. This involves:

(a) Nationalization of some sectors of the defence industry, including the profitable civilian sectors, such as offshore engineering.
(b) Workers' participation in the nationalized industries, in order to generate ideas for alternative products and shift the central direction of government policymaking; i.e. wave-power is unlikely to be adopted by a government committed to oil and nuclear power. (An additional consideration is safety. For example, the need to improve the safety of submersibles and to reduce the problem of

fuel extraction in shipbuilding might receive greater
emphasis.)
(c) Planning agreements with all private armaments companies,
in order to ensure that capacity freed from armaments
production and that profits from armaments (which are
currently high because of the practice of cost-plus
contracting) and compensation monies from nationalization
are invested in suitable projects in Britain rather than
abroad.

Notes and References

1. This paper was written in March 1976 and has not been
updated.

2. British Shipbuilding 1972. A report to the Department of
Trade and Industry by Booz-Allen and Hamilton International
BV. HMSO, 1973.

3. Final Report on Industrial Conversion Potential in the
Shipbuilding Industry. For US Arms Control and Disarmament
Agency, Mid-West Research Institute Contract No.ACDA/E-66.
MRI Project 2833-D. 18 March 1966.

4. Vickers Limited Annual Report and Accounts 1974.

(The author is grateful to the shop-stewards of Barrow and the
Vickers Shop-stewards Combine Committee who provided the basic
material for this paper.)

THE SHIPBUILDING PROCESS

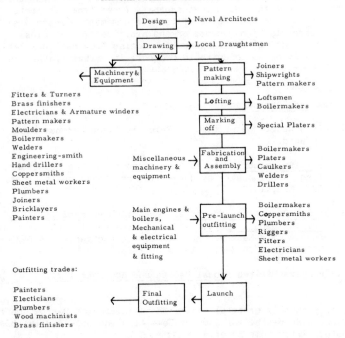

Design → Naval Architects

Drawing → Local Draughtsmen

Machinery & Equipment

Pattern making → Joiners / Shipwrights / Pattern makers

Fitters & Turners
Brass finishers
Electricians & Armature winders
Pattern makers
Moulders
Boilermakers
Welders
Engineering-smith
Hand drillers
Coppersmiths
Sheet metal workers
Plumbers
Joiners
Bricklayers
Painters

Lofting → Loftsmen / Boilermakers

Marking off → Special Platers

Miscellaneous machinery & equipment → Fabrication and Assembly → Boilermakers / Platers / Caulkers / Welders / Drillers

Main engines & boilers, Mechanical & electrical equipment & fitting → Pre-launch outfitting → Boilermakers / Coppersmiths / Plumbers / Riggers / Fitters / Electricians / Sheet metal workers

Outfitting trades:

Painters
Electicians
Plumbers
Wood machinists
Brass finishers

Final Outfitting ← Launch

Skilled Trades, March 1970

BARROW SHIPBUILDING WORKS

	Journeymen	Apprentices	Percentage of Total
Anglesmith	1	–	.03
Caulkers	169	10	6
Coppersmiths	78	31	4
Drillers	57	3	2
Electricians	297	83	13
Fitters & Turners	480	95	20
Brass Polishers	6	–	.2
Joiners	210	32	8
Loftsmen	51	17	2
Painters	133	14	5
Platers	151	35	6
Plumbers	194	–	.2
Riggers & Sailmakers	8	–	.03
Riveter	1		
Sheet iron workers	148	14	6
Shipsmiths	6	–	.2
Shipwrights	185	28	7
Welders	296	67	12
Woodcutting machinists	18	2	.7
	2489	431	100 *

* Figures do not add up to
total due to rounding.

Skilled Trades, March 1974

BARROW ENGINEERING WORKS

	Journeymen	Apprentices	Percentage of total
Fitters & Turners	631	110	50
Brassfinishers	20	-	1
Electricians) Armature Winders)	106	46	10
Patternmakers	10	2	1
Moulders	18	-	1
Boilermakers	117	29	10
Welders	138	28	11
Engineering Smiths	6	2	1
Hand drillers	20	4	2
Coppersmiths	28	15	3
Sheet metal workers	32	12	3
Plumbers	15	13	2
Joiners	39	-	3
Bricklayers	3	-	.2
Painters	36	-	2
	1219	261	100 *

* Figures do not add up to total due to rounding.

INDUSTRIES SUITABLE FOR SHIPBUILDING CONVERSION*

Fabricated Structural Metals

3441 Fabricated structural metals
3442 Metal doors, sash, frames, et.
3443 Fabricated plate work
3444 Sheet metal work
3446 Architectural metal work
3449 Miscellaneous metal work

Railroad Equipment

3741 Locomotives
3742 Railroad cars

Construction & Mining Machinery

3531 Construction machinery & equipment
3532 Mining machinery & equipment
3533 Oil field machinery & equipment

Special Industrial Machinery

3551 Food products machinery
3552 Textile machinery
3553 Woodworking machinery
3554 Paper industries machinery
3555 Printing trades machinery
3559 Other special industry machinery

Material Handling Equipment

3534 Elevators
3535 Conveyors
3536 Hoists, cranes, monorails
3537 Industrial trucks, etc.

Trailers & Miscellaneous Transportation Equipment

3715 Truck trailers
3791 Trailer coaches
3799 Miscellaneous transportation

General Industrial Machinery & Equipment

3561 Pumps, compressors & equipment
3564 Blowers & fans
3565 Industrial patterns
3566 Mechanical power transmission equipment
3567 Industrial process furnaces & ovens
3569 Other general industrial machinery

Service Industry Machines

3581 Automatic merchandising machines
3582 Commercial laundry machines
3585 Refrigerators and air conditioners
3589 Other service industry machines

*Numbers refer to SITC classification.

Farm Machinery

3522 Farm machinery and equipment

Metalworking Machinery

3541 Metal cutting machinery
3542 Metal forming machinery
3544 Special dies, tools, jigs
3545 Machine tool accessories
3548 Miscellaneous metalworking machinery

Furniture & Fixtures

2531 Public building furniture
2541 Wood office furniture
2542 Metal office furniture

Others

3599 Machine shops, jobbing and repair
3391 Iron and steel forgings
3479 Coatings and engravings

3511 Turbines and steam engines
3519 Internal combustion engines
3611 Electrical measuring intruments

3433 Nonelectrical heating equipment
3494 Steel springs
3499 Other fabricated metal products
3451 Screw machine products
3452 Bolts, nuts screws

Vickers Limited

ESTABLISHMENTS, PRINCIPAL DIVISION, SUBSIDIARY AND
ASSOCIATED COMPANIES

Engineering Group

Crabtree-Vickers Division, Leeds, Gateshead, Otley & London
Crayford Works, Crayford.
   Kirby's (Engineers) Limited, Walsall.
Vickers-Dawson Division, Crayford, Gomersal, nr. Leeds
and Thetford.
   Vickers-Vandergeeten S.A. (99.7%) Brussels.
Elswick Works, Newcastle-upon-Tyne.
   Armament Division
   Foundry Division
   Non-Ferrous Metals Division
   Pressing Division
Medical Engineering, Basingstoke.
Michell Bearings, Newcastle-upon-Tyne.
Scotswood Works, Newcastle-upon-Tyne.
South Marston Works, Swindon.
   Automated Systems Division
      Four Plus Limited, Oldham.
   Design and Procurement Division
   Hydraulics Division
   Nuclear Engineering Division

Shipbuilding Group

Vickers Shipbuilding Group, London.
Barrow Shipbuilding and Engineering Works
   Compact Oribital Gear Works, Rhayader.
   Ship Model Experiment Tanks, St. Albans and Dumbarton.

Roneo Vickers Office Equipment Group

Roneo Vickers Limited, Croydon, Dartford, Crayford, Hemel
Hempstead and Romford.
   Fanfold Limited, London (with effect from 29 March 1974)
   Hirst Buckley Limited, Scissett.
   Langdon Precision Engineers Limited, London.
   Roneo-Neopost Limited, London.
   Roneo-Vickers Partitions Limited, Liverpool.
   Antonio Corona S.p.A (70%) Como, Italy.
   Compagnie du Roneo, S.A. (98.41%) Paris, Les Lilas & Noyon.
      Columbia S.A., Paris and Issy-les-Moulineaux.
   S.A. Compagnie Roneo, Brussels.
   Roneo (Suisse) S.A. Geneva.
   Societe d'Etude et de Fabrication d'Articles Modernes
   d'Organisation, Paris.
   Societe SEFAMO Bureoeinrichtungen G.m.b.H. Frankfurt.
   Societe Immobilier de la Sente Giraud, Paris.
   Societe SPES, Paris.
   Societe Iviorienne de Distribution et d'Equipement
   de Bureaux, (49.85%) Abidjan, Ivory Coast.
   Societe Europeenne de Realisation d'Organisation et
   de Participation & Perth.

Ertma S.A. Geneva.
"Hadewe" B.V. Drachten, Holland.
Roneo Vickers Canada Limited, Toronto.
Roneo Vickers Holland N.V., Rotterdam.
Roneo Vickers India Limited (85.71%), New Delhi.
Roneo Vickers Norge A/S, Oslo.
Roneo Vickers Svenska A.B. Stockholm.
  Berg Bolinder A.B. (95.33%) Stockholm (with effect
  28 May 1974).
Roneo Vickers Limited, Pakistan, Karachi.

Howson-Algrphy Group

Leeds, Thetford, Orpington and Margate.
Howson-Algraphy Limited.
  Howson-Algraphy A.B. Stockholm.
  Howson-Algraphy B.V. Soest, Holland.
    Howeson-Algraphy (Europe) B.V. Soest, Holland.
  Howson-Algraphy (France) S.A. Paris.
  Howson-Algraphy S.A. (75.09%) Barcelona.
  Howson-Algraphy S.A.R.L. (76%) Lebanon.
  Howson-Algraphy S.p.A. Milan.

Offshore Engineering

Vickers Oceanics Limited (63.16%) Barrow in Furness, London
& Leith.
Brown Brothers & Company Limited, Edinburgh.
  John Greig & Sons Limited, Edinburgh.
Slingsby Sailplanes, Kirkbymoorside.

Other U.K.

Hebburn Works, Hebburn.
Vickers Instruments, York.
Vickers (Insurance) Limited, London.
Vickers Limited Leasing Division, London.
Vickers Properties Limited, London.
  Inter-Vickers Limited, (50%) Scotswood.

Australia

Vickers Holdings Pty. Limited, Melbourne.
    Vickers Australia Limited (61.62%) Melbourne.
    Vicker-Hadwa Division, Perth.
    Vickers Hoskins Division, Perth.
      Vickers Keogh Pty. Limited (80%), Kalgoorlie.
      W.A. Mining Engineering Services Pty. Limited
      (50%) Perth.
    Vickers Ruwolt Division, Richmond, Moorabbin,
    Ipswich, Waterloo and Barry Beach.
      Vickers Research Pty. Limited, Richmond.
  Vickers Cockatoo Dockyard Pty. Limited, Sydney.
  Howson-Algraphy Pty. Limited, Melbourne, Sydney,
  Adelaide, Brisbane & Perth.
  Crabtree-Vickers Pty. Limited, Melbourne, Sydney,
  Adelaide, Brisbane & Perth.
  Middows Brothers Pty. Limited, Sydney and Melbourne.
  Roneo Vickers Pty. Limited, Melbourne, Sydney, Adelaide,
  Brisbane & Perth.

Canada

Canadian Vickers Limited (72.04%) Montreal.
Industrial Division
Marine Division
Montreal Ship Repairs Limited.
Crabtree-Vickers (Canada) Limited, Toronto.

India

Vickers India Private Limited, New Delhi.

South Africa

Vickers Southern Africa (Pty.) Limited, Johannesburg.
Cooke, Troughton & Simms (Pty.) Limited, Cape Town.
Pinalex Investments (Pty.) Limited, Boksburg, nr.
Johannesburg.
Roneo Vickers (Manufacturing) (Pty.) Limited, Boksburg,
nr. Johannesburg.
Roneo Vickers South Africa (Pty) Limited, Johannesburg.
Vickers (Engineering) (Pty) Limited, Johannesburg.
Vickers Instruments (Pty) Limited, Johannesburg.

United States of America

Vickers America, Incorporated, New Jersey.
Medi-Computer Corporation, Connecticut and New Jersey.
Vickers Medical Products Corporation, Delaware.
Crabtree-Vickers Incorporated, New Jersey.
Howson-Algraphy Incorporated, New Jersey.
Vickers Instruments Incorporated, Woburn, Mass.

Associated Companies

Barber-Greene England, Limited (49%) Bury St. Edmunds.
British Aircraft Corporation (Holdings) Limited (50%) London.
International Research & Development Company Limited (50%)
Newcastle-upon-Tyne.
Oyster Lane Properties (Holdings) Limited (50%) London.

Note: Except where otherwise stated the companies names are
wholly owned.

14 TORNADO : CANCELLATION, CONVERSION AND DIVERSIFICATION IN THE AEROSPACE INDUSTRY

by Dan Smith

Part 1:   Background to the Project

MRCA, the aircraft known as Tornado, is a collaborative
project involving Britain, Italy and the Federal Republic of
Germany (FRG).   When MRCA is finally produced it will make
up about half the RAF's combat capability in the 1980s;   it
is intended to replace Canberra, Buccaneer, Vulcan and,
eventually, Phantom.   It is the major military aerospace
project for Britain at the present time.

Organization

In July, 1968 Belgium, Britain, Canada, the FRG, Italy and
the Netherlands initialled a Memorandum of Understanding
declaring their interest in the development of a multi-role
variable geometry aircraft.   Within a year, Belgium,
Canada and the Netherlands had withdrawn from the project,
apparently because of the expense and complications of the
project, both technically and organizationally.

   In March 1969, Panavia was formed.   It is a consortium of
BAC, MBB and Fiat;   BAC and MBB each now have a 42.5 per cent
share in Panavia, Fiat has 15 per cent.   Panavia is subordinate
to the NATO MRCA Management Organization (NAMMO), which over-
sees the whole project.   NAMMO's instructions are carried out
by the NATO MRCA Management Agency (NAMMA), which functions as
NAMMO's executive body and also pays the sub-contractors.
Panavia and NAMMA each have a staff of about 140.

   NAMMO and NAMMA function on behalf of the three governments;
Panavia is a German company, subservient to German company law,
with an international board of six directors appointed from
member companies. (1)

   The engine - the RB199 - is being produced by Turbo-Union,
a consortium of Rolls-Royce (40 per cent share), MTU (40 per
cent), and Aeritalia (20 per cent).

   Panavia is responsible for the airframe, and for coordinating the activities of Turbo-Union and Avionica (the company
responsible for the avionics).

   Contracts for work have been carefully divided into systems
and sub-systems, and in some cases sub-sub-systems, and allotted
on a basis reflecting the interest of each government in the
project   In the airframe, BAC is responsible for the nose and
rear fuselage, MBB for the centre fuselage including the wing
joints (despite its lack of previous experience in this the
most sensitive part of a swing-wing aircraft), and Fiat for
the wings.   Responsibility for the engine has been divided up
in a similar fashion.   Each contractor sub-contracts out;
many of the sub-contractors are involved in work-sharing
arrangements.

   Each country will have its own production line and assemble
aircraft from the components provided by the various sub-
contractors.

   British, German and Italian firms are participating in the
development of each and every system in the aircraft. (2)

The British will be developing an Interceptor variant alone
(see below).

Timetables

Project definition was completed in 1970 and development
begun.   The intention was that the first prototype should fly
in September 1973; the nine pre-production models would all
have flown by September 1975, and the aircraft was expected
to enter service around the end of 1977. (3)
By May 1972 the delay on sending out contracts (because of
the large number of permutations possible) had delayed the
programme six months and the service entry date was accordingly
deferred into early 1978. (4) In September 1972, the maiden
flight was put back three months to December 1973.   There
were several more delays before the maiden flight finally
took place on 15 August 1974, nearly a year behind schedule.
In October 1974, after a review of the project, the
British MoD announced approval for the project through to
December 1975, when production orders would be placed.   It
is unlikely that the pre-production models will fly for some
time.
When the White Paper on the Defence Estimates was published
in March 1975, the Interceptor variant was in the project
definition stage. (5)

Numbers

In 1970, Mr B.O. Heath, the MRCA project manager at BAC,
believed that total orders would lead to a likely production
figure of 'over 1,000 aeroplanes, nearly 1,200'. (6)
In 1972, after various other estimates had been eliminated,
the generally agreed figure was a total of around 900 — a
maximum of 420 for FRG, 350 to 400 for Britain, and around 100
for Italy. (7) In August 1972, the FRG reduced its order to
320 because of rising costs; (8) the order now stands at
320 for the FRG, 385 for Britain and 100 for Italy.
The RAF's 385 is made up of about 165 of the Interceptor
variant, and 220 of the so-called 'common MRCA'. (9)

Roles and Characteristics

MRCA will be required to fulfil five roles for three air
forces:
(a)   Strike and interdiction (including naval strike) — for
      RAF and Luftwaffe.
(b)   Close air support — for all three.
(c)   Reconnaissance — for all three.
(d)   Air defence interception — for RAF.
(e)   Air superiority — for Luftwaffe and IAF. (10)
To perform all these roles, the aircraft requires seventeen
basic characteristics (see annex to this paper, p.436),
plus a flexible armaments-carrying ability.

The Interceptor is regarded as a separate aircraft from the 'common MRCA' which will perform the other roles; however, it will have the same basic airframe and the same engine. Aerodynamic modifications, and the removal of certain avionic systems to reduce weight, are the major changes envisaged for the Interceptor. (11)

The requirements of the different versions are in some cases conflicting. For example, the Interceptor needs an engine capable of high thrust, while the others need long range/duration, i.e. low fuel consumption. The short take-off and landing (STOL) necessary for the 'common MRCA' calls for an unswept wing, while low-level supersonic speed needs a swept wing; STOL needs a low wing loading while low-level speed needs high wing loading, but maximum manoeuvrability calls for a moderate wing loading. (12)

Different versions of MRCA will have abilities they do not need - 'wasted characteristics'. For example, the RAF will have a strike aircraft capable of high manoeuvrability and cab-rank loitering; it will have a close support bomber capable of supersonic speeds at low level, and a reconnaissance aircraft with a fast rate of climb

The swing-wing airframe and the turbofan RB199 engine appear to be the way to reconcile the conflicting requirements, but it does seem somewhat profligate to waste characteristics in this way, especially since each one costs money and real resources. From the annex (p.434) it will be noted that there is only one characteristic shared by all versions of MRCA; even within the 'common' version, sharing is not particularly high.

It is the technical sophistication and innovation required to reconcile the conflicts and provide all the characteristics which have resulted in the high costs of the MRCA. In addition, real increases in the unit cost probably stem from this factor. Anything innovative on such a large scale is a risky project, and the increased costs resulting from hitches is multiplied by the size of the bureaucracy controlling the project, the committee method of taking development decisions (itself likely to cause delay), and the inflexible development strategy employed. (13)

Conclusions

(1) The three governments set out on a project to succeed where others had failed, and with a necessity for sophistication and innovation. This, together with the collaborative method, has pushed costs up.

(2) Given the economic circumstances and the limited progress made thus far on the Interceptor variant, there seems every possibility that it will be cancelled - which will increase unit costs on the other versions.

(3) Cancellation of production in toto seems less likely but
     possible.

(4) The problems encountered in this project should inspire
     extreme caution about other collaborative projects.

Annex to Paper: Roles and Characteristics
Characteristics required, together with the roles to which
they attach, are as follows:

All-weather operation: All roles
Long range/duration: All except Interceptor
Transonic low-level speed: Strike/interdiction
Large weapon load ability: Strike/interdiction
High navigation accuracy: Strike: Close support; Reconnais-
sance
Accurate weapon delivery: Strike; Close support
Good subsonic handling: Close support
Recce equipment carriage: Reconnaissance
Rapid data evaluation: Reconnaissance
Good manoeuvrability at low and medium level: Air superiority
Rapid acceleration: Interceptor
High rate of climb: Interceptor
Supersonic cruise: Interceptor
Maximum manoeuvrability: Interceptor
Target acquisition and identification: Interceptor
STOL: Strike; Close Support; reconnaissance
Low speed for 'cab-rank' loitering: Air superiority

In some cases, the characteristic will be acquired for the
aircraft by the addition or removal of certain avionic
systems. In other cases, where the characteristics apply
to the engine or the basic airframe (as in cases of speed,
manoeuvrability and load), various versions will have 'wasted
characteristics'; the engine will be capable of all the
various abilities, and the airframe able to stand up to all the
different types of stresses to which it will be subjected in its
various roles.

Notes and References

1. W.B. Walker, 'The MRCA: a Case Study in European Collabora-
    tion', Research Policy, 2 (1974).

2. As n.1.

3. As n.1; and Flight International, 4 May 1972.

4. As n.3.

5. 1975 Defence Estimates, White Paper, ch.VII, para 12.

6. Quoted in Walker, op.cit.

7. Interavia, April 1972.

8. Aviation Week and Space Technology, 11 September 1972.

9. Flight International, 28 March 1974.

10. See Interavia, April 1973; Flight International, 28
    March 1974; and White Paper, March 1975, ch.VII,
    para 11.

11. Flight International, 28 March 1974.

12. Interavia, April 1973.

13. See Walker, op.cit., for a fuller discussion.

(The author is indebted to the Institute of Strategic Studies
for use of their facilities in the preparation of the paper.)

Part 2: Conversion and Diversification in the Aerospace Industry in the Event of Tornado's Cancellation

The discussion in this paper outlines recent developments in the Tornado saga, and, although cancellation is viewed as extremely unlikely, considers the conversion potential embodied in the resources devoted to that project.

During the discussion, two terms are used which should be understood separately. Conversion refers to the process by which part of our military aerospace capacity would move into a different field of manufacture — a once and for all change; diversification, relating to unconverted capacity, implies that other work could be alternated with aerospace development and production. The conditions under which diversification might be possible are discussed in the paper.

Tornado

Approval for production of the 'common' version (IDS) MRCA Tornado was announced on 5 March 1976; 220 will be produced, and the intention to order 165 copies of the British-only Air Defence Variant (ADV) was confirmed, and full development approved.

Tornado costs remain a matter for controversy. The government quotes £6.34 million as the IDS unit cost, and £7.72 million for the ADV (1977/8 estimate prices), but it is not clear that the cost of R & D is included in either figure; it has also been suggested that the ADV's air-to-air missiles, without which it is hardly a weapon system, are excluded from the figures.

Briefly, the 'life-time' cost per copy for the IDS is calculated at £16.81 million, and for the ADV at £20.22 million; this gives a 'life-time' cost for the whole project and the full order of 385 aircraft in the region of £7,000 million. This estimate is higher than estimates generally made by even the severest critics of Tornado — estimates above about £4,000 million have not normally been seen. Of course, these estimates are in 1975 prices, and actual cash outlay will be vastly higher over the period of fifteen years

On the basis of these estimates, the budget outlay on Tornado for the years 1978/9 to 1980/81 will be somewhere over £250 million for each year. Should ADV production begin in 1980/1, which seems a little early given present progress, then the outlay for the second of those years might be around £300 million, and for the third closer to £350 million These figures are more or less compatible with those suggested by David Greenwood. (1)

Estimating cancellation costs is extremely difficult because of the paucity of data. Nor is it possible to use costs for cancelling previous projects (e.g. TSR-2) as a guide, because work on Tornado is so much further advanced. Tooling-up has begun at the BAC plant in Preston where the aircraft will be assembled, with full production likely to get under way in 1978.   This advanced stage alone makes it virtually impossible that Tornado will be cancelled - or the IDS version, at any rate; it might still be possible to stop the ADV, not only because it is less advanced but because it is a British only project.   As a case-study, the survey of conversion prospects arising from cancellation of Tornado remains useful, but barring something unforeseeable, the RAF will be equipped with the IDS version at least in the 1980s.   Of the eventual outlay on the ADV, only the development investment is irretrievably committed;   ADV cancellation would probably mean 'losing' a little over the total development cost (i.e. somewhere above £225 million), but with important resources saved, although the longer one waits the more expensive cancellation will become.

Recently there has been growing criticism of Tornado in the press and elsewhere, dealing not with the project's basic desirability or suitability, but with its alleged deficiencies in two categories - weaponry and performance.

The former was the subject of a short debate in the House of Commons, raised by Ronald Brown, MP, concerning contract procedure for the main weapons computer, euphemistically known as the 'stores management system'. (2) He complained that Marconi-Elliott Avionics Systems Ltd (MEASL), while prime contractor for the system, was unable to fulfil the specifications, but, when the contract was re-offered for tender, the British and Italian governments reselected MEASL, although the FRG opted for an American company, Base Ten.

Whatever the propriety of MEASL getting a second chance, the failure of its first effort has meant a delay of four years;   in addition, Base Ten and MEASL are reported to be working on two completely different and incompatible systems. If production and delivery schedules are maintained, both contractors have extremely tight schedules for their work. Replying to the debate, the Minister of State placed the blame for the failure not on MEASL, but on the government's insufficiently detailed and over-ambitious specifications.(3) Indeed, given Tornado's small size and multiplicity of roles (resulting in a wide range of tasks for the computer), MEASL's job was like fitting a quart into a pint pot. One might, however, argue that MEASL should have assessed the task more carefully before taking it on.

It is alleged that the aircraft's RB-199 engine is not producing the thrust it should, although the reported scope of the failure depends upon the source.   Air et Cosmos, a French journal, suggests Tornado will fly no faster than Mach 1.3, while Interavia has estimated a maximum speed of Mach 1.7 (4)   Other sources have suggested that the engine's present thrust can produce an airspeed of Mach 1.4, but that continued work should get it up to Mach 1.8.   All these

speeds are below BAC's claim of Mach 2, and below the Mach 2.2
originally advertised for the ADV.

The relatively low speed is not a worry in relation to
Tornado's close support or battlefield interdiction roles,
where the emphasis is on pinpoint accuracy rather than speed.
It is, however, a matter of concern for the long-range
strike, deep interdiction, reconnaissance and interceptor
roles. By way of comparison, the Mig-25 Foxbat, which
recently sneaked into Japan, and which performs as both
interceptor and reconnaissance aircraft, is reportedly
capable of speeds of Mach 2.8. (5)

Should the problems with the engine be ironed out, there
is yet another snag, brought to light by Stephen Thornley,
an aviation scientist sacked for his criticisms of Tornado.
Apparently, the variable air-intake on the fuselage has been
mal-designed, so that whatever thrust the engine produces,
the aircraft will be incapable of speeds above Mach 2, and
probably not even that high.

With Tornado far from completed, it seems that the sequel is
ready to run. Air Staff Target (AST) 403, generally discussed as a replacement for Harrier and Jaguar, may eventually
result in another collaborative project on the scale of Tornado.

To the Expenditure Committee, MoD officials commented that
plans for a new fighter aircraft are 'under consideration';
without being drawn, they implied that the aircraft would be
neither a Tornado derivative, nor the American F-15 Eagle, but
probably a completely new aircraft. (6) It has been suggested
that AST-403 is conceived of primarily as a bomber which could
also fulfil various fighter roles. (7) If this is so, and if
the MoD officials' remarks concerned AST-403, and if it is not to
be bought off the shelf from the USA, then, even at this early
stage, and with such incomplete evidence, there are grounds for
concern. It is the view of General Dynamics, prime contractors for the F-16, that 'it is possible to adapt an air-to-air
fighter for ground attack, but not vice versa'; it could be
that AST-403 is starting back to front. (8)

There must also be severe doubts about the efficiency both
of the multi-role concept, when taken to the lengths it has
been in Tornado, and of the collaborative framework for
development and production. Leaving aside the more profound
political implications of collaborative projects, it does
appear that costs are increased by collaboration. It causes
a proliferation of roles, requiring differing and sometimes
conflicting performance characteristics, which all have to be
fitted in at great cost and with the consequences we have
considered in MEASL's inability to fulfil the specifications;
this causes delay, and extra cost. It also makes the
project extremely hard to keep under effective control — by
many, this last criticism is seen as the most powerful
argument in favour of collaboration, since cancellation
becomes harder. The multi-role concept, adopted because of
continuing and staggering cost increases for advanced

424

weaponry, aims at having one aircraft perform a variety of
tasks, which results in a product unnecessarily sophisticated
for each task, often performing less effectively than a
specialized aircraft would, and still enormously expensive.
Tornado has been sarcastically but relevantly described as
'the egg-laying, wool-producing, milk-giving sow' - it may be
nice to have sows laying eggs, but hens do the same thing more
easily and cheaply. (9) And it is no longer certain that all
the roles that aircraft now fulfil are relevant, desirable or
necessary or possible in the 1980s and beyond; there are
strong reasons for thinking that in a rapidly changing strategic environment aircraft can and should have a more limited
range of tasks in the future. (10)

Before the decision is taken to embark upon a new multi-
role collaborative project which will again develop almost
unstoppable momentum, there should be a much closer and more
critical examination of the experience and issues involved in
Tornado. Its one potential benefit is as a warning against
such gross waste of resources in the future. Government
assertions that Tornado is a wonderful aircraft, and that
collaboration works beautifully, just will not do.

Tornado Contractors

This long list of firms involved in Tornado work causes its own
problems for a consideration of alternative work. It becomes
next to impossible to quantify the workforce involved in any
detail. The Government either does not have the data or is
unwilling to provide them. An answer to one Parliamentary
Question could provide only a limited list of contractors, while
another could not provide the division of skills in the workforce, the regions in which they are employed, or the numbers
presently employed, apart from 5,500 at BAC and 4,000 at Rolls-
Royce. (11)

It is thus impossible to define with any precision the
details of the conversion and diversification task consequent
upon cancellation of Tornado. Accordingly, our discussion
deals in terms of possibilities and suggestions, rather than a
blueprint. Since the Corporate Plan of the Lucas Aerospace
Shop Stewards Combine Committee provides such an exceptional
range of alternatives for the aerospace equipment industry, the
discussion is limited to airframe and aeroengine - to BAC and
Rolls-Royce.

Conversion and Diversification and the Aerospace Industry

British aerospace, along with the rest of West European aerospace, is in a profound crisis. Since 1968, BAC's civil
aviation division has announced mass redundancies on no less
than seven occasions. In civil aviation in particular,
European firms suffer blow after blow at the hands of their
American rivals who have tremendous advantages through economy

425

of scale and the size of their domestic market (even though US
aerospace is far from trouble-free). The EEC Action Plan for
European Aerospace asserted that without decisive collective
action by European aerospace firms and governments, European
aerospace would wither away to nothing. (12) But although the
distinctly federalist tinge of the proposals might have the
profoundest implications for government relations, its effects
on civil aerospace at least might be little more than cosmetic,
which may explain the delay in implementing the plan's
proposals, and the apparent low interest in projects such as
the European Airbus. (13)

Unfortunately for the industry, no long-term solution to
its problems can be found in military aerospace manufacture,
which would, anyway, leave civil aerospace workers redundant.
The problems of the industry are not so much to do with bad
management or selling techniques, as with the difficulties
created by expanding capacity and escalating costs, in a
strategic environment in which their military wares' roles
are increasingly threatened and may not be viable at costs
which are anything like acceptable.

Even if Tornado survives, conversion of military aerospace
capacity remains important - for the workers in the industry,
for allocation of resources, and for the development of more
coherent defence policy better tailored to our economic
capacity.

However, a programme of conversion should not merely
consider alternative development production, but also the
extent to which parts of the industry should be maintained
in their present form, or even perhaps expanded, for two main
reasons.

There will certainly be a continuing requirement for military
aircraft to perform certain missions, and the appropriate
development and production capacity would therefore need to be
maintained (barring a decision to purchase all military aircraft
from abroad). Additionally, it should be noted that Rolls-
Royce maintains a dominating position in the EEC aeroengine
industry - it is larger than all its EEC rivals put together.
While this is not an argument for excluding Rolls-Royce from
conversion planning, it would clearly be foolish to discard
an industry in which Britain has such a strong advantage.

What is needed is therefore not only conversion of capacity
superfluous to our requirements, but also a measure of diversification to provide the retained industry with greater stability
and security, with the added advantage of potentially reducing the
cost of maintaining development and production capacity for
military aerospace, in periods when there is no major procurement under way. This possibility is discussed in more detail
below.

Despite spirited opposition from the House of Lords during
the debate on airframe nationalization, the new nationalized
airframe corporation does have the statutory possibility of
diversifying manufacture, if the order by the Secretary of
State is approved by both Houses of Parliaments. (14) Given

the Lords' opposition to this during their attempts to mutilate
the Bill, one may fear the corporation would not be permitted
to diversify; however, the motive behind the Lords' amendment
to this part of the Bill was apparently the fear of the Society
of British Aerospace Companies that the nationalized industry
would compete with or absorb ancillary manufacturers, and the
most desirable diversification would not be into that area,
but into completely new product fields. (15) The Act also
contains clauses obliging the corporation 'to promote
industrial democracy in a strong and organic form', which
could be most important in helping to ensure wide involvement in the planning process which the Study Group has
recognized as so important. (16)

Conversion and Diversification at BAC and Rolls-Royce

Cancellation of Tornado now would release in excess of 5,500
workers at BAC and 4,000 at Rolls-Royce, plus plant space and
machinery, for alternative work. (17) It would also affect
the job prospects of a further 4,500 at BAC, and 2,000 at
Rolls-Royce, who can expect to be employed on Tornado when
work is at its peak. (18) If associated development capacity
were also converted or diversified, the number affected would
be much greater.

We can be fairly sure that the alternative products discussed below are far from constituting a complete list of
conversion and diversification options for these workers.
Although the bulk of the suggestions come from the workforce,
at neither BAC nor Rolls-Royce has there been the kind of
systematic search for alternatives which was undertaken at
Lucas Aerospace. While the BAC Combine Committee has considered
the question of diversification, largely as a fall-back
should redundancies loom, it has not done so in detail; in
Rolls-Royce, suggestions are circulating for discussion, but
the process has gone no further. We can be reasonably sure
that a much larger range of alternatives would be produced were
the task to be taken on systematically.

However, there are in BAC suggestions that a strategy be
adopted, similar to that which underlies the Lucas Aerospace
Corporate Plan, to develop Job Protection Agreements rather
than fighting each redundancy as it emerges. This would be
based on a five- to ten-year plan, with full information
available about investment intentions and possibilities, aimed
at providing alternative production during dips in military
aerospace production (as between Jaguar and Tornado). Thus,
production would be alternated between one type of manufacture
and another. This plan, possible in a nationalized industry,
but only with effective industrial democracy and advance
planning, would aim at providing security where now there is
only the fear of redundancy, whenever production of an aircraft is wound down.

However, it would not be possible to alternate production
in this way if it were suggested that the entire plant be

switched away from military aerospace, then back to it, then away, and so on. As these ideas are developing, therefore, they envisage the maintenance of a skilled 'pool' of labour constantly involved with whatever types of production were undertaken. Production lines would need to 'tick over' whenever they were not operating at full pressure; this already happens at BAC (on Canberra and Jaguar, for example) so it is nothing new. The weight of the workforce could then be directed wherever necessary, but the basic capability would not be lost. To some extent, this does already happen at BAC: teams have, for example, worked on helicopters or submarines for periods instead of on fixed-wing aircraft.

While this kind of alternating production would provide greater security and stability, and while the concept has potential not only for BAC but also for Rolls-Royce, there are two particularly important factors which limit the scope of the proposal.

Non-military aerospace production would have to be in fields employing closely related skills or the 'learning' period at the start of new aircraft production would be lengthened and costs accordingly increased; at the extreme, the skills might have to be totally relearned unless the other types of work were in compatible fields.   While it is not particularly difficult to identify work with the appropriate compatible skills, this is a factor limiting the range of options.

More importantly, alternating production in this way would appear to rule out the non-military part of the enterprise being in a field which involves competition on the open market.   In a competitive and rapidly changing market, it is, first, difficult to get a foot in the door; once there, a full effort is necessary to maintain the position.   Therefore, the non-military part of the alternating production should be making things for which the Government is the market, and the fields which offer themselves for consideration are such as energy, health, transport or education.

Should it be decided, as it is reasonable to suppose it will, that a continued military aerospace industrial capacity is required, the Government will either come under pressure to keep that capacity in full business by constantly procuring newer and 'better' aircraft (and this would be so whether or not the industry were nationalized), or it will, more sensibly, find cheaper ways of maintaining that capacity.   It could do this by going against all precedent and deciding not to go for completely new aircraft, concentrating instead on continued marginal improvements on already established types.   This appears to be happening to some extent with the Chieftain tank, and with the French main battle tank, the AMX-30. But the two main weaknesses of this plan are that, if it started from Tornado, it would be beginning on a very unsound basis, and that limits of improvement might quite frequently be reached; however, it remains worth considering whether

each replacement generation of aircraft has to be so utterly
different from its predecessor.   Alternatively, in times
when the Government did not require new or further aircraft,
it could wind production on aircraft down and provide other
kinds of work.   Thus the costs of maintaining the capacity
would be offset by also using it to produce other things we
really need.

    Essentially, then, we are talking about a three-part
process:

(a) establish a level of development and production capacity
     needed for military aircraft, compatible with our
     economic capacity and strategic requirement;

(b) convert the rest of the present capacity to other kinds
     of work;

(c) diversify the remaining aerospace capacity, creating a
     basic minimum for aerospace, a minimum for other types
     of production, and, so to speak, a 'floating' workforce.

    At first consideration, there do not seem insurmountable
barriers to development staff being 'mobile' in the same way
as production workers.   The need to have compatible production
skills in all parts of the alternating production has already
been discussed;   this would mean that the underlying technology
of the products need not change — the ends to which it would be
put are different.   Development staffs' experience in such
fields as stress in materials and structures, hydraulics,
components, electronics, instrumentation and production and
assembly techniques, could be applied equally to the military
and the other types of production.   Where parts of the
development tasks on the non-military projects demanded
different skills, there could be a measure of occupational
conversion, or, more simply, new staff could be brought in.
    It will be readily admitted that all of this involves a
major planning task, with government prepared to co-ordinate
its procurement policy and timetables between departments.
However, the alternative, with or without defence cuts, is
the threat of unemployment, often coming true, a continued
waste of resources, and the creation of surplus capacity
which will continue to confuse weapons procurement decisions.
If, in the event of defence cuts, the unconverted capacity
were not diversified, we should, by reducing capacity, have
reduced the problems generated by capacity expansion, but it
would remain a problem creating constant pressure for higher
military budgets;   we would be better off than we are now,
but, by diversifying, we could be better off still.   The
ideas emerging in BAC along these lines, while not as
developed as those which have emerged from Lucas Aerospace,
do thus provide a reasonable basis on which to proceed.

Alternative Products

In seeking alternative products, we start from the basis of
looking for work which is socially and/or economically
valuable, representing a sensible use of resources, contributing
either to government social programmes, or to the general
strength of the economy.   To fulfil the latter aim we should
be thinking about products which can be import substitutes or
have export potential (or both).   While converted capacity may
need support from government over a possibly quite lengthy
period, it could and should aspire to competitiveness.   As
suggested above, diversified capacity will need to find safer
and less volatile markets - essentially, the government
through its various programmes, and possibly local authorities.

The process of conversion and diversification cannot be
viewed in isolation, nor should it be seen as a purely
technical issue.   In following sections, the discussion
raises related areas of policy - adopting some of the alternative products below would require important decisions over
the National Health Service, energy or transport, for example
- and the need for industrial restructuring, so that
managerial and development methods and organization are
appropriate to the new fields of enterprise.   It should also
be realized that making converted capacity competitive
involves public investment in areas of profitable manufacture, along the lines already worked out for Labour's
industrial strategy. (19)   Support from the Government while
competitiveness is established does not just mean financial
aid;   it could also require steps such as preferential
purchasing by nationalized industry or selective import
controls (on machine tools, for example).   A more vigorous
implementation of the planning agreement system could also
be required for both conversion and diversification.   These
are political decisions which could be part and parcel of a
successful transfer of resources.   The technical issues are
important, but they only have real meaning against a political
background.

Full use of resources involves identifying work which most
benefits from those resources.   In the airframe industry, to
quote an early study, 'The primary resource of the industry is
its ability to design, develop and manufacture new and
advanced products.' (20)   Our inquiries will be most fruitful if directed towards advanced and advancing technologies.
Rolls-Royce has tremendous experience in the production of
sophisticated engines of all kinds and uses, and this,
together with its high potential for skilled precision work,
provides us with useful guidelines for assessing alternative
products.

The following examples of alternatives are drawn from
three sources - one document from each of the workforces at
BAC and Rolls-Royce, and the earlier paper on aerospace
conversion for the Study Group. (21)

Machine tools: The decline of the British machine-tool

industry is reflected not just by the fact that the majority of
new tools are imported, but also by noting that virtually all
the most advanced tools come from abroad (particularly from
the FRG and Italy). Conversion at either BAC or Rolls-Royce,
or both, could be an opportunity for boosting British manufacture of machine tools, particularly if it were decided to
concentrate on the sophisticated types - with computer controls
and servo-hydraulic operation. Buns Before Gutter (21)
points out that Pratt & Whitney, the US aeroengine corporation,
has converted some of its capacity to machine-tool manufacture.
Rebuilding the machine-tool industry seems a basic condition of
revitalizing British industry as a whole, and conversion
provides an opportunity; by reducing imports it would also
help the balance of payments, and might eventually provide
exports.

Processing plants: Again, this work would suit both BAC and
Rolls-Royce. The Recovery Works at Avonmouth has been able
to reclaim, from sewage, water for industry, gas for heating
and power generation, and soil conditioners. The 1976
drought makes the first of those a topical goal, but the
others are equally important. Future developments are likely
to provide processes for recovering proteins from fine
chemicals; ethylene recovery from oil is already under way
and could be further expanded. Recovery and re-use of parts
of waste clearly helps to get the most out of resources, and,
if development and production capacity can be made available,
it would be pointless to continue largely to rely on imports
from the USA. With a fairly small numerical requirement and
long lead times, this could be an option for diversification;
should it be decided to export the products, conversion would
be more suitable. Either way, Rolls-Royce could make the
engines, with BAC responsible for components and assembly.

Energy: The primary market here is the Government, but,
before conversion or diversification into this field, basic
decisions are needed. Two particular types of production
have emerged as suitable: (a) Barrage schemes: further
development of tidal-power schemes, such as the Severn
Barrage, would require engines which Rolls-Royce could
manufacture. This has also been raised as a conversion
option for Vickers Barrow, and the two places could dovetail
their operations; (22) And (b) Nuclear material disposal:
if we were committed to relying on nuclear power for energy
generation, then every effort should be made to provide
safe methods of waste disposal; airframe experience in stress
technology, materials and structures could all go some way
towards this.
    In addition, some of the ambitious alternative energy
schemes, such as solar-panel generators or windmower
generators, would require assembly which could be done at
BAC, particularly because, as has been shown by the Lucas
Plan, aerospace technology and experience can help to turn

these from visionary ideas into concrete realities.

Marine engines: This is an option for conversion at Rolls-Royce; gas-turbine engines are already successfully adapted for use in naval shipping, and it has been suggested that commercial shipping will also utilize them. One of the problems about this is that turbine engines may be too sophisticated, providing advantages which are really illusory for merchant shipping, as suggested by Mary Kaldor. (22) On the other hand, Rolls-Royce was some time ago approached by Cuba for gas turbines for its trawler fleet; it is suggested in Buns Before Gutter (23) that the approach led to nothing because of American pressure. There has also been considerable interest, not least in Japan, in using gas turbines in oil tankers. There might therefore be possibilities in either or both of those types of shipping, although the retraction of world trade (and thus the excess shipbuilding capacity which exists world-wide) raises obvious problems. It is therefore an option to be approached with some care, but one worth detailed consideration of the possibilities and potential advantages, including a full survey of market potential; it would be foolish and frustrating for Rolls-Royce to fail to identify possibilities if they do exist, and thus miss an opportunity which would have tremendous export potential.

Freighter aircraft: It has previously been suggested that a short take-off and landing (STOL) freighter/passenger aircraft might open up a market in the Third World, where previously inaccessible parts could be reached more cheaply and easily than in such methods as building railways. (24) Naturally, such a project could be suitable for both BAC and Rolls-Royce, as well as aerospace equipment firms; it might also have military applications - the USA is now in the process of developing a STOL military transport. (25) Perhaps precisely because the USA industry is already into this field, it is an option which should be approached with care. It may be, however, that a civil STOL transport could be less sophisticated, and therefore cheaper, with advantage. Again, this requires detailed study.

Prefabricated parts for construction: This has long been regarded as one of the airframe industry's major options for conversion, but it is full of problems, some of which were discussed in Aspects of Conversion of Arms Industries. (26) It would mean entry into a field which is already under critical examination in the Labour Party, with fundamental reforms now being discussed. (27) Not only would there need to be an infusion into BAC of architectural talent, but also, particularly if prefab housing were taken up, an entirely new promotion and marketing organization would have to be created. For this reason, prefabrication of parts for industrial structures and, for example,

bridges, might be a more attractive proposition.   Any moves
taking up this option would need to be planned alongside
other possible reforms in the building industry.

Rolling stock: The idea of developing new types of lightweight railway rolling stock, replacing the century-old
concepts now utilized, emerged from the Lucas Aerospace
Combine Committee's Corporate Plan.   It remained a possibility,
utilizing airframe experience together with traditional
coach design expertise, to develop cheaper and safer
rolling stock (the present rigid-structure and extremely heavy
stock tends to exacerbate the effects of accidents);
the hybrid road/rail vehicle which has been developed also
raises attractive possibilities.   It opens the possibility of drive-on drive-off rail transport;   it could make
tracklaying cheaper by being able to take steeper gradients
(and this could be suitable for use both in Britain and the
Third World).   With the Government as the sole purchaser,
in the initial stages at least, there would be plenty of
scope for planning production, so, although a good deal of
R & D work would be necessary, this could be a possibility
for diversification at BAC, depending upon the layout of
the plant.

Medical manufacture: This is an option for either conversion
or diversification at BAC, and would particularly involve
production of the kind of sophisticated monitoring equipment
used in intensive care units, but also of aids for the elderly,
crippled and handicapped, making use of vastly simplified
aerospace controls technology.   This would demand a basic
decision for greater investment in the National Health Service.

It has also been suggested that 'compact and efficient
agricultural equipment' could be manufactured at BAC, although
this would depend on numerous factors, not least the fact
that the government may well be unwilling to compete against
British companies already involved in the field. (28)   The
same source suggests that the expertise in all kinds of
fields available at BAC could be used to establish a testing
and advice centre for British industry;   were such a centre
established, it should involve other types of manufacture as
well, or else be one of a series of centres.   It is a useful
idea, the demand for which should be assessed, but it is
neither necessarily dependent upon a conversion programme,
nor necessarily best suited in Preston.

Problems of Conversion and Diversification
The products above, while probably not a complete list,
represent a reasonable range of conversion and diversification options - as Peter Ward puts it, 'a good starter for ten'.
(29)   When this list is added to others which have been
reported to the Study Group together with those in the
Corporate Plan for Lucas Aerospace, and even when the over-

lapping projects are eliminated, the list of possible alternative products for conversion and diversification in arms industries looks pretty impressive. The problems are not to do with finding alternatives.

The first problem is to do with selecting between the various alternatives. The selection must be made on the basis of suitability to existing facilities and skills, and on the basis of social and economic value. American experience in base conversion has shown the value at a community level of a proper strategy, identifying assets, requirements and constraints; (30) the same principle holds for a national conversion effort. Obviously, a good deal of preliminary work is necessary as a basis from which to plan the details.

The second problem is one of co-ordination, avoiding duplication and phasing the changeover from one kind of production to another.

Converted capacity will have to be supported during a transition period. While capacity can be converted relatively quickly, with occupational conversion for production workers in particular being quite straight-forward, more time would be required for some development staff and administrative staff. (31) Support will also be necessary during the period needed to penetrate or create new markets; there must be an awareness of this, and explicit willingness to make proper provision for it.

In general, converted capacity should not enter competition with established British industries. This can be avoided by concentrating on fields of advanced technology in which British performance is now poor (or, in some cases, virtually non-existent). Alternatively, employees no longer required in military aerospace could be transferred to existing firms, who could receive development contracts from the Government through the NEB, to enable them to expand their expertise and markets. Established firms involved in this should be brought within the planning agreements system to establish full accountability.

It has already been pointed out that some of the alternative products involve basic decisions about major areas of government policy, decisions about, for example, boosting the National Health Service, developing new transport policy, entering profitable manufacture, how best to support newly converted capacity, relying on or turning away from nuclear power. There is a clear two-way relationship between these areas of policy and the potential for success in conversion and diversification of the arms industry.

In the nuclear power debate, for instance, a decision to rely on nuclear power would invalidate some of the most exciting conversion options available, although it could lead to another decision to utilize airframe experience to develop safer means of waste disposal. On the other hand, it is not permissible for the decision about nuclear power to be taken without consideration of other means of power

434

generation. Successful experiments with solar-powered
generators, the development of the Nodding Duck wave-power
system (see pp.418-19) and the interest in both types of project
shown by the South-West Energy Group and Harwell Energy
Technology Support Unit make it impossible to dismiss 'alternative energy', (32) particularly when one realizes that the
industrial capacity to develop these systems could be made
available following arms cuts.

At no stage can consideration of conversion and diversification be divorced from other policy areas. The process will
be anarchic and uncertain unless other policy is considered,
and the proper planning undertaken.

Attitudes of the Workforce

A problem many people have pointed to is the attitude of the
workforce. This, of course, brings us in a sense back to
square one — views attributed to workers and trade unions are
regularly used as an argument against arms cuts, normally as
a way of avoiding rather than entering debate.

In fact, there does not generally seem to be opposition to
conversion and diversification per se. Naturally enough
there is opposition to redundancy, and to incompletely planned
and conceived schemes which might go off at half-cock, causing
redundancies in the long run. There is opposition to purchasing arms from abroad while capacity lies idle or under-used in
Britain, and there is opposition to taking on work which is
less well paid, less skilful and less interesting.

Clearly, however, with so many of the suggestions for
alternative products coming from the workforce, the argument
that workers are all against conversion or diversification
just does not hold water.

There will, however, be opposition to arms cuts unless and
until there are concrete alternatives to provide jobs. There
is a great deal of suspicion of proposals for arms cuts at
present; this should cause us no surprise in view of the
present unemployment total, at a time when investment is
falling in Britain, and when it seems that the Treasury has
been allowed to get away with a policy of creating unemployment. This should provide us with a warning, first that arms
cuts are only acceptable to many if they are linked with
provision of other work. Secondly, we should be clear that
what has emerged in the work of the Study Group is a set of
proposals and possibilities which adequately demonstrate the
feasibility and desirability of conversion and diversification,
but do not yet add up to a blueprint. We can say that such a
plan is possible, and that it would make a valuable contribution to sorting out defence policy and many of our industrial
and economic problems — but we cannot claim to have developed
that plan. The discussion above has shown that a good deal
of preliminary work is necessary — market surveys, systematic
search for products, collection of fuller data, etc. — before
a government could draw up such a plan. The conclusions and
recommendations of the Study Group must include urging that
preliminary work be now undertaken.

Restructuring the Industry

A consideration that frequently escapes the attention of those working in this field is the question of the kind of framework within which converted capacity would function. The preference here is for a complete overhaul of the structure of the industry to   this would include taking it out of the hands of present management, changing management structures, rejigging development teams, introducing a limited number of new personnel at key points, and developing new methods of operation. We can identify two major reasons which make this task essential.

In military aerospace, and, indeed, throughout military industry, it is a basic tenet that each product should be better than the last. The main rationale for the introduction of replacement weapon systems is that the existing systems are obsolescent. Improvements are seen in terms of better perform-ance, and not, significantly, in terms of cost-reducing produc-tion techniques or other economies. Improved performance means increased sophistication. Of course, there have been sound reasons for this; an attempt to fight the Battle of Britain with Sopwith Camels would have been ludicrous as well as unsuccessful, and the Spitfires and other fighters of the Second World War are now, rightly, museum pieces. Sophistication, however, carries with it the penalties of expanding industrial capacity and increased costs. (32)

This drive for improvement is not a recent phenomenon – it begins with the first development of military aircraft. It has, however, now created practices and industrial structures which confuse and distort decisions about defence policy, and which are incompatible with competitive civil manufacture. This is not to say that each new civil product benefits by being worse than its predecessor, but an improved product may be one which is simpler to use, cheaper to produce and cheaper to buy, not one which is more sophisticated and so necessarily more expensive. The kind of cost increases we have seen in major weapon systems are not supportable in the civil field. (33) Furthermore, increased sophistication does not always produce a better final product: Tornado is an example of this, as is the American B-1 bomber – benefits can be illusory.

This leads us to the second point. Military industry is not really profitable or competitive in the normal sense of the term, particularly in Britain, where the aerospace industry in particular has been rationalized by reducing the number of firms involved, to reduce the cost of maintaining military industrial capacity. With military industry, government fixes profit levels; development is funded so that the corporations themselves need not make irretrievable investment; purchase of the completed product is guaranteed if it comes up to scratch, and often if it falls far short of scratch. If a government decides it needs a military industry, it has to support and maintain the capacity involved – especially the development capacit the technological 'core' of the industry – and provide it with

436

work. In this sense, whether or not private profit is
gained as a result, military industry is really a public
industry, maintained by the Government according to its
definition of the public good. It may be argued that
industry has to compete in the export field; even there,
the Defence Sales Organization (and foreign counterparts
for other countries' industry) helps out in every way
possible.

Neither development nor management techniques in
military aerospace are really suitable for competitive
manufacture on an open market. On a very simple level,
it cannot be assumed that people whose whole experience
has been in fields where cost is little or no object and
where increased sophistication must always be provided,
can easily adapt themselves to a field where cost is all-
important, and increased sophistication of the final
product is but one of many relevant considerations. We
are demanding different methods and techniques - a different
ethos.

In diluted form, these problems also apply to diversified
capacity. Although the Government would remain the main or
sole purchaser of the non-military products, we would certainly
not want to see costs of, for example, medical equipment
spiralling upwards through unnecessary extra sophistication
for marginal or illusory benefits.

There should, therefore, be a clear separation between
converted and retained capacity. Ideally this separation
should be both administrative and physical - converted capacity
should function in different buildings, or even on a different
site if one were available. As well as enabling people to
have more control over matters directly affecting their
prosperity and well-being, and providing a fertile new source
of ideas, industrial democracy would be a way of enforcing
change in industrial structures, and introduces a sense of
urgency into marketing and cost reductions. A body such as
the NEB would need to supervise and check on much of the
functioning of the new industry; industrial democracy
could also provide a kind of internal check. It is not that
management or development methods are inefficient in military
aerospace (though by one set of criteria they certainly are),
but that they are inherently unsuited in many respects to the
kind of manufacture in which conversion would involve them;
therefore, they must be changed and replaced by ones more
suitable.

Conclusions

Despite the unlikelihood of Tornado cancellation, and despite
limited data, it is valuable to consider other uses to which the
resources involved in the project could be committed. We
have concentrated on work which can utilize present skills and
technology, and this particularly means fields of advanced and
advancing technology.

There will undoubtedly be a requirement to retain some

military aerospace capacity; maintenance of this capacity could be made cheaper through diversification, with the important benefit of producing worthwhile things during periods when aerospace capacity is not required to be in full operation. In addition, this would provide the workforce with greater security.

For both conversion and diversification we can identify useful work which could contribute, variously, to the balance of payments, to the basic strength of our economy and industry, and to public spending programmes. For several of the products, certain fairly basic decisions are required about other areas of policy. Adequate planning is a crucial part of the process, and this must be based on yet more preliminary work. If these alternatives are presented in a concrete and practical form, there is no reason to expect opposition from the workforce involved in military aerospace. We shall, however, face real barriers to success unless conversion planning includes restructuring in the industry which is converted.

We can conclude that this planning task is possible, and that the opportunity for a successful programme of conversion and diversification exists.

Notes and References

1. David Greenwood, 'Defence Programme Options to 1980/81', Study Group Paper No.2.

2. Hansard, 18 March 1976, cols. 1627-53.

3. William Rodgers MP, in ibid., col. 1650.

4. Both quoted in 'More Attacks on NATO Plane', Sunday Times, 3 October 1976.

5. Various estimates in the region of Mach 3 had been made before the Mig-25 was closely examined; it now appears that the machometer is 'redlined' at 2.8; see 'The Mig-25 Saga', Air International, November 1976.

6. See Defence and External Affairs Sub-committee of the Expenditure Committee, Minutes of 27 April 1976, House of Commons, 236-v, Questions 407-12.

7. Geoffrey Pattie, Towards a New Defence Policy, Conservative Political Centre, 1976, p.16.

8. See 'Europe's F-16 Plans Unfold', Flight International, 23 October 1976.

9. 'The End of the MRCA?', in Ulrich Albrecht and others. The Anti-White Book, FRG, 1974.

10. W.B. Walker, 'The Multi-Role Combat Aircraft (MRCA): a Case-Study in European Collaboration', in Research Policy, 2, 1974, discusses the disadvantageous consequences of collaboration; William D. White, U.S. Tactical Air Power, Brookings Institution, Washington DC, 1974, discusses the relative merits of multi-role and specialized aircraft (pp. 55-9), and the changing strategic environment. The implications of this for British defence policy, and Tornado in particular, are discussed in Dan Smith, 'Strategic and Political Implications of Reduced Defence Programmes', Study Group Paper No.5.

11. Hansard, 5 November 1975, and 17 December 1975.

12. See Financial Times, 10 October 1975.

13. For example, most recently, see 'Aerospace Decisions Now, EEC tells Britain', Flight International, 16 October 1976; it would appear that as far as there is British interest in civil aerospace collaboration, it is directed towards further bilateral collaboration with France, not towards EEC-wide projects.

14. Aircraft and Shipbuilding Industries Act 1977; see Part 1, Clauses 2(5) and (6), and 3(1).

15. 'Lords Modify Nationalization Bill', Flight International, 23 October 1976.

16. Part I, Clause 2(8); see also Part I, Clauses 5(1), 7(1), and 8(2). (This paragraph has been drafted to replace two in the original paper which discussed the then uncertain future of the Bill.)

17. Hansard, 17 December 1975.

18. Roy Mason, MP, Hansard, 31 March 1976, col. 1334; at peak, a further 8,000 are expected to be employed on avionics and equipment, giving a total of 24,000 workers directly employed, with another 12,000 employed 'indirectly'.

19. See Labour's Programme for Britain 1976, chs. 2 and 3.

20. James J. McDonagh and Steven M. Zimmerman, 'Mobilization for Peace: A Program for Civilian Diversification of the Airframe Industry', unpublished thesis at Columbia University, 1961, p.181.

21. Buns Before Gutter, discussion document for aerospace shop stewards, 1976; Peter Ward, Alternatives to Arms Production, paper for Preston Trades Council Day School, April 1976; Dan Smith, 'Aspects of Conversion of Arms Industries', Study Group Paper No.12.

22. See Mary Kaldor & Albert Booth, 'Alternative Employment for Naval Shipbuilding Workers: A Case-Study of the Resources Devoted to the Production of the ASW Cruiser', Study Group. Paper No.13.

23. See n.21 *supra*.

24. Dan Smith, 'Aspects of Conversion of Arms Industries'. Study Group Paper No. 12

25. The USAFAMST programme for a STOL military transport involves two competing prototypes (Boeing's YC-14 and McDonnell Douglas's YC-15), both of which have flown; selection between them is expected in September 1977 for entry into service in 1983: see 'YC-14, All Blow and No Puff!', *Air International*, November 1976.

26. As n. 21 *supra*, pp.13 and 15.

27. See *Labour's Programme for Britain 1976*, Ch.3.

28. Peter Ward, paper op-cit.

29. *Ibid.*

30. See Dan Smith, 'Community Planning and Base Conversion', Study Group. Paper No.15.

31. Dan Smith, 'Aspects of Conversion of Arms Industries'. Study Group Paper No. 12

32. See 'Million-Watt U.S. Solar Boiler is Tested in French Pyrenees', *International Herald Tribune*, 20 October 1976, and 'Second Chance for Solar Energy', *Guardian*, 1 November 1976.

32. For a full discussion of this process, see Mary Kaldor, 'European Defence Industries - National and International Implications', ISIO Sussex University, 1972, and 'Defence, Industrial Capacity and the Economy', March 1976, for the Study Group.

33. See Norman R. Augustine, 'One Plane, One Tank, One Ship: Trend for the Future?', *Defense Management* Journal, April 1975; the discussion there shows a remarkably steady increase in costs of all major weapons systems, and includes the calculations that if the trends continue, the entire US Department of Defense budget will be able to purchase only one aircraft in the year 2036!

(For information and other help in preparing this paper the author is grateful to the following: Bob Crook, David Griffiths, Brian Hesketh, Mary Kaldor and Peter Ward. Naturally none of them is responsible for the contents or conclusions of the paper.)

Annex I — MRCA Tornado Costs

The basis of the calculations is the government estimate of
£6.34 million unit production for the IDS, and £7.72 million
for the ADV; I have used the convention that production
accounts for 35 per cent of the total cost, Research and
Development (R & D) for 15 per cent, and Operation and
Support (O & S) for the remaining 50 per cent.
    The IDS R&D cost has been reduced to account for the
effects of sharing. It is estimated by the government that
collaboration adds 25 per cent to the total cost of R&D but
the government's share of these overall costs was 42½ per
cent.
    ADV R&D costs are not reduced by sharing. However, the
government estimates that 65 per cent of the IDS R&D is
applicable to the ADV, so the R&D figure has been reduced by
an appropriate amount.
    O&S is not made cheaper by collaboration, or by experience
on previous development.   Therefore O&S figures are calculated
on the basis of notional R&D costs — what IDS would have cost
to develop without collaboration, and what ADV would have cost
had it not been preceded by IDS work.
    Naturally, these figures are _estimates only._ They are
higher than normally seen. Such figures as have been released
for development costs, however, lead me to believe I am on the
right lines in that category, and with such a sophisticated
aircraft, maintenance costs in particular will be very high.
In view of that I am fairly confident that these estimates are
in the right order of magnitude. It could be pointed out
that even if they were to err by an average of £2 million per
aircraft, the cost of the whole project over the full life of
the aircraft would still be over £6,000 million at 1977/8
estimate prices, which is still far higher than any previous
estimate.

(£m 1975 autumn prices, rounded figures)

	Per copy				Full order	(220 IDS,	165 ADV)	
	R & D	Prod	O & S	Total	R & D	Prod	O & S	Total
IDS Tornado	1.44	6.34	9.06	16.84	325	1400	2000	3700
ADV Tornado	1.54	7.72	11	20.22	250	1275	1825	3350
Combined Totals					575	2675	3825	7000

(Note: totals may not add up due to rounding)

Annex II — MRCA Tornado Contractors

Firms marked with an asterisk have been described by the
government as 'first-level' contractors or 'sub-contractors',
though not necessarily for all the systems upon which they are
working.

I have been unable to obtain certain information from a
significant number of the contractors of where the work is
being done. The information that is available suggests what
might anyway seem likely, that almost all regions of the
country have some Tornado work.

FIRM	SYSTEM
* BAC	Nose, aft-fuselage, tail. Assembly. Martel missile sub-systems.
* Chelton	Instrument Landing System (ILS) glideslope, localiser and marking beacon. UHF/VHF antennae.
* Cossor	ILS. IFF Transponder.
* Computing Devices Co.	Navigator's weapon aiming mode selector.
* Davall & Son	Head-Up Display Camera.
* Decca Navigator	Doppler navigation radar.
* Dowty-Rotol	Undercarriage. Engine re-heat control.
Dowty Boulton Paul	Air intake hydraulics.
* Dunlop	Wheels, brakes, tyres, anti-skid system.
Epsylon	Voice recorder.
* Fairey Hydraulics	Secondary control system.
* Ferranti	Laser Range Finder. Inertial navigation system. Combined map and radar display. Transformer rectifier
Flight Refuelling	Flexible couplings.
* (GEC) Marconi Elliott Avionic Systems	Stores Management System. Main control system. Tactical navigation. TV tabulated display.

442

FIRM	SYSTEM
	Triplex Transducer unit.
	Terrain following E-scope display.
	Fuel flow metering.
	Stability Augmentation System.
	Intercept radar.
* (GEC) Easams	Avionics systems integrator.
Graviner	Fire detection and extinguishing.
Hawker-Siddeley Dynamics	Martell Air-to-Surface missile.
	Air intake control.
Hymatic Engineering	Anti-G valve.
* Lucas Aerospace	Windscreens.
	Main engine control unit.
	Integrated Drive Generator.
	Pitch Feel System.
	Wheel-brake and fuel control systems.
	Auxiliary power unit.
	Constant Frequency electrical generating.
* Marston Excelsior	Intercooler.
* Martin Baker	Ejector Seat.
Miles Engineering	Weapons release actuators.
Normalair Garrett	Cold Air unit.
	Pre-cooler.
	Oxygen system and pressure reducing valve.
Plessey	UHF/VHF
	Electric power system controller.
Rolls-Royce	RB-199 engine.
Smiths Industries	Head-Up Display.
	Horizontal situation actuator.
	Digital guidance equipment.
Sperry Rand	Crash recorder.
	Hydraulic pumps.
Ultra Electronics	Communication control system.
Uniroyal	Fuel cells.

lso engaged in work, under work-sharing arrangements, are:-
ge Engineering; Aeroquip; Simmonds Precision Products.

urces: Interavia, September 1972;
S Market Report 1975 'MRCA';
lliam Rogers MP, answer to written question, Hansard,
  November 1975.

ANNEX III US and EEC Aerospace Industry

(a) Output and employment 1974

	UK Output $m	UK Jobs	Other EEC Output $m	Other EEC Jobs	USA Jobs[1]
Airframes and missiles[2]	1222.6	93,992	3356.1	126,013	588,900
Aeroengines	1051.8	93,992	869	35,317	226,600
Equipment	710.3	22,116	612.8	40,898	149,500
Total	2387.7	210,100	4837.9	202,228	965,000

(b) Leading Airframe and Missile Corporations 1974

		Final sales[3] $m	Jobs
Boeing	(USA)	3730.6	75,400
Lockheed	(USA)	3279	62,100
McDonnell Douglas	(USA)	3075	70,739
General Dynamics	(USA)	1967.5	n.a.[4]
Aerospatiale	(France)	976.1	40,242
Dassault-Breguet	(France)	760.1	15,161
BAC	(UK)	636.3	34,994
MBB	(FRG)	578.1	19,978
VFW/Fokker	(FRG/Netherlands)	527.9	17,978[5]
Hawker Siddeley	(UK)	440	35,000
Dornier	(FRG)	202	7,000
Westland	(UK)	192.8	11,904[6]

(c) Leading EEC Aeroengine Corporations 1974

Rolls Royce (1971)	(UK)	1097.5[7]	61.924
SNECMA	(France)	368.8	19,095
MTU	(FRG)	205.3	11,333

Source:
The European Aerospace Industry: Position and Figures Commission
of the European Communities, SEC (76) 2657, May 1976, Tables 35, 36, 43,
57 and 58; all monetary totals converted from European units of account.

1. US aerospace output in 1974 was 22,253m, amounting to 68.8% of the capitalist world's aerospace output, compared to an EEC total of 7,823m, amounting to 24.2% of the total.
2. The space sector in the EEC employs upwards of 3,000, and in the USA employs 140,900; space employment figures (and output in the EEC's case) are included under airframes and missiles.
3. 'The level of a firm's sales gives no indication of the nature of its activities—in particular, the volume of subcontracting, production under licence, and R&D.' SEC (76) 2657, p. 45.
4. In 1973 General Dynamics employed 62,400.
5. Approximately one-third of VFW/Fokker employees work in the Netherlands, the remainder in FRG.
6. Approximately half of Westland employees work on helicopters.
7. There is no explanation in the source document of the disparity between UK aeroengine output and Rolls Royce output; this may be due to Rolls Royce's other industrial activities.

15 COMMUNITY PLANNING AND BASE CONVERSION

by Dan Smith

In our interim report we commented:

'The details of plans for conversion would have to involve
the firm, the industry and the government, as well as the
community in the locality where arms factories are converted.'
(1)

It is an important principle that the planning necessary after
defence cutbacks should involve all those affected, including
those indirectly affected. We stress this both because we
wish to see the extension of democratic participation in
matters of crucial importance to people's livelihoods and
well-being, and because extending participation in the
planning could lead to some extremely valuable inputs to the
planning process. This paper discusses the effects on
Preston of reductions in military spending, and indicates
the kind of contribution which local involvement in planning
could make to the process of conversion; community response
to local defence cutbacks over the past thirteen years is
examined, for the light it throws both on the ways in which
communities have been able to organize to overcome the
apparent problems posed, and on the possibilities for conversion and joint use of bases.

Preston Area

Preston and the surrounding area is particularly dependent
upon arms contracts. BAC employs 10-12,000 people at its
military division; close by is the British Leyland Truck and
Bus division, at Leyland, where a small proportion of the
13,000 strong workforce is engaged in production of the L-60
tank engine; also in the area is ROF Euxton, employing some
$2\frac{1}{2}$-3,000 workers, probably as a filling factory. In addition
there is at least one local components firm (Attwater & Sons)
which includes MOD contracts in its work, and there may be
others. Related to our concerns about conversion and
diversification, given the number of energy projects which
have been proposed as alternative work for arms industries, is
the joint British Nuclear Fuels Ltd/Atomic Energy Authority site,
employing 3-4,000 people.
    A good deal of local prosperity is bound up in these enter-
prises. Cancellation of contracts without provision of
alternative work would lengthen dole queues, reduce spending
power, badly affecting local commerce and the whole basis of the
area's prosperity.
    It is however worth noting that the contract for the L-60
engine at British Leyland is being reduced and may shortly be
eliminated. Apparently this has caused no concern whatsoever
in the plant. The order was fairly small, and the engines
were assembled individually, not on an assembly line; the
skilled workers released from this production will be very
welcome elsewhere in the same factory which has plenty of work

producing trucks and buses.   So there is no problem about conversion there.

Nevertheless, this fortunate pattern may not be repeated at BAC.   The sizeable workforce there will be called upon to make major adjustments, every effort should be made to ensure that the transition away from arms manufacture is as smooth as possible, with no loss of income for the workers.   Apart from aid and planning from central government, a good deal could be done to minimize the seriousness of the impact, by a local planning body, constituted perhaps from the Trades Council, the local business and community organizations.

Their contribution could be particularly valuable because the area is already the subject of a major development plan, for the Central Lancashire New Town, for which the designation order was made in 1970, and which envisages a more than two-fold increase in population in the area together with the establishment of much new industry.   Although full approval is still awaited, and may obviously be delayed given the public spending situation, some work is already in progress, described in Annex I, p.454.

In the event of arms cuts, a local planning body might usefully modify the existing development plans.   The envisaged population increase might be lessened, or delayed slightly, so that the incoming industries could provide work for dis-placed arms workers.   An effort could be made to attract those industries which could best utilize the skills of the released workforce.   A new enterprise with central government support might be established in some of the factory space under construction by the Central Lancashire Development Corporation (CLDC).   New factories are already being built 'on spec' by CLDC, so that the present configuration of the local BAC plant need not be a constraint upon the choice of alternative types of production. (2)

Thus planned local development could aid in the redeploy-ment of arms workers, and keep wider options for conversion, while the skills of the workforce would mean that incoming industry has tremendously valuable resources to tap.

It should be noted in passing that ROF Euxton lies right in the middle of the designated area.   The workforce there at peak production was 39,000 it is now less than 8 per cent of that figure.   The plant and surrounds - which at first sight seem about as big as Hyde Park - must contain plenty of idle space, but CLDC's plans are not permitted to involve use of this resource.   This really is a sad waste, particu-larly in view of experience of similar situations in the USA discussed below.

At present, Preston's eggs are very much in the arms basket, especially the Tornado basket.   With the develop-ment planned and the work already in progress the community is in an ideal position to establish a more diverse and stable base for its prosperity;   the release of a good proportion of the skilled workforce now engaged in arms work would aid developing that base tremendously.

Community Planning Experience in the USA

The Office of Economic Adjustment (OEA) in the Department of Defence was established to help American communities overcome the consequences of closures of bases, withdrawal of contracts, and other local military cutbacks; in other words it helps American communities deal with exactly those problems that Preston might face. It works as the permanent office of the President's Economic Adjustment Committee, with the membership listed in Annex II. Its experience, gained over the years, is useful to us for a number of reasons.

The employment consequences of arms cuts inevitably seem at their most intractable when viewed nationally; in most cases, when seen case by case, they become more manageable, and it is this kind of experience that the OEA has. It has shown what a community can do, and has articulated various rules for a successful conversion effort which are relevant for us. These conversion efforts have all taken place in an atmosphere far less sympathetic to planning than exists in Britain, and particularly in the Labour Party: planning has been almost entirely at a local level, even when there has been State or Federal funding, and the OEA's role is to provide expert assistance when requested. While we view the importance of planning within a wider social and economic context than just industrial or base conversion following arms cuts, it should be evident that adequate planning at a national level (where the main planning task would be co-ordination, to prevent unwanted duplication) will make the tasks more straightforward than they have been in the USA.

Finally, as has been said before to the Study Group, conversion 'can be done, because it has been done'. (3) It has been done in Britain, and in the USA, and the OEA experience can help answer certain specific problems we might expect to encounter here, as well as give further evidence to support Frank Blackaby's assertion.

The prevailing attitude at the OEA can be summed up as follows: 'as a matter of fact, the imagined disasters that a base closing portends can actually become catalysts for community improvements never before thought possible.' (4)

In general, despite often serious transitional problems and uncertainties, communities seem better off after the bases have gone. There have been failures - the record is not 100 per cent successful; one potential failure is briefly considered below but the record does seem remarkably good. In particular by diversifying the basis of their prosperity, by attracting several employers to an area where there was previously but one base or installation, many communities have achieved greater stability, in exactly the way that was discussed with relation to the Preston area.

The ingredients for a successful conversion effort are identified as follows:

The creation of a mutually agreed upon organization is essential for the proper development of a workable base use plan. The <u>second</u> ingredient for a successful conversion effort is the need for definitive development strategy.... (<u>This</u>) incorporating specific goals and objectives provides a solid foundation for delineating a base-use plan - the <u>third</u> ingredient. (5)

In the view of OEA officials, virtually anyone or anything will do for the organization, including a workers' committee, as long as it has adequate resources, sufficient local backing, and enough initiative and drive to make a success of the project. According to one report a common factor for success has been 'an all-out commitment from all segments of the community, including labor, business, government'. (6)

The important point about achieving the widest participation in the planning process is that it increases the chances of success. Conversion plans, however good they look on paper, which do not have the backing in the affected community will not work.

The development strategy should identify the assets in the community and area and aim to utilize them fully. These assets can be understood in terms of plant, space and facilities, including both the factory affected by arms and cuts and others in the area which may have spare capacity, or areas which could be developed as industrial estates, and in terms of what Mike Cooley has described as 'our most precious asset - people, with their skill, ingenuity and creativity'. (7) The strategy should also involve a market survey, involving the requirements of the community, or spreading more widely to the national and international market; here, expert advice can often make a most useful contribution. Finally, possible constraints upon development must be considered. A general survey of the topic provides examples of these which may limit the ability to attract and service industry; there may be lacks such as, 'an adequate water and sewer system, the ability to provide the kind of skilled worker needed by a specific industry, adequate transportation, sufficient housing in desirable price range, recreational and cultural facilities, a hospital or a good educational system'. (8)

Many of these deficiences can be remedied in the course of the conversion process, and the plans include this aspect as a matter of course.

The final ingredient, the base-use plan itself, obviously varies from place to place. It is here that the process moves away from general principles and into details specific to the locality. With the general principles identified it may be useful to provide a limited number of examples.

Conversion Efforts in Progress

a) <u>Long Beach, California.</u> Conversion here was necessary due to 'realignment' of the naval complex. Although the

loss of 780 civilian jobs at the complex was offset by a gain
of 1040 jobs at the nearby shipyard, the relocation away from
the area of 17,824 military personnel could have had the gravest
effects on local prosperity. The Long Beach Economic Development Corporation was established to co-ordinate the utilization
of a 355 acre site close by for development as an industrial
estate. The development plan involves the creation of about
6,500 jobs over ten years. Incoming industries are screened
on such criteria as adequate job provision and environmental
effects. (9)

b) Newport, Rhode Island. 1,442 acres of a 2,162 acre
naval base were declared superfluous to the USN's requirements. Despite the problems of base-sharing (further
considered below), a detailed development plan has been drawn
up, envisaging the creation of 3,085 jobs over the next few
years. The available acreage will be used for industrial
and commercial areas, health and education facilities,
recreation and housing. (10)

c) Glasgow, Montana. This was unofficially described to me
as 'an absolute bummer'. It is an air base, 18 miles from
Glasgow (1973 population, 4,877), in a sparsely populated
area of North-East Montana. The main problem has been to
find a body — State, local or private — to take the lead in
development. Despite the obvious problems of isolation, the
general intention is to make the most logical use of the
facilities, centering on aviation and support facilities,
providing a method of transporting livestock from east
Montana. (11)

Past Successes

The following four examples of conversion efforts begun in
the '60s and now successfully completed are taken from
Economic Recovery (see note 4); this document is essentially
an advertising plug for the work of the OEA (with a forward
from President Ford), and an encouragement to communities who
may be affected by future base closures. It provides
details of 22 successful conversion projects, and sketches
the results of a further 24; the results of all 46 are
summarized in Annex II.

d) Benecia, California (1964). The closure of the arsenal
and depot at Benecia (which is fairly close to San Francisco)
meant a loss of 2,318 jobs, a staggering blow in a community
of only 6,450 persons. Industrial development was initiated
reasonably quickly, and has created 3,000 new jobs in the
locality.

e) Sidney, Nebraska (1967). Sidney is the county town of
Cheyenne County whose 1960 population was just less than
15,000. The Sioux Army Depot at Sidney was the region's

largest employer, and its closure affected 20 per cent of the country's families. The initial impact was very serious, a blow to both the well-being and morale of the region, but part of the 19,360 acre site was developed for industry (the rest being turned to agricultural use) and became a magnet for manufacturing activity. All lost employment has been replaced, and extra employment has been created; since 1970 Sidney's small population has increased by 1,000.

f) <u>Neosho, Missouri (1968).</u> In a working population of approximately 50,000, 1,200 were made redundant following the closure of Air Force Plant, 65 at Neosho. The plant and a near-by disused army base were developed. All jobs lost at the plant were replaced within a year of closure; there are now 2,034 jobs on the site of the former plant, and other development in the town has generated a further 822 jobs. The former base is now the site of a municipal airport, a college and a technical school.

g) <u>Springfield, Massachusetts (1968).</u> The Springfield Armoury (established 1794) was the third largest employer and a traditional industry in a city of 164,000; it provided 2,400 civilian and a few military jobs. Closure came when the city was already experiencing a severe decline. The armoury site was incorporated into development plans which the closure has catalysed. On the site, 9 firms now provide 1,000 jobs, and the development plans brought a further 37 new firms into Springfield, providing an extra 9,200 jobs.

Joint Base-use

Discussing the economic effects of closure of the Clyde Submarine Base (CSB) David Greenwood and Timothy Stone point out that total closure of CSB is less likely than a partial run-down of base activity. (12) The kind of defence programme options considered by the Study Group, including a possible phasing out of the Polaris fleet would probably not result in total closure of CSB. The facilities there would remain extremely attractive to the RN - it is an ideal area for a submarine base and support facilities. When Scottish CND produced its report on alternative employment at CSB, it was partially predicted on total closure, and use of the complete site. (13) Would a partial run-down necessarily invalidate the report's conclusions, by keeping the site unavailable for the establishment of new local industry?
    The Ministry of Defence has a habit of erecting walls and barbed-wire fences, and then keeping all-comers out, whether or not the space and building inside are being fully utilized. ROF Euxton, previously referred to, is a case in point. The MoD's instinctive reaction would undoubtedly be to rule out sharing a site.

It may be that the particular conditions at CSB or ROF
Euxton, previously referred to, is a case in point. The
MoD's instinctive reaction would undoubtedly be to rule out
sharing a site. It may be that the particular conditions at
CSB or at ROF Euxton would eventually be seen to make sharing
impossible. But experience in the USA shows that that is not the
only response, nor should it be an automatic response.

Apart from anything else, commonsense indicates that fences
and walls can be taken down and erected around smaller areas,
releasing land and possibly buildings for other use.
Utilities can be shared quite easily, and the arrangements for
them so made that both civil and military authorities are
happy. 'It can be done, because it has been done' - several
times.

An air base at Topeka, Kansas, is now both a civil and a
military airfield; Annex III summarizes the joint-use
agreement. The Hunters Point Shipyard in San Francisco is
the subject of an agreement now being negotiated. At the
airfield, the Metropolitan Topeka Airport Authority is the
landlord, and the USAF is the tenant; at Hunters Point, the
government will be the landlord, and it seems likely that a
private ship repair firm will be the tenant. Both examples,
two among many, show that mutually agreeable arrangements can
be made, providing both parties to the agreement with the
facilities they want and need to operate.

Thus, there is no technical reason why sharing at CSB or
ROF Euxton should be dismissed out of hand. The details
would have to be worked out according to how much of the
installation is superfluous to the requirements of the
industries to be sited there, and so on, but precedents
exist to show it is worth trying to work these details out.

One should add that the decision on whether or not the
installation can be utilized should not be left in the sole
hands of the MoD, whose instinctive reactions might make an
objective assessment rather difficult to achieve. Interest-
ingly, in the case of Hunters Point, and perhaps in other
cases, the OEA has intervened against the government side of
the agreement to make the conditions of the agreement less
stringent and more favourable to the civilian party; that
is done because it has been found to be necessary, and it may
perhaps provide a note of caution about how joint-use agreements should be drawn up.

Conclusion

A community in which a base is closed or partially run down,
or in which arms contractors curtail their activities and
provide less employment, faces a serious challenge and the
threat of losing its prosperity. It also faces a tremendous
opportunity for development, and for increased local prosperity. In the event of local or national inertia, the

consequences are potentially disastrous; this survey shows, on the other hand, that it is possible, even in small communities, to replace employment and add to it.

Community involvement in the planning process is not just a token gesture towards democratic participation. It makes capitalizing upon the opportunities more possible. Local knowledge can mean more effective and imaginative planning, more decisive and swift action. It can ensure that real needs and assets are clearly identified, that the former are adequately met and the latter efficiently utilized.

Developing a coherent planning strategy involving all the relevant factors can make the final details of the plan straightforward. Government assistance can help the plan become reality, and can ease the problems of the transition period. Sound planning can provide the basis for economic use of both local and national resources; a relatively small amount of investment, well directed, will create more jobs, more prosperity, and more worthwhile products than a fortune thrown in haphazardly.

DEVELOPMENT IN PRESTON AREA

The Chorley/Leyland/Preston area of Lancashire is the site
for the Central Lancashire New Town, with a proposed population ultimately of 430,000, together with industrial and
other development. Present populations are:-

Chorley	31,000
Leyland	23,000
Preston	135,000

The Central Lancashire Development Corporation (CLDC) was
established by a designation order in 1970 and is now fully
staffed. The necessary preliminary planning has been carried
out.

While approval from the Department of the Environment to proceed fully is still awaited (with a decision to be given 'as
soon as possible'), some development projects are already
under way, including 16 miles of sewage system in three areas,
on which £5.6 million had been invested by September 1976,
five residential development areas and one industrial.

The residential areas will eventually consist of 5,580 dwel-
lings. 2,880 of those are under contract or licence, and
most of these are already under construction; a further 236
had been completed, and all but a few occupied, by September
1976.

In the industrial area the CLDC has constructed and let 21
factories, 18 of which are occupied; the first of these
factories only became available at the beginning of 1976;
5 more have been completed but are unlet, 11 are under
construction of which 3 are earmarked for tenants, and a
further 22 are planned. 242,000 square feet of factory
space has been built; there is 52,000 square feet under construction and a further 137,000 planned. Private construction adds a further 6,000 square feet in operation (with a
planned extension of 3,000 square feet) and 80,000 under
construction.

Capital expenditure so far exceeds £50 million.

Sources:-

Chorley Official Guide & Industrial Review;

Leyland Official Guide & Industrial Handbook;

The Borough of Preston Official Guide;

CLDC, A third review of progress to date (September 1976)
and

Guide to Development in Progress - No, 4 (September 1976)

BASE CONVERSION IN THE USA

a) Membership of the President's Economic Adjustment
    Committee is made up of representatives of the
    following:-

    Department of Defense (in the Chair)

                        Agriculture
                        Commerce
                        Health, Education & Welfare
                        Housing and Urban Development
                        Interior
                        Justice
                        Labour
                        Transportation
    Office of Management & Budget
    Civil Service Commission
    Federal Energy Administration
    General Services Administration
    Small Business Administration
    Environmental Protection Agency
    Council of Economic Advisors
    Arms Control & Disarmament Agency
    Domestic Council
    Community Services Administration

b) Use made of 46 former defence installations:-

    Utilised for Industry   in 40 cases
    Education   33
    Airport & aviation   27
    Recreation facilities   25
    Housing   21
    Municipal & government   20
    Commerce   13
    Health   11
    Agriculture   9
    Other   2

Source:-

Economic Recovery, Office of Economic Adjustment, 1976

Summary of a Joint-use Agreement

The agreement is between the USA and the City of Topeka, Kansas, acting through the Metropolitan Topeka Airport Authority (MTAA), relating to Forbes Air Force Base.

The Government conveyed all land to MTAA, retaining rights to joint use of specified areas, exclusive use of specified areas, and rights of entry and exit.

MTAA has exclusive use of certain terminal buildings, hangers, the air freight terminal, and non-military parking ramps. Runways, taxiways, lighting systems and navigational aids are jointly used.

MTAA is responsible for the operation and upkeep of the jointly used areas. It provides the fire-fighting capability for the whole field, and the Government shares in the costs. The Government provides its own crash/rescue facilities, over which MTAA and the Government can come to a separate agreement for co-operation if they desire.

The Government is MTAA's tenant, paying an annual rent; it may terminate the agreement on 6 months notice, or by mutual agreement. The Government must gain MTAA's approval for alterations to areas to which it has exclusive rights. MTAA is not to be held responsible for any injuries arising from Government activities at the field.

The transition from use as an Air Force Base, to the new shared use, is catered for in the agreement on terms aimed at facilitating the changeover and demonstrating 'a tangible and long-range Government contribution to the operation of the Airport'.

The agreement, dated 1 December 1975, is for three years' initial duration, with open options for renewal beyond that time.

Source:-

Office of Economic Adjustment, US Department of Defense.

Notes

1. _Labour's Programme for Britain 1976_, Chapter 16 (emphasis added).

2. The possibility that options for alternative production could be limited by details of the plant space available is mentioned briefly in another paper to the Study Group, as is a related concern – namely, ensuring that converted industry is free to operate under different procedures from those in aerospace; both problems are lessened, of course, if industrial development is providing plenty of new plant, as in the Preston area. See Dan Smith, Tornado – Cancellation, Conversion and Diversification in the Aerospace Industry. Study Group. Paper No.14.

3. Frank Blackaby, 'Note on the employment consequences of a £1,000 million cut (at 1974 prices) in military expenditure over 5 years', Study Group Paper No.10.

4. _Economic Recovery_ (OEA, 1975), p.4.

5. _Newport Naval Base, Rhode Island_ (OEA, April 1974), pp. 2 & 3 (emphasis in original); the formula is repeated throughout the OEA's reports.

6. _Long Beach, California_ (OEA, June 1974), p.2.

7. Mike Cooley in David Griffiths, ed. _Defence Cuts and Labour's Industrial Strategy_ (CND Labour Committee, September 1975).

8. _The Impact of Defence cutbacks_ (OEA, July 1973), p.11.

9. _Long Beach, California_ (OEA, June 1974).

10. _Newport Naval Base, Rhode Island_; the report does not indicate any civilian unemployment directly resulting from the base closure, although there would obviously be indirect effects without a conversion plan; in this case the base itself is converted.

11. _Glasgow AFB_ (OEA, December 1975).

12. David Greenwood and Timothy Stone, _The Clyde Submarine Base and the Local Economy_ (February 1975), paper for Scottish CND conference.

13. _Replacing Employment at the Nuclear Bases_ (Scottish CND, February 1975).

16   THE LUCAS AEROSPACE CORPORATE PLAN

by Steve Vines

In January 1976, the Lucas Aerospace Combine Shop Stewards
Committee published their proposals for the production of
alternative products. The committee, representing shopfloor trade unionists throughout the company, hoped to
suggest ways to secure employment and establish their right
to work on socially useful and needed products. The
Corporate Plan which emerged provided a tremendous stimulus
to those who were campaigning for cuts in defence expenditure,
yet were concerned that such cuts might cause unemployment.
A detailed alternative to weapons manufacture, coming from
the workforce of one of Britain's major military producers
showed that cutting defence expenditure should not necessarily
bring about unemployment for defence workers.

For this reason the Labour Party's Defence Study Group paid
considerable attention to this initiative; in fact one of the
Combine Committee's members, Mike Cooley, was an active
participant in the group. This summary is an edited compilation of some of the many published essays about the Corporate
Plan. The Study Group examined the Corporate Plan itself in
some detail but as this is readily available from the Secretary
of the Combine Committee (Mr E. Scarbrow, 86 Mellow Lane,
East Hayes, Middlesex, UK), we have tried to present a more
general picture, showing what the plan is about, how it was
received and discuss some of the questions, raised by the
response.

The main sources for this compilation are:
The Corporate Plan - by the Lucas Aerospace Combine Shop
    Stewards Committee (LACP).
'Design, Technology and Production for Social Needs - An
    Initiative by the Lucas Aerospace Workers' - by Mike Cooley
    in The New Universities Quarterly, Vol.39, No.1.
'The Lucas Aerospace Alternative Corporate Plan' by Dave
    Elliott in Alternative Work for Military Industries,
    Richardson Institute for Conflict and Peace Research,
    London, 1977.

The Genesis of the Plan (1)

The idea of preparing an overall Corporate Plan arose at a
meeting in November 1974 with Tony Benn, then Minister of
Industry, requested by the Combine Committee to discuss the
possible nationalization of Lucas Aerospace. (2) As the
introduction to the Corporate Plan reports:

    In the course of the meeting Mr. Benn suggested that
    there was the distinct possibility of further cutbacks
    in certain aerospace and military projects. Even if
    this did not occur the rate at which new projects would
    be started was likely to be reduced. Accordingly he
    felt that the Combine Committee would be well advised
    to consider alternative products, not excluding inter-
    mediate technology on which our members could become

engaged in the event of a recession.

The Plan also quotes Roy Mason, then Defence Secretary, who pointed out in the House of Commons 'that with few new projects coming along there would be a marked reduction over the next decade in the level of activity in military aerospace projects, particularly on the design side.' (Financial Times, 7.5.75.)

Although they viewed these reductions 'as both inevitable and desirable', pointing out that 'it is the national policy of almost all the unions the Combine represents that there should be cuts in defence expenditure', they were only too aware that,

> when these cuts are made our members are placed in the position of being made redundant or fighting for their continuation. We ourselves have done this in the past and will support our colleagues in the rest of the aerospace industry in doing so in future. Indeed, recently when the campaign to protect the H.S.146 (3) was at its height our members at the Wolverhampton plant seized drawings in support of their colleagues at Hawker-Siddeley's. (LACP)

Likewise, they realized 'that the traditional method of fighting for the right to work has not been particularly successful. Between 1960 and 1975 the total number in the aerospace industry has been reduced from 283,000 to 195,000 workers.' (LACP)

Apart from the specific problems there was the more general problem of structural unemployment:

> Industries are tending to become capital intensive rather than labour intensive. Over the past 8 or 9 years there has been some 5,000,000 people permanently unemployed in the United States. The same sort of structural difficulties are now manifesting themselves even in West Germany where there are some 1,000,000 people out of work and some 700,000 on short time working. (LACP)

The Combine Committee feared that these structural problems would be 'further compounded by the rationalization of the European aerospace industry within the Common Market' and further anticipated that Lucas Aerospace would 'attempt a rationalization programme with its associated companies in Europe'. (LACP)

The Committee also felt a commitment to diversification was necessary as a consequence of government's plans to nationalize the aerospace industry. Lucas Aerospace and other component firms were not to be included in the nationalization programme and there would consequently be a changed relationship between Lucas and the nationalized firms.

460

There were clear indications from our fellow trade unionists in those bodies that they will not be prepared to see the lucrative parts of the industry hived off by the component manufacturers; in this we fully understand their motives and support them. Rolls-Royce workers have already forced a project to be pulled back from Lucas so as to prevent redundancies in Rolls-Royce. (LACP)

In particular, the Committee did not want to permit aerospace component firms like Lucas to 'cream off the research and development which was paid for by the taxpayer':

As trade unionists we do not wish to see a relationship between the aerospace component firms and the nationalized sector of the industry which would be similar to the relationship of the equipment manufacturers to the National Coal Board. Such a relationship would provide the opportunity for those forces in society hostile to nationalization to point out that nationalized industries were economically unsuccessful. (LACP)

Thus, the creation of non-aerospace jobs was vital.
Against the background of all these pressures, the Combine Committee saw the Corporate Plan as having two basic objectives:

Firstly, to protect our members right to work by proposing a range of alternative products on which they could become engaged in the event of further cut-backs in the aero-space industry.
Secondly, to ensure that among the alternative products proposed are a number which would be socially useful to the community at large. (LACP)

The Committee realized both that the company's existing policies, product range and market could not guarantee jobs, and that many pressing but unmet social needs could be satisfied using the workforce's skills. As the Committee pointed out,

The desire to work on socially useful products is one which is now widespread through large sectors of industry. The aerospace industry is a particular glaring example of the gap which exists between that which technology could provide, and that which it actually does provide to meet the wide range of human problems we see about us. There is something seriously wrong about a society which can produce a level of technology to design and build Concorde but cannot provide enough simple urban heating systems to protect the old age pensioners who are dying each winter of hypothermia. (LACP)

461

The Right to Useful Work

Ken Coates has recently edited a book bearing this title. (4)
It contains articles examining this issue in a variety of
industries (shipbuilding, the energy industries and the arms
industries). Starting from the problem of unemployment as a
growing threat to workers' livelihoods, the book examines
responses to this threat and shows that there is a possibility
of transforming this problem into a creative possibility for
effecting changes in both the running of industry and determination of the goods produced. Workers should no longer be
considered as merely passive participants in the production
process, they have a right to both secure their own employment
prospects and have a say in what they produce. A society
which is unable to guarantee employment for those who want to
work and to produce needed goods has no right to object to the
extension of collective bargaining into this area. Indeed
the notion that the trade unions should play a wider role in
industry than their former emphasis on questions of pay and
conditions is gaining widespread credence.

The idea of a right to useful work has a long tradition in
the Labour movement. It is part of the broader concept of
the submission of the productive process to the needs of mankind
rather than the submission of mankind to this process. The
pioneer socialist, William Morris, spoke about this problem
when he described some labour as being a curse rather than a
blessing. Speaking to the Hampstead Liberal Club in 1884
he suggested that 'it would be better for the community and
the worker if the latter were to fold his hands and refuse to
work' if the work he were doing were a 'mere curse, a burden
to life'. He continued, 'It remains for us to look to it
that we do really produce something, and not nothing that we
want or are allowed to use.' (5)

The Lucas Aerospace Plan therefore is both about alternatives
to arms production and in many ways more fundamentally, a
challenge to some deeply held assumptions about work, management and industrial democracy. It is unsurprising that this
challenge has come from people working in an industry concerned
with producing better means of destruction; working towards
this end product must have provided a considerable stimulus to
think about alternative uses for the resources of the military
industries.

Technological Development versus Social Progress?

Mike Cooley from the Lucas Shop Stewards Combine has pointed
to the contradictions found in technologically advanced
societies who have failed to match technological achievements
with social needs. He saw this as a substantial motivating
force for the Corporate Plan; four contradictions in particular
were identified:

First, there is the appalling gap which now exists
between that which technology could provide for society
and that which it actually does provide. We have a
level of technological sophistication such that we can
design and produce Concorde, yet in the same society
we cannot provide enough simple heating systems to
protect old-age pensioners from hypothermia. In the
winter of 1975-6, 980 died of the cold in the London
area alone. We have senior automotive engineers who
sit in front of computerised visual display units
'working interactively to optimise the configuration'
of car bodies such that they are aerodynamically stable
at 120 miles an hour when the average speed of traffic
through New York is 6.2 miles an hour. It was in fact
11 miles an hour at the turn of the century when the
vehicles were horsedrawn. In London at certain times
of the day it is about 8.5 miles an hour. We have
sophisticated communication systems such that we can
send messages round the world in nano seconds, yet it now
takes longer to send a letter from Washington to New
York than it did in the days of the stage coach. Hence
we find the linear drive forward of complex esoteric
technology in the interests of the multinational
corporations and on the other hand the growing deprivation
of communities and the mass of people as a whole.

The second contradiction is the tragic wastage our
society makes of its most precious asset - that is the
skill, ingenuity, energy, creativity and enthusiasm, of
its ordinary people. We now have in Britain 1.6 million
people out of work. There are thousands of engineers
suffering the degradation of the dole queue when we
urgently need cheap, effective and safe transport systems
for our cities. There are thousands of electricians
robbed by society of the right to work when we urgently
need economic urban heating systems. We have, I believe,
180,000 building workers out of a job when by the government's own statistics it is admitted that about 7 million
people live in semi-slums in this country. In the London
area we have about 20 per cent of the schools without an
indoor toilet, when the people who could be making these
things are rotting away in the dole queue.

The third contradiction is the myth that computerization,
automation and the use of robotic equipment will automatically free human beings from soul-destroying, back-breaking
tasks and leave them free to engage in more creative work.
The perception of my members and that of millions of workers
in the industrial nations is that in most instances the
reverse is actually the case.

Fourthly, there is the growing hostility of society at
large to science and technology as at present practised.
If you go to gatherings where there are artists, journalists
and writers and you admit to being a technologist, they

treat you as some latter day Yahoo, to misquote Swift. They
really seem to believe that you specified that rust should be
sprayed on car bodies before the paint is applied, that all
commodities should be enclosed in non-recycleable containers
and that every large-scale plant you design is produced
specifically to pollute the air and the rivers. There
seems to be no understanding of the manner in which
scientists and technologists are used as mere messenger
boys of the multinational corporations whose sole concern
is the maximization of profits. It is therefore not
surprising that some of our most able and sensitive sixth
formers will now not study science and technology because
they correctly perceive it to be such a dehumanized activity
in our society.

All these four contradictions - and indeed many others -
have impacted themselves upon us in Lucas Aerospace over the
past five years. We do work on equipment for Concorde,
we have experienced structural unemployment and we know day
by day of the growing hostility of the public to science
and technology.

It seemed absurd to us that we had all this skill and
knowledge and facilities and that society urgently needed
equipment and services which we could provide, and yet the
market economy seemed incapable of linking these two.

The Corporate Plan

Initially the Combine Committee tried to obtain advice and
assistance for formulating the plan from a wide range of
specialists, trade unions and other organizations. With
a few significant exceptions, none of these individuals or
organizations responded.

Mike Cooley describes the Combine Committee's next move:

We then did what we should have done in the first instance:
we asked our own members what they thought they should be
making. I have never doubted the ability of ordinary
people to cope with these problems, but not doubting it
is one thing, having concrete evidence is something different.
That concrete evidence began to pour into us within three or
four weeks. In a short time we had 150 ideas of products
which we could make and build with the existing machine tools
and skills we had in Lucas Aerospace. We elicited this
information through our shop stewards committees via a
questionnaire. I should explain that this questionnaire
was very different from those which the soap powder companies
produce where the respondent is treated as some kind of
passive cretin. In our case, the questionnaire was dialec-
tically designed. By that I mean that in filling it in the
respondent was caused to think about his or her skill and
ability and environment in which he or she worked and the
facilities which they had available to them. We also

464

deliberately composed it so that they would think of themselves in their dual role in society, that is both as producers and as consumers. We were therefore quite deliberately transcending the absurd division which our society imposes on us, which seems to suggest that there are two nations, one that works in factories and offices and an entirely different nation that lives in houses and communities. We pointed out that what we do during the day at work should be meaningful in relation to the communities in which we live. We also deliberately designed the questionnaire to cause the respondents to think of products not merely for their exchange value but for their use value.

When we collected all these proposals we refined them into six major product ranges which are now embodied in six volumes, each of approximately 200 pages. They contain specific technical details, economic calculations and even engineering drawings. We quite deliberately sought a mix of products which on the one hand included those which could be designed and built in the very short term and those which would require long-term development; those which could be used in metropolitan Britain mixed with those which would be suitable for use in the third world, products incidentally which could be sold in a mutually non-exploitative fashion. Finally we sought a mix of products which would be profitable by the present criteria of the market economy and those which would not necessarily be profitable but would be highly socially useful.

The Proposals (6)

The end product of the Combine Committee's work was a detailed plan for the manufacture of 150 products. The proposals are of varying stages of elaboration; at the time of writing work continues at individual plant level and nationally in co-operation with the newly formed Centre for Alternative Industrial and Technological Systems.

The proposals for products fall into six major areas of technological activity: oceanics, telechiric machines, transport systems, braking systems, alternative energy sources and medical equipment.

We examine below some of the products proposed and their possible uses.

Alternative Energy Technologies.

Lucas already had considerable experience of heat pump technology and used to manufacture a small wind-electric machine. The company's expertise in aerodynamics would, the stewards considered, make it well suited to developing a series of

465

large scale windmills for both electricity generation and
direct (friction) heating (e.g. for greenhouses).   Mike
Cooley cites one of the proposals:

> Drawing on our aerodynamics knowhow, we have proposed a
> range of wind generators.   In some instances these
> would have a unique rotor control in which the liquid
> which is used as the media for transmitting the heat
> is actually used to achieve the breaking and is thereby
> heated in the process itself.

It was clear that in considering these various technologies,
the Lucas workers were concerned to develop systems relevant
on the community scale - i.e. for complete housing estates -
rather than individual domestic units which, as one steward
put it, could be just 'gimmicks for individual architect
built houses' and 'playthings for the middle classes'.
    In the field of solar power, they proposed development of
switching and control circuits for solar heating systems and
fluid dynamics.
    Mike Cooley describes why the plan focussed on the crucial
issue of energy storage:

> One of the proposals for storing energy was to produce
> gaseous hydrogen fuel cells.   These would require
> considerable funding from the government but would
> produce means of conserving energy which would be
> ecologically desirable and socially responsible. We
> also designed a range of solar collecting equipment
> which could be used in low energy houses and we worked
> in conjunction with Clive Latimer and his colleagues
> at the North East London Polytechnic in producing
> components for a low energy house.   I should add that
> this house was specifically designed so that it could
> be constructed on a self-build basis.   In fact some of
> the students working on the Communications Design Degree
> course at that polytechnic are now writing an instruction
> manual which would enable people without any particular
> skills to go through a learning process and at the same
> time to produce very ecologically desirable forms of
> housing.   One can now see that if this concept were
> linked to imaginative government community funding it
> would be possible in areas of high unemployment where
> there are acute housing problems to provide funds to
> employ those in that area to build their own housing.

Among the energy proposals are some which would be of
particular interest to Third World countries.   The emphasis
here is on self-sufficient products rather than the supply
of the sort of goods which increase the dependence of the
Third World countries on products from the industrialized
world.   Instead of the usual policy of supplying these
countries with single purpose products for fixed tasks the

466

Lucas plan proposes a power pack which is capable of providing a
wide range of services.   This power pack could operate on a
range of indigenous fuels and methane and which, by means of a
variable speed gear box, would be capable of alternatively
pumping water, compressing air, providing high pressure
hydraulics and generating electricity.   This would, therefore,
be a sort of universal power pack which could provide a small
village or community with a range of services.   This is quite
contrary to the present design methodology which seeks to do
actually the reverse.

Medical Technologies

Much of the interest in the Lucas Plan has focussed on the
medical products suggested by the Combine Committee.   With
the increasing concern shown about the state of the National
Health Service, the Lucas proposals suggested some ways in
which shortcomings could be avoided.   It should be noted
however that the emphasis on advanced and therefore relatively
expensive technological products has been criticized by those
who feel that the limited resources of the NHS would be better
devoted to more basic provision of health care and preventive
medicine.
    Mike Cooley explains how the Combine Committee proposes to
extend the existing interest of the company in medical products:

    In the medical field Lucas already makes pacemakers and
    kidney machines.   About three years ago the company
    attempted to sell off its kidney machine division to an
    international company operating from Switzerland.   We
    were able to prevent them doing so at that time both by
    threats of action and the involvement of some MPs.   When
    we checked on the requirements for kidney machines in
    Britain we were horrified to learn that 3,000 people die
    each year because they cannot get a kidney machine.

    The plan therefore proposes an increased output of these
machines.

    Before we even started the corporate plan our members at the
    Wolverhampton plant visited a centre for children with
    Spina Bifida and were horrified to see that the only way
    they could propel themselves about was literally by
    crawling on the floor.   So they designed a vehicle which
    subsequently became known as Hobcart - it was highly
    successful and the Spina Bifida Association of Australia
    wanted to order 2,000 of these.   Lucas would not agree
    to manufacture these because they said it was incompat-
    ible with their product range and at that time the
    corporate plan was not developed and we were not able
    to press for this.   But the design and development of
    this product were significant in another sense: Mike

Parry Evans, its designer, said that it was one of the most enriching experiences of his life when he actually took the Hobcart down and saw the pleasure on the child's face. — It meant more to him, he said, than all the design activity he had been involved in up to then. For the first time in his career he actually saw the person who was going to use the product that he had designed. It was enriching also in another sense because he was intimately in contact with a social human problem. He literally had to make a clay mould of the child's back so that the seat would support it properly. It was also fulfilling in that for the first time he was working in the multi-disciplinary team together with a medical doctor, a physiotherapist and a health visitor. I mention this because it illustrates very graphically that it is untrue to suggest that aerospace technologists are only interested in complex esoteric technical problems. It can be far more enriching for them if they are allowed to relate their technology to really human and social problems.

Some of our members at another plant realised that a significant percentage of the people who die of heart attacks die between the point at which the attack occurs and the stage at which they are located in the intensive care unit in the hospital. So they designed a light, simple, portable life support system which can be taken in an ambulance or at the side of a stretcher to keep the patient 'ticking over' until they are linked to the main life support system in the hospital. They also learned that many patients die under critical operations because of the problem of maintaining the blood at a constant optimum temperature and flow. This, it seemed to them was a simple technical problem if one were able to get behind the feudal mysticism of the medical profession. So they designed a fairly simple heat exchanger and pumping system and they built this in prototype. I understand that when the assistant chief designer at one of our plants had to have a critical operation they were able to convince the local hospital to use it and it was highly successful.

The plan also proposed the development of artificial limb control systems, drawing on expertise in aerospace control engineering; for research into sight-substituting aids for the blind, again drawing on experience in aerospace; and for specialized aids for operating theatres, including body cooling systems and improved surgical equipment.

Telechiric Devices

The Plan proposes a range of equipment which would mimic the motions of a human being (telechiric devices) thus preserving human skills, although these skills would be used at a control

point away from the more dangerous environment of say, a
mine or underwater vessel, or pipe line. Similar applications could be found in fire fighting and other circumstances
where human skills can cope with the problem but human life
should not be endangered in the process - 'Thus,' writes
Mike Cooley,

> human beings would continue to be involved in that
> precious learning process which comes about through
> actually working on the physical world about us and
> it would also mean that we would be countering
> structural unemployment. We would, in a word, be
> very creatively linking a relatively labour-intensive
> form of work with a reasonably advanced and responsible
> technology.

Transport Systems

The emphasis in these proposals is on energy conservation,
mass transportation and increasing the scope of transport
provision over difficult terrain.
   In view of the recurrent discussions about the so-called
energy crisis, one of the plan's most exciting proposals is
for a hybrid power pack which could be used in cars, coaches,
lorries or trains. Mike Cooley explains:

> There is now a growth in the use of battery driven
> vehicles. This is clearly ecologically desirable
> but has the great disadvantage that in a stop-start
> situation they have to be charged every forty miles
> and on a flat terrain about every hundred miles.
> We are proposing a power pack in which we have a
> small internal combustion engine running at its constant
> optimum reve: this will mean that all the energy which
> is lost as one accelerates, decelerates, idles at
> traffic lights, starts cold and so on, is put in as
> useful energy through a generator which charges a
> stack of batteries which then operates an electric
> motor. Our initial calculations (which have subse-
> quently been supported by work done in Germany),
> suggest that this would improve specific fuel consump-
> tion by 50 per cent; it would reduce toxic emissions,
> since the unburned gases are not going out into the
> atmosphere, by about 80 per cent. Further, since the
> whole system would be running at constant revs one
> could calculate all the resonance of the system and
> effectively silence it: our calculations suggest that
> a power pack of this kind would be inaudible against a
> background noise of about 60 or 70 decibels at 10
> metres.
>   It may be asked, of course, why such a power pack
> had not been designed and developed before. The

simple answer, it seems to us, is that such a power
pack would have to last for about ten or fifteen years
and this is absolutely contrary to the whole ethos of
automotive design, which has as its basis the notion
of a throwaway product with all the terrible waste of
energy and materials which that implies.   We are
convinced that Western society cannot carry on in
this wasteful and arrogant fashion much longer.

Another proposal is for:

a unique road-rail vehicle which is capable of driving
through a city as a coach and then running on the
national railway network.   It could provide the basis
for a truly integrated, cheap, effective public transport system in this country.   It uses pneumatic
tyres and is therefore capable of going up an incline
of 1:6 - normal railway rolling stock can only go up
an incline of 1:80.   This meant in the past that
when a new railway line was laid down it was necessary
literally to flatten the mountains and fill up the
valleys or put tunnels through them.   This costs about
£1 million per track mile - this was the approximate
cost of the railway in Tanzania which the Chinese put
down.   With our system a track can be put down at
about £20,000 per track mile since it follows the
natural contours of the countryside.   This vehicle
would therefore not only be of enormous use in metropolitan
Britain but would also be of great interest in developing
countries and even in areas such as Scotland and some of
the less densely populated areas in Europe.

Both the Highlands and Islands Development Board in Scotland
and the Tanzanian and Zambian Governments have expressed
considerable interest in this vehicle.
    The Combine Committee also investigated possible joint
production of a new type of airship, using vectored thrust
jet engines for fine control over its loading/unloading
position, drawing on experience gained with the control
system for the Harrier 'jump jet'.
    In the transport field there is also considerable emphasis
on the development of braking systems.   Lucas already has
experience of producing retarder braking systems for
vehicles.   The Plan envisages production of auxiliary or
fail safe braking systems for cars, coaches and trains.
Systems would also be produced to cope with the varying
needs of heavy road and rail vehicles and the hybrid road/
rail vehicle described above; as well as anti-skid systems
and an integrated braking system incorporating both mechanical disc brakes and dynamometers.

470

Other Proposals (7)

Most of these proposals suggest an expansion or modification
of existing product ranges to be found within the Lucas
Aerospace group. They include:

* Work on fuel cells technology, particularly the development of a 30 kw fuel cell power plant, using gaseous
  hydrogen and oxygen.

* Application of micro-processor technology for the systems
  and products proposed elsewhere in the Plan.

* Standby power units for the computer industry using
  automatic sensing and starting systems.

* Development of complete systems for submersibles and
  the application of telecheiric devices for underwater
  work.

Other proposals under consideration for the future are:

* Linear motors to operate pumps and compressors.

* A new range of applications for the existing '60 and 90
  Gas Turbine' engines.

* A robot helicopter using Lucas gas turbine for crop
  spraying.

* High speed motors.

* Ballscrews for converting rotating to linear motion or
  vice-versa, with wide application to machine tools and
  other products in the Plan. A new type of screw has been
  long awaited by industry, such a development has
  considerable potential.

The Production Process (8)

In addition to these and other product proposals not made
public, the Corporate Plan includes proposals for radically
restructuring work organization and control, to reverse the
tendency towards fragmentation, de-skilling, and an ever
increasing work pace. The plan calls for new forms of
autonomous work groups and project teams, so that

> the skill and ability of our manual staff worker is
> continually used in closely integrated production
> teams, where all the experience and common sense of
> the shop floor workers would be directly linked to

the scientific knowledge of the technical staff. (LACP)

Although the alternative products were selected to
utilize existing skills, the plan's authors recognize that
for such a dramatic programme of diversification considerable
retraining would be needed for both blue and white collar
workers. They were highly critical of existing plans for
employee development:

> Very little is being done to extend and develop the
> very considerable skills and ability still to be
> found within the workforce ... there is little indication
> that the company is embarking on any real programme of
> apprenticeships and the intake of young people ... The
> company is making no attempt to employ women in technical
> jobs, and apart from recruiting these from outside there
> are many women doing routine jobs well below their
> existing capabilities. (LACP)

Consequently the plan calls for a major appraisal of the
employee development programme, arguing that: 'The entire
workforce including semi-skilled and skilled workers are
capable of retraining for jobs which would greatly extend
the range of work they could undertake.'

It was also argued, that in the event of any cutbacks,
redundant workers should have the opportunity to opt for
training, transforming the potential redundancy into 'a
positive breathing space during which re-education could
act as an enlightened form of work sharing'.
Developing this proposal it was argued that the government could use money which would normally be spent on
earnings related unemployment benefits (and perhaps also
on Temporary Employment Subsidies) to fund this type of
part-time education, thus both saving people from the dole
queue and making a positive investment for the future with
taxpayers' money.

Markets

Central and local government would clearly be a major consumer of many of the products proposed in the Corporate Plan.
Ultimately, central government is the only domestic consumer of
current Lucas defence products, and in theory at least it
would be possible to find a way of shifting procurement
expenditures in order to buy alternative products designed
for public sector consumption. In fact, reallocating
expenditure is a complex process, but proper planning and
liaison between the relevant government departments and local
authorities should be up to solving any problems.
Some of the products suggested are already in production
and others have a proven market judged by conventional

472

commercial standards. A central concern of the plan however
is to produce goods which directly meet people's needs. The
shop stewards combine have made considerable efforts to identify what these needs might be. Representatives were despat-
ched to local community and environmental groups, to other
trade union bodies and, in some cases, to individual potential consumers. Although an important exercise, this kind
of ad-hoc survey by the Combine Committee cannot be regarded
as an exhaustive market assessment.

Many of the objections to the plan have revolved around
the question of markets for the proposed products.
Government ministers, for example, pointed out in the course
of the Labour Party Study Group's meetings that central
and local government were, as a result of public spending
reductions, in no position to buy many of the products
intended for their consumption. It was also suggested that
viable markets simply did not exist for some of the more
original products and that if they did the usual commercial
considerations would surely have persuaded the company to
make the products. These arguments have never been
elaborated to a sufficient degree of precision so that a
judgment can be made of their validity. The reason is that
the company, who are the only body which can do this, are
unprepared to carry out market feasibility studies on behalf
of the Combine Committee. However, they did apparently
carry out an internal market assessment of one of the
suggested products, the heat pump, which indicated the
existence of a considerable market for this technology.(9)
The proposal to expand production of kidney machines, was
given indirect encouragement in the government's April 1978
budget when the Chancellor announced the allocation of
sufficient funds to purchase 400 additional units and associated
facilities. These indications as well as the interest shown
by potential consumers has been such to suggest that the plan
is pointing in the right direction. Even so marketing
considerations seem to be the weakest aspect of the plan, but
in fairness it should be stated that the Combine Committee is
in no position to undertake a marketing operation. However,
work is currently underway at the Centre for Alternative
Industrial and Technological Systems to produce a market
survey for some of the products.

The Plan challenges the established idea of what constitutes
a market, denying that the criteria of profitability is the
most important aspect. The combine committee say: 'progress
can only be minimal so long as our society is based on the
assumption that profits come first and people last.' (LACP)

The criterion of profitability is therefore replaced by
that of social need.

The Response

Although the prime purpose of producing the Corporate Plan
was to have firm proposals ready for the advent of redundancy
schemes, the Committee initially hoped to be able gradually to

negotiate the implementation of their proposals in advance of redundancies.   Not that sweeping changes were expected:

> It is not suggested in this report that Lucas Aerospace is suddenly going to cease to be deeply involved in the aerospace industry.   We recognise, whether we like it or not, that the aerospace industry is going to remain a major part of the economic and technological activity of the so-called 'technologically advanced nations'.   The intention is rather to suggest that alternative products should be introduced in a phased manner such that the tendency of the industry to contract would firstly be halted and then gradually reversed as Lucas Aerospace diversified into new fields. (LACP)

Even with this more modest strategy, problems were envisaged:

> There is obviously the danger that the discussion with the management about the implementation of the Plan (if it were agreed that such discussions should take place) could gradually degenerate into a form of collaboration. There is also the danger that even if collaboration were carefully avoided the company might simply take parts of the Corporate Plan and have all this technology on the free.   The Plan has taken a very considerable length of time to prepare and involved many evenings and weekends of work.   It has also meant that outside experts have been prepared to give generously of their detailed knowledge in order to help the development of the Corporate Plan.   In these circumstances the greatest care will have to be taken to ensure that the company does not succeed in drawing off the 'money spinners' from the Plan, and perhaps even having these produced abroad, whilst declining those products which would be socially useful.   It is even conceivable that whilst the company would take sections of the Plan, our members may still be confronted with the perennial problem of redundancy.   Because of these dangers it is suggested that the correct tactic would be to present only part of the plan to the company, and then to test out in practice the manner in which the company will attempt to deal with it.   (LACP).

Accordingly, the Committee initially presented only 12 of the plan's 150 product proposals to the company, in January 1976.   The Labour government were approached in the hope that the Plan could be adopted as a 'planning agreement' between government, trade unions and the company. If the Combine Committee found the initial response to requests for assistance in drawing up the plan disappointing they must have found the reaction to the finished product

474

positively disheartening. The company are adamant in their refusal to give serious consideration to the proposals, the government have refused to intercede between the company and the Combine Committee, and the main trade union body in the industry, the Confederation of Shipbuilding, and Engineering Unions (CSEU) has, at best, been ambivalent in its attitude towards both the Combine itself and its proposals. The response in other words is a damning indictment of the inability of British industry and those associated with it to think creatively about the problems it faces, and accept that many of these problems can be actually overcome. The plan has demonstrated that those best equipped to tackle these problems are often those employed at the point of production and that these people often have a clearer idea of possibilities than some others who are commonly regarded as 'experts'.

In the following extract, Mike Cooley describes some of the disappointments and encouragement received by the Combine Committee since the publication of the Plan:

We have of course approached the government and we have had every sympathy short of actual help! We have been enormously impressed at the ability of the various ministries to pass the buck; indeed we have experienced at first hand the white heat of bureaucracy. Although the Company has centrally rejected the Corporate Plan and is now refusing to meet the Combine Committee to discuss it, no Minister has been prepared to insist that the company should meet us to do so. In fact junior ministers, like Les Huckfield, continuously write to us saying 'In my considered view those best suited to deal with this question are the company and the trade unions involved'. It is absolutely clear that the company will not and has not met us to discuss the plan. However, support from the trade union movement is growing – large shop stewards committees at Chryslers, Vickers, Rolls Royce and elsewhere are now discussing corporate plans of this kind. One of our colleagues from Burnley, Terry Moran, has made a tour of trades councils in Scotland discussing these matters. The Combine Committee is now itself organizing a series of meetings in the towns in which Lucas Aerospace has sites. We believe it is arrogant for aerospace technologists to think that they should be defining what communities should have. We are seeking through the local trade unions, political parties and other organizations in each area to help us to define what they need and to begin to create a climate of public opinion where we can force the government and the company to act. At the national level the TUC has produced a half an hour television programme on BBC 2 dealing with our corporate plan: this is part of its Trade Union training programme for shop stewards. The Transport & General Workers Union has just come out with

a statement indicating that its shop stewards throughout
the country should press for corporate plans of this kind.
At an international level the interest has been truly
enormous. In Sweden, for example, they have produced
six half-hour radio programmes dealing exclusively with
the Corporate Plan and have made cassettes which are now
being discussed in factories throughout Sweden. They
have also made a one hour television programme and a
paperback book has been produced dealing with the corporate
plan. Similar developments are taking place in Australia
and elsewhere and the interest centres not merely on the
fact that a group of workers for the first time are
demanding the right to work on socially useful products,
but that they are proposing a whole series of new
methods of production, where workers by hand and brain
can really contribute to the design and development of
products and where they can work in a non-alienated
fashion in a labour process which enhances human beings
rather than diminishes them.

   The more recent history of the Lucas Aerospace has been
marked by a series of redundancy plans announced by the company.
A 'get tough' attitude by management towards leading members of
the Combine Committee and growing evidence that the firm would
like to move as much of its production outside 'troublesome'
Britain as possible. (10) On the positive side, some moves
have been made towards implementing the plan at a local level,
the Burnley plant took up particular proposals in 1976 and, at
the time of writing, the company have agreed to start discussions
at two plants, threatened with redundancies. These discussions
will clearly centre on the corporate plan which is probably
the reason why management are keen to exclude from the talks
those who were most active in drawing up the proposals. On
the union side there has also been some movement, the CSEU has
had meetings with the Lucas management to urge that serious
consideration should be given to the plan; some of the member
unions are in fact pushing for the Confederation to take a
firmer stand in this matter. As for the government position,
this is a matter of concern because the Department of Industry
maintain that they are constrained from acting for purely
procedural reasons, that is, that they can only act on a request
from the CSEU and that that request has not been forthcoming.
They are being urged by both the Labour Party National Executive
Committee and a large group of Labour MPs and trade unionists to
intervene and use their considerable influence to persuade the
Company to embark on serious negotiations about the plan.
   Mike Cooley writes:

   Our society in the past has been very good at technical
   invention but very slow at social innovation. We have
   made incredible strides technologically but our social
   organizations are virtually those which existed several
   hundred years ago. One of the Swedish television

interviews said 'when one looks at Britain in the past it has been great at <u>scientific and technological invention</u> and frequently has not really developed or exploited that. The Lucas workers corporate plan shows <u>great social invention</u> but it probably is also the case that they will not develop or extend that in Britain' (my emphasis). If this were true it would be very sad indeed.

This is a very pessimistic conclusion but it should be remembered that whatever the eventual fate of the plan its very production has been of tremendous importance and there is therefore no possibility that the plan could be regarded as a waste of time. Both the Combine Committee themselves, and others who have studied their work have drawn various conclusions from the plan and its reception. For the purpose of the work of the Labour Party study group with its emphasis on alternatives to military production, the Lucas initiative can be seen as a breakthrough. It brought alive the years of discussion about alternatives to military production and showed that the alternatives were real. During the work of the study group we looked at other workforce initiatives and experience of industrial conversion abroad, some of which has been stimulated by the Lucas Plan itself.
In Britain the Lucas Aerospace Combine Committee have been the pioneers, historical experience seems to indicate that pioneers often get a rough deal even though the value of their work is widely acclaimed after they have ceased to be in a position to benefit. If the workers at Lucas were to suffer this fate it would not be as a result of the Laws of nature or some other mystical process, but because those in a position to take decisions have failed to carry out their responsibility to society. A failure to make use of one of the most imaginative products of British industry in recent history will have wider repercussions than those affecting the Lucas Aerospace workforce. That the initiative should have come from workers in the arms industries is unsurprising but probably more problematic than if it should have arisen elsewhere.
The arms industry has been particularly dependent on safe markets, easy access to finance and a lack of penalties for rising costs. (11) Arms production is characteristically known to be carried out according to marketing criteria which would be the envy of other civil producers. Few other industries can consistently turn to the government asking for orders which, as an end in themselves, merely fill otherwise surplus capacity, rather than, as a prime consideration, fulfilling a particular requirement for goods. The owners of the arms producing industry therefore have a tremendous vested interest in remaining safely insulated within the sphere of military production, they less than others, will wish to venture out into the vagaries of the free market, the principle of which they profess to hold so dear. That

is why the Combine Committee realized at an early stage in
their work that the plan would not gain acceptance by gentle
persuasion or indeed without vigorous governmental action.
Mike Cooley offers this conclusion:

> We in Lucas Aerospace are trying ... (to form a) ...
> political and ideological view of what we want technology
> to do for us and the courage and determination to fight
> for its implementation. We hope that in that fight we
> will be supported by widespread sections of the community
> because we will not be able to create an island of
> responsibility in Lucas Aerospace in a sea of national
> depravity.

Notes

1. Taken from Dave Elliot, 'The Lucas Aerospace Corporate
   Plan'.

2. This was before plans had been completed for the nationalization of the airframe and shipbuilding industries.

3. The HS 146 is a shorthaul aircraft; the Government have
   now announced plans for it to go into production.

4. Ken Coates (ed.), The Right to Useful Work (Spokesman,
   Nottingham 1978).

5. William Morris, 'Useful Work Versus Useless Toil' in
   Political Writings of William Morris, A.L. Morton (ed.),
   (London, 1973).

6. Taken from Dave Elliot, 'The Lucas Aerospace Corporate
   Plan', and LACP.

7. Ibid.

8. Ibid.

9. New Scientist (16 Feb.1978).

10. See Lucas - A British Company, survey prepared for APEX
    by TURU (1978).

11. This point is developed in Study Group Paper, No.8.

17 THE VICKERS PROPOSALS : BUILDING A CHIEFTAIN TANK AND THE ALTERNATIVE USE OF RESOURCES

by The Vickers National Combine Committee of Shop Stewards

The mechanical engineering industry is the largest in the UK. It accounts for nearly 5 per cent of all civilian employment, 20 per cent of all UK exports and 40 per cent of all UK plant and machinery. Since the war, the industry has stagnated; the British share of the world market has declined, fixed investment has been low, employment has fallen since the early 1970s and the value per ton of British engineering exports, a measure of their technical content, is lower than the value per ton of the engineering exports or imports of any of Britain's main competitors. A recent NEDC report concluded that a major requirement of the British mechanical engineering industry was improvement in the quality of detailed design. (1)

In contrast, the export of tanks has risen substantially in recent years and the value per ton of the Chieftain Tank is higher than most of its main competitors — the American M-60, the West German Leopard, the French AMX-30, or the Soviet T-54. (2) Some of the finest engineering skills and talents are involved in building tanks and guns. While ordnance and other military equipment represents a rather small proportion of the total output of the mechanical engineering industry, 60 per cent of the design engineers and scientists working in the industry are engaged in military work. It is evident that if those skills and talents were devoted to civilian work, to improvements in the quality of engineering exports as well as the plant and machinery needed to manufacture a wide range of civilian products, this could not only improve Britain's economic position but it could also contribute to the fulfilment of many basic needs in British society today.

This is particularly relevant for West Newcastle, where the Chieftain Tank is built However, competitive British armaments may be, the market is, by its nature, erratic and unpredictable. It depends on wars and arms races and shifting political allegiances. In peacetime, it can never be expected to grow as fast as civilian markets. For a hundred years, Vickers, or its predecessor Armstrong's, has dominated employment in West Newcastle. The arms business was built upon the experience of two world wars. Older people remember the days when 'tides of men' trudged up the streets from the Elswick Works at the end of their shifts. (3) Since then, employment has dwindled; from 12,250 at Elswick in 1939 and 20,850 at the height of World War II, to 5,700 in 1963, to 1,850 in October 1977. And this, despite the growth of arms exports and the success of the Chieftain tank, Elswick's staple, in the Middle East.

This was the reason why the Vickers Combine Committee undertook to examine the resources involved in the construction of a Chieftain Tank and the alternative uses to which they could be put. But there was also another more general reason. If unions are to challenge the widespread redundancies now occurring in the engineering industry and if victories won in wage struggles are not to

be overtaken by unemployment and inflation, it is important
that they be able to demonstrate at the level of the
individual plant, the concrete alternative strategies which
could ensure future work and social well being.
   This paper is divided into four sections. The first
section examines the market for tanks and the future
prospects for employment. The second section describes
the process of building a tank. The third section is
about the past experience of Vickers' Elswick Works in
the manufacture of commercial products and tries to explain
the failures. The last section considers the principles
on which an alternative strategy should be based and the
suggestions that have been made for alternative products.

The Future of Tanks

The Chieftain Tank that the Elswick workers are building for
Iran is claimed to be the most advanced in the world. It
is protected by the new Chobham plastic reinforced armour,
described by the Ministry of Defence as 'the single most
significant development in the design of tanks since World
War II'. (4)   Its 120mm gun is the most powerful tank gun
in the world. The 105mm gun on Centurion, Chieftain's
predecessor which was also built at Elswick, was widely
reported to have proved superior to Soviet guns in the Arab/
Israeli war of October 1973; the Chieftain gun is said to
be even better. Moreover, it uses an extremely sophisticated
and efficient fire control system, including infra-red
devices for night time operations and as a range finder. The
main weakness of earlier Chieftains, their lack of power and
speed, is said to have been corrected by the new Rolls Royce
engine which will replace the problematic Leyland L.60. The
Chieftain also has various little extras, like a collapsible
fabric screen on which it can float and facilities to enable
men to remain under the hatch for 48 hours in the event of
nuclear, chemical or biological attack. Its main disadvantage stems from its very sophistication - the fact that
complexity may entail a loss of ruggedness and reliability.
According to the latest Annual Report, 'prospective orders'
for this splendid weapon of war should 'ensure continuity of
production ... well into the 1980s'. (5)
   Beyond the 1980s, however, the future for tank orders is
uncertain. First of all, by that date the Americans and
Germans will have developed competing tanks, both using
Chobham armour. The German tank is likely to have an
advantage in speed and mobility. The American tank has
been designed, at least in part, to operate in desert
conditions and some versions may have a missile in place of a
gun. The British Government has made it clear that there
will not be a new British Battle Tank before the late 1980s.
Secondly, the current market for tanks largely consists of
oil-rich regimes in the Middle East (despite continued

Israeli interest, the British Government has vetoed the supply of Chieftains to direct participants in the Arab/ Israeli war). The repressive nature of these régimes is associated with political instability. Quite apart from the risks of British involvement in a war in the area, it is quite possible that such régimes may be overthrown by people who would not look kindly upon the arms suppliers of their oppressive predecessors. Thirdly, and perhaps most importantly, it is widely believed that tanks are obsolete. Lord Robens, for instance, considers that the future of armaments lies in electronics and not mechanical engineering and that tanks are unlikely to be used for much other than 'internal' policing in the future. The electronics revolution has led to a vast improvement in the accuracy of guided weapons so that all tanks, even those protected by Chobham armour, have become much more vulnerable. Except in the desert, where concealment is difficult anyway, the case for these big sophisticated vehicles which are hard to hide and expensive to replace is questionable. During the three weeks of the 1973 Middle East War, more than one tank was destroyed every 15 minutes. A British Government, operating under severe financial restraint, will have to take this argument seriously when deciding upon a replacement for Chieftain. Its logic has already been accepted, at least implicitly, by Vickers management when the tank design team was disbanded.

But even if it were possible to anticipate continued growth in tank orders, this would not provide any kind of employment guarantee. Contrary to the popular impression employment in defence work is just as insecure as civilian work, if not more so. Because individual plants are dependent on one or two major contracts, fluctuations in sales and jobs can be extreme. At Elswick, such fluctuations have been much greater on the military side than on the civil side. There have been major gaps in contracts which led to heavy redundancies; for example, in the early 1960s when work had ended on Centurion and not begun on Chieftain and in the late 1960s when the British Army's order for Chieftain had been completed and before export orders were received. This is shown in Table 1.

Moreover, despite the real increase in the volume of arms sales, employment has steadily declined. In 1939, tank orders amounted to £8m. In 1951, Centurion orders were £11m. The latest Chieftain order for Iran, which is to be shared with the Royal Ordnance Factory Leeds, is worth more than £500m. And yet, as we saw above, the labour force has fallen dramatically. On the last order, Elswick were producing 4 vehicles a month with 570 productive workers. On the new order, production is to be stepped up to 14 vehicles a month with only 520 productive workers. There are several reasons for this. One reason, probably the most important, is the utilization of spare capacity. As we shall see below, the organization of tank production tends towards underutilization of people and space. The bigger the contract, the easier it is to organize production

482

in such a way as to minimize this underutilization. Another reason is the introduction of numerically controlled (NC) machines, which can do several operations that were once undertaken by a number of different individual machinists There is one NC machine at Elswick which does 48 operations and it only requires a setter and an operator. On the new Chieftain order, for example, painters will be replaced by aerospray guns. Very often, the effect of new NC machines is not immediately apparent in redundancies. This is partly because there is currently an acute shortage of machinists and partly because the redundancies occur on the run down of a previous order before the machines are introduced. Thus the painters were declared redundant on the run-down of the last vehicle order. In fact, they demanded short time instead of redundancy and organized a stoppage. But in the end there was work going in the shipyards and so they accepted redundancy.

Finally, and perhaps most importantly, reduced employment can be explained by an increase in outside work. Before the war, Vickers built the entire tank - the gun, the armour plate, etc. Now Vickers only make 18 per cent of the value of the total tank. To some extent this is offset by the increased complexity of successive tanks. Chieftain, for example, has many more individual parts than did the Centurion. But there has been a substantial increase in the amount and sophistication of electronic equipment incorporated into tanks. Outside work takes the form of 'free issue', parts supplied free by ROF Leeds, the prime contractor for Chieftain, or subcontracting, including bought-in components like the engine or fire control equipment, by Vickers.

In the future, one can expect the introduction of more machines, and more improvements in capacity utilization. It is probable that, given the very low levels of employment at Elswick, the decline in jobs, assuming continued arms orders, is likely to be slower than before. But it is also most unlikely that a new Middle Eastern war or similar event is going to stem the gradual decay of West Newcastle.

Building the Chieftain Tank

As of October 1977, there were 1,846 people employed at Elswick, in the three divisions - Defence Systems, which used to be known as armaments, Pressings, and Non-ferrous Metals. The foundry, which employed about 200 people, was recently sold. Defence systems is the heart of the factory. It employs 1,517 people of which 608 are white collar workers. The proportion of overhead workers is higher than for any other division. Formerly, Elswick used to build naval armaments. Today, the entire work-force of defence system division is engaged in the construction of armoured fighting vehicles: the Chieftain Tank, the Vickers Main Battle Tank, developed by Vickers for export, an Armoured Recovery Vehicle, based on the Chieftain, and an armoured vehicle launched bridge (the Mexi Bridge based on Centurion suspension). A list of these is contained in Table 2. On the current order, Vickers expects to produce 14 tanks a month, although work is already somewhat behind schedule because of delays on the engine.

Building a tank is rather similar to building a ship. Armstrongs, as the Elswick plant was known before the merger with Vickers in 1927, built a reputation on its ability to construct a complete warship in the latter years of the last century. Many of the concepts of tank manufacture were drawn from the experience of warship building. The chassis is known as the 'hull' and the early tanks were classified as 'cruisers' and 'battle tanks'.

Formerly, Vickers was at the forefront in the design of tanks. Indeed, during the interwar period, the tank department at Elswick was the only institution in Britain which maintained facilities for the design and development of tanks and the Vickers tanks provided a model for tank developments in many countries. (6) The official historian of British war production commended the Vickers-Armstrong designers for their 'solitary and pioneering efforts'. (7) During the Second World War, the Government began to take a much more important role in design and development and it was then that the establishment at Chobham, now known as Military Vehicles and Engineering Establishment (MVEE), was set up. Both the Centurion and the Chieftain were designed at Chobham, although in the case of Centurion, Vickers took over responsibility for co-ordinating design and production. Nevertheless, after the war, Vickers maintained a considerable design team. They designed the Vickers Main Battle Tank, which is cheaper and lighter and, in Lord Robens' words, 'nippier' than the Chieftain, especially for export. It has been sold to Kuwait and Kenya and is built under licence in India. They also designed a version of the Abbott self-propelled gun, the Falcon anti-aircraft system consisting of an Abbott chassis and Hispano Suiza 30mm anti-aircraft guns, and a 'retro-fit' package to bring old Centurions up to modern tank standards. (8) There was close co-operation with the Ministry of Defence and, at one time in the early 1960s, there were as many as 90 people from MOD working on quality control. But following the armoured fighting vehicles lull in the late 1960s, the design team was run down. Many of the designers were made redundant. Many of those who stayed were hived off into the International Research and Development Company Ltd., formed jointly by Vickers and Parsons in 1973. There is still a small research laboratory but the two empty floors of Head Office bear testimony to the disappearance of design.

The production of tanks was streamlined during World War II. To the experience of shipbuilding was added the new example of automobile manufacture. A new tank shop was built on the site of the old shipyard in 1937 and on the outbreak of war, new

methods of production were introduced.   According to J.D. Scott,
the official historian of Vickers:

> Production started in one corner of the large shop,
> where plates were placed together and secured in
> position by a few bolts through rivet holes in order
> to form a skeleton of the hull.   This fabrication
> was then rivetted.   Then, stage by stage, the hulls
> advanced down the shop until they were ready to receive
> the suspension and the wheels.   In the early days this
> took place in the same shop, but as production grew it
> became necessary to take them into an adjacent shop for
> 'suspension and wheeling'.   After getting its suspension
> unit and wheels, the hulls went back to the erecting
> shop to receive engine, transmission, fuel tanks and
> electrical and other equipments.   Meanwhile, construc-
> tion of the turrets had been going forward and the
> turrets and hulls met and were fitted together.   Some
> of the units involved weighed several tons and mass
> production on this scale had never even been attempted
> before. (9)

At the height of World War II, they were producing 60 tanks a
month.
     Next to the tank shops were the naval armament shops.
They were huge shops with nooks and crannies where you could
hide an army.   No.24 shop was purpose built for the 14 inch
guns of George V and other big ships and it was said to be
the best shop in the North East.   It had lifting gear and
gun pits and, later, they installed handling gear for missiles.   People from Barrow who saw it were very impressed.
But when naval armaments moved to Barrow in 1968 the big
shops were knocked and a year or so later, the tank shops
were sold to Parsons.   You can still see the ruins of the
gun shops, with their imposing arches, from the river.
     Today, the methods of flow production have been abandoned
and production is dispersed in several different shops.
The tanks or parts are moved backwards and forwards on tank
transporters from shop to shop.
There are three main stages in tank production.   The first stage
is fabrication   Armour plate is purchased from the old Vickers plant at
Sheffield which is now part of British steel and the turret casting is
received from one of the Royal Ordnance Factories.   The
steel is cut and shaped and then fitted, tacked and finally
welded together to make the hull.   The skills involved are
Burners, Sheet Metal Workers, Whitesmiths, (10) Platers,
Welders and Grinders.
     The second stage is machining,This consists of big machining on the
hull and turret (making holes, etc., in which to fit various

parts and apparatus) and small machining on parts. The skills are related to the machines. There are borers — vertical and horizontal — millers, plano-millers, slotters, turners — including combination turret lathe operators, grinders, drillers, capstan operators etc. As mentioned above Vickers has introduced a number of NC machines and these have replaced much labour. Because machining work has increased, the number of machinists has not fallen by much. The unions at Elswick have a policy that a skilled man should operate the machines if they replace jobs that were previously skilled or a mix of skilled and semi-skilled. On NC machines, heavy machining takes about 40 hours per operation. There are 8 operations on the turret, 3 on the hull, and 3 on the glacis or panniers (protected storage space above the tracks). Then there are numerous smaller items where the machining times vary from 5 minutes to 40 hours. There are hundreds, perhaps thousands, of parts and as tanks grow more sophisticated, the number of parts increase. Some machining is subcontracted to other Vickers plants at Scotswood, Crayford, Team Valley and occasionally Thetford.

The last stage is fitting and final erection. This stage involves final assembly of the hull and turret with the various machined parts, the gun and the engine, and the electrical equipment, finishing — i.e. painting etc. and testing. Nearly all the electrical equipment is provided by the Ministry of Defence as 'free issue'. The gun comes from RUF Nottingham and the engine comes from British Leyland or Rolls Royce. Some of the people who have fabricated or machined equipment are involved at this stage. In addition, there are fitters, coppersmiths, joiners, electricians, painters and rough painters, and leather workers. Some of the fitting jobs have also been affected by NC machines.

In all three manufacturing sections, there are also ancillary workers — handymen, craners, slingers and labourers — who do odd jobs such as carrying parts from one section to another and so on. There are also 'hangers on', overhead workers, without whom, many believe, the work might proceed much faster. These include the progress people, who are supposed to make sure that equipment is in the right place at the right time which usually means that a worker asks for the equipment and a progress persons goes and finds it, the store people, the rate fixers, the inspectors, the foremen, and, of course, the management.

The production process takes about 18 months; testing starts at 12-14 months. But these times are subject to great variations. There are enormous problems in program-ming the work and because these problems are not overcome, the process involves a good deal of waste — waste of space, capital equipment and above all people. It is not just the inefficiency resulting from the dispersal of shops, though this presumably adds to the problems. Workers are always

waiting around. Sometimes it's waiting for orders. Sometimes it's waiting for work from other sections; or new
orders, the fabrication shops may be working flat out while
the finishing trades are on short term. Sometimes, it's
waiting for parts, like the engine, from outside. Sometimes,
big delays can be caused by minor events; for example, two
slingers off work can prevent the crames from operating and
halt production in an entire shop. The problems are easier
to manage if the contract is big. Then there is enough
work for all sections and subcontracting can be used to
smooth the schedule. Subcontracting can also be used to
undermine the unions, as when work is subcontracted to firms
employing cheap non-unionized labour. This happened with
the petrol tank for the Chieftain and, for a time, the tank
was blacked.

Nevertheless, the primary problem is planning and this is
all the more difficult if management do not understand the
production process and workers are excluded from management.
Given the current framework of capital/labour relations,
increased efficiency in management terms tends to mean
increased control over workers. If workers are to ensure
a certain degree of autonomy over their work and an adequate
rate of pay, they need to maintain a monopoly on the information they possess. A skilled craftsman has considerable
autonomy in determining the rate for a job. On average,
machinists at Elswick earn £12 less than other skilled
workers because their jobs are easier to time. At Elswick,
there is a method of batch costing, whereby each piece worker
books his time, together with his rate, against an order
number. Theoretically, this ought to facilitate costing and
scheduling. In practice, the workers often book against
different contracts in order to spread the work. The matter
is complicated by the heirarchical nature of management so
that the line management have an interest in confusing the
higher-ups since they do not wish to be seen to be far behind
on any one contract. The consequence is that it is very
difficult to obtain correct information on the work involved
in any particular product, so that there is no reliable basis
for efficient planning.

Apart from the boredom, the waste may not matter. The
alternative is increased output of tanks for the repressive
Iranian régime. It was different in World War II when the
tanks were needed for the war against fascism. But when one
considers how the time could be spent, the socially useful
work that could be done, then this waiting around, the
antagonistic social organization that impedes co-operative
production, is unacceptable.

Commercial Experience at Vickers

Since the war, Vickers Elswick has undertaken the production of
a wide range of commercial products. Apart from non-ferrous
metals, Vickers never established a basis for sustained

production in any of these products and none of them served
to stem the decline in employment.   In 1939, the two main
armament plants in West Newcastle, Elswick and Scotswood,
employed 18,000 people.   In 1963, they employed 7,000 people
and in 1976, they employed 2,800 people.   There used to be
50 pubs along the Scotswood road, with names like the Rifle,
the Gun, or the Ordnance Arms.   Today, there are only six.
Being on the dole has become part of the way of life.

The commercial products that have been manufactured at
Elswick include moulds for prefabricated houses, electric
furnaces and other equipment and plant for British steel,
car presses, car dies, cranes, deck machinery, printing
presses, newspaper presses, mining machinery, the variable
speed gears and delivery pumps, rubber processing equipment,
pulverizing plant, tripwound pressure vessels, experimental
tunnelling equipment and subcontract work on the Sulzer
engine.   These are shown in Table 1.   Most of them were
undertaken under licence from other companies or under sub-
contract to other plants in Vickers.   The great engineering
works in the North East are widely regarded by management
as jobbing shops.   Commercial products tended to be treated
as 'filler-ins' to employ spare capacity between armament
orders.   Those products that proved successful were generally
hived off to other parts of the Vickers empire.   Thus printing
presses were transferred to the old Crabtree Plant at Leeds
while the Variable Speed Gear, a highly successful gear system
based on hydraulic pressures, was transferred eventually to
the Hydraulic Division at South Marston.   The car presses and
marine equipment were transferred to Scotswood.

Vickers also had a forge and a foundry at Elswick.
Originally the forge and the foundry formed a single division.
The foundry had a heat treatment plant which made a profit.
It also had iron casting and pattern making facilities.   The
forge was situated on the riverside between the old tank shops
and the old naval armament shops;   it was where the bombs
had been forged and it was the only forge in the North East.
When the old tank shops were taken over by Parsons, the
forge had to be moved to higher level.   It needed stronger
foundations to take the heavy hammers.   The management
decided instead to close the forge and modernize the foundry.
But they made mistakes in the modernization and, without the
forge, the foundry could not make a profit.   The plan for
modernization had depended on the demand for printing
machinery;   but the new printing venture at Crabtree failed and
other markets had to be found.   They sold the foundry in
1977 for a song;   it might have been because of the expenditure
needed to bring poor health and safety conditions up to
standard.

There were two examples, during this period, of attempts by
Vickers to build up an indigenous capability in a new
commercial product in the North East.   One was the die shop at
Elswick.   This was a spin-off from the Clearing presses.

The die shop, which began about 1957, employed about 250
people and supplied dies to Vauxhall, Saab, Volvo, the
Polish State Car Company, Renault and Volkswagen. It also
supplied dies for other products, like fridges and baths,
and research was undertaken at Vickers into a new kind of
explosive forming die. According to the shop stewards,
'the work was terrible in the die shop but a great solidarity
grew up there you know ... in the shop itself we established total control on flexibility and overtime ... The
solidarity was good – it came out of the situation we were
in'. (11) Management challenged the control over the die
shop and the result was a strike which lasted ten weeks.
The workers won the strike but the die shop was closed down.
At just this time Ford opened a new tool-making shop outside
Cologne and Ford had been the biggest customer. Some of
the men went south to join Hall brothers in Rugby.
    The second example was the tractor scheme at Scotswood.
Immediately after the war, Vickers were engaged in converting
Sherman tanks for use as tractors known as 'Shervik'. In
order to break into the tractor market which was then an
American monopoly, Vickers went into partnership with Jack
Olding & Company, who were the agents for selling the American
Caterpillar tractor in Britain, and Rolls Royce, who were to
manufacture the engine. With the aid of a designer from
Ferguson, Vickers developed a heavy earth moving tractor, which
had a Centurion suspension. There were two main versions,
the VR 180, named Vigor, with a 200 hp Rolls Royce engine,
and the VR.X, with a 150 hp Rolls Royce engine, named Vikon.
Vigor went into production in 1952. Among other things, it
was used in the construction of the M1, for clearing and deep
ploughing large areas of land in the south of Yugoslavia and
by the British and Pakistani armies. It was said to cost
about £10,000, of which labour costs at Scotswood were only
£700. And it had a larger work capacity than its main
competitor, the Caterpillar. At its peak it employed about
a thousand people. It is said that Jack Olding pushed the
tractor into the market place before it was properly proved
and tested and also that the engineering was too fine and
complex – there were many people who felt that it was wrong
to have higher engineering standards for machines to kill
people than for machines designed to benefit people.
According to Jim Hendin, head of the Vickers engineering
group,

    we ran into quite a lot of problems because a tractor
    is a different animal to a tank. A tank is very well
    maintained, it has a low service life, they change
    the tracks very frequently and the rest of it, whereas
    a man buying a construction machinery tractor, he
    expects the tracks to last a fair long time, I should
    think from memory four or five thousand hours,
    against four or five hundred for a tank. (12)

After the tractor sold to a rather large number of customers
at home and overseas, snags began to develop.   In particular,
the powerful engine proved too much strain for the gearbox.
Even though the gearbox was hardened and most of the snags
ironed out, the tractor gained a reputation for unreliability.
The scheme finally folded in 1960.

Right up to 1959, the company appears to have been optimistic about the future of the tractors.   In 1957, a special
tractor division was established, combining the tractor
production at Scotswood with Onions and Sons (Leveller) Ltd,
which produced large scrapers and had been acquired by Vickers.
The distribution of tractors was taken over from Jack Olding,
additional space was made available at Scotswood, and subsidiary companies were set up in Canada and Australia.   There
were plans for an ultimate turnover of £10m and for mass
production at Long Benton.   As late as 1959, the company were
reporting hopefully on export prospects, particularly in South
America.   The following year, tractor production was discontinued because 'markets for the crawler type tractor narrowed
and the future did not hold sufficient promise to justify
continuing production'. (13)   It was around this time that
Caterpillar established a plant at Birtley.

The failure of the tractors and other commercial products
is often attributed to the nature of Vickers as an armaments
company - the lack of cost and market consciousness, the view
of itself as a 'national institution' which can always survive.
The interwar experience, when Vickers attempted to enter the
electricity and automobile industries, is often cited as an
example.   Wolseley, for instance, was very good at producing
luxury cars, including 'high speed' cars for staff officers and
scouting duties.   But it was not so good at marketing or mass
production.   According to J.D. Scott: 'Wolseley had built
costly showrooms in Piccadilly before finding that the police
would not let them bring cars across the pavement, and the
incident was not unrepresentative.' (14)   Likewise, Armstrongs
turned down a plan to mass produce automobiles in 1906 because
the profit was less, according to the directors than on a single
river gunboat. (15)

But today, Vickers has succeeded in moving into civilian
industries outside the North East.   After the successive
waves of nationalization - steel in 1965, aircraft and shipbuilding in 1977 - armaments account for a small proportion
of Vickers total output.   The hiving off of successful
commercial products from Elswick was part of a policy of
streamlining and rationalization, in which various spheres
of production were concentrated in different locations.   It
also involved the separation of design, production and
distribution so that in the development of products like the
tractors, the designers were often remote from the customer
and unfamiliar with the production process.   It was a policy
of financial diversification, in which Vickers entered new
fields through the takeover of existing plants, often overseas,
rather than through new investment.   Increased profits have

490

been achieved in the short run through the elimination or run down of older plants rather than through modernization and innovation. The consequent expansion of Vickers' output was thus merely fictitious representing an increase in the share of output controlled by Vickers rather than an absolute increase. This meant that, over the last ten years, trading profits have increased threefold as a percentage of sales and sixfold in absolute terms, while employment has fallen by 10 per cent, a reduction of 6,000 people. If one takes into account the fact that acquisitions during this period have added around 9,000 people to the labour force - a period in which there has been no nationalization - the total loss to the labour force is probably around 15,000. This strategy is evidenced in the composition of foreign sales as shown in Table 3. In the highly competitive advanced industrial markets of West Europe, North America and Australia, increases in sales have been achieved through the establishment of overseas subsidiaries. Indeed, the entire increase in sales of overseas subsidiaries has occurred in these regions. In contrast, exports have been stagnant in real terms. Such increase as there has been probably consisting of Elswick tanks and Barrow warships has mostly gone to the underdeveloped markets of Asia and South America - areas where technically backward, cheaper products are more likely to find a customer.

There may have been gains in employment and output in some areas in Britain; Vickers has undertaken some new investment in offshore engineering, office equipment and bottle washing machinery. But its solution to the declining rate of profit on all but armament work at Elswick has been abandonment rather than innovation. That this is a long term policy is evidenced in the run down of designers and the low apprentice intake. The great weakness at Elswick and Scotswood was the failure to maintain independent teams for design, development and testing. In 1970, only 30 apprentices were accepted at Elswick and some people say they would be better off apprenticed to estate agents. For the main asset at Elswick now is space and land. Hence the growth of warehousing where there were once productive factories.

The attachment of Vickers to the criterion of short term profit has to be understood as a consequence of its growth. Over the years, the management of Vickers has become hierarchical and traditional. As one person has put it, there is a kind of notion of the 'divine right to manage, a premium apprenticeship based on heredity. It leads to terrible inertia'. The heads of the Engineering Group are always Barrow men. They believe in waiting for orders and not counting costs. They think they should 'make a crane which lasts for sixty years' which is admirable no doubt but it does not lead to markets. Even more important perhaps is the fear of taking risks which stems from the nature of the hierarchy. The line management are unwilling to accept new ideas and they tend to oppose new investment for fear of penalization in case of failure. They never put decisions on paper and information rarely gets passed up through the hierarchy. The aim is to preserve things as they are until retirement. (In fact, the current management of the Engineering Group was responsible for huge fiascos.) The growth of Vickers and the consequent need for increased financial control has led to the replacement of engineers by accountants. They do not understand the production process or may not even be able to tell the difference between a drill and a lathe. They judge a new venture by the number that appears on a calculator and not on a comprehensive view of the long term. It was probably this kind of reason, the un-willingness to anticipate the future, the fear of greater losses that led to the decision to abandon the tractor scheme. Vickers has been described by one worker as a 'huge animal that has come out of the prehistoric age. It wallows around in the mud and occasionally comes up for sun. It survives because it has got a thick skin.' Of course, the cautiousness of management, as we have seen, is compounded by the cost of making the necessary structural adjustments at Elswick. Elswick's past has become a liability to the management. Road access is limited to a tunnel under the railway. River access is limited by the swing bridge built at Elswick in 1876. And the degree of control established over the production process after years of struggle is probably greater than is some of Vickers newer plants overseas.

The risks of innovation and new investment should not, of course, be minimized. In many of the commercial fields they entered, they were caught in the fierce competition between established multinational monopolies, which militated against the intrusion of newcomers. The market was the market for capital goods and heavily dependent on booms and slumps in investment. To survive a recession in investment like the present one and to build up a new capability in a particular product line in anticipation of the next recovery might well require resources and an ability for risk-taking would be that beyond the capacity of a smaller, more dynamic company.

From the standpoint of the Vickers' shareholders, it was probably true that the decision to pursue rationalization instead of innovation made sense. But this kind of sense looks altogether different from the standpoint of the people who lost their jobs and the people who could have benefited from the products of their skills.

Alternative Products

Out of the stock of Vickers' past experience, it is possible to pick a number of products which may well have a growing market in the future and which could be built at Elswick. It is also possible to think of additional products that are

492

technically feasible and satisfy some currently unfilled
social need.   But if such products are to succeed where
earlier products failed, they would have to entail an
entirely new form of organization for production.   The
problem, as we have seen, is rooted in the basic contradiction of Vickers as a large corporate organization.   On
the one hand, only an organization of similar scale can
provide sufficient resources and market protection.   On
the other hand, the very size of Vickers and its profit-
seeking nature lead to inertia and timidity.   The
alternative is a system of national planning in which
government support is provided, on a long term basis, for
the build-up of a truly indigeneous capability.   This
would have to combine design and manufacture with intimate
knowledge of user need.   To ensure that the structure of
the state is not patterned by the structure of the corporation, that real needs are identified and social priorities
shifted accordingly, and that the productive and creative
potential of workers is fully utilized, the system would
have to be based on grass roots participation of workers
and consumers.   (A proposal for Workers Participation at
Vickers has already been put forward by the North East
Shop Stewards Committee and discussed by the Combine
Committee.)(16)
    It should be stressed that the list of alternative
products presented here does not represent a fully worked
out plan.   Given the information available to workers, it
would be surprising if it were.   Rather, it represents a
basis on which workers could begin to extend the range of
issues subject to collective bargaining.   Currently,
choice of product, investment, organization of production
are not thought to be the workers' business.   And yet
decisions on these issues have far-reaching effects on
workers' lines.   Furthermore, the workers' knowledge about
production and also about social needs is currently wasted;
yet it could shape positively the future direction of
social and economic change.   Once it is accepted that
these issues are a subject of collective bargaining, then
it would be possible for workers to create links with user
groups, in hospitals, schools, old people's homes, etc., and
build up a stock of experience and a set of alternative
planning units, so that effective democratic planning could
become a reality.   Nowadays society allocates resources
through the market mechanism which is rather inefficient
at meeting social needs because it depends on the distribution
of income and is biased towards the rich.   Democratic
planning, involving direct contact between the people who use
products and the people who make them, represents a socialist
alternative to the market mechanism.
    The products suggested here are, then, a basis for
discussion and for developing an alternative process of
democratic planning as part of a process of widening
collective bargaining and increasing workers' power.

Annex Two provides a list of alternative products. The
most feasible are described below. They are predicated on
some such structural change. Some of the products suggested
here are theoretically feasible within the current organizational
framework. The money received by Vickers in compensation for
the nationalization of aircraft and shipbuilding provides an
opportunity to put the suggestions into practice. Indeed,
some of them have already been put to the management at Scotswood by members of AUEW Tass. But organizational change may
still be required because of the natural conservatism of
Vickers. The management do not encourage bright ideas - many
people have been deterred from participation in the suggestions
scheme because of the lack of response and the puny remuneration.
To ask Vickers to change, as one person said 'is like asking a
brewer to make jam'.

The products concentrate on heavy engineering. The assets
at Elswick are large spaces, heavy engineering skills and a
custom-built approach. The fact that there has never been
assembly line production and the plant is not now geared up
for specific production could prove an advantage in conversion.
Where possible, the products have been related to local needs
in Newcastle just as were the original hydraulic cranes built
by Armstrong in the mid-nineteenth century.

1. Manufacturing Machinery

The main proposal is the design of car presses. Elswick used
to make clearing presses until these were transferred to
Scotswood. Currently, Scotswood make Schuler presses but
there are very few orders at the moment because of the strong
European competition. To stay in the market, Vickers need
to have their own press. The Scotswood draughtsmen have
been thinking about a new economy press. The experience
with Schuler presses has given people ideas about how the
design could be improved and how cheaper methods could be
adopted. They undertook a design exercise for about 18
months and found that it was not as easy as they had
thought and that there were reasons for apparent complexities
of design. Nevertheless, they concluded that there are real
possibilities. This is probably the most immediately
feasible project and management are considering whether to
recruit seven designers to undertake the project. Another
suggestion is the manufacture of plant and machinery needed
at Elswick which would be designed around the kind of
conditions of work which workers consider to be most
appropriate for productive and satisfying labour.

2. Mining Machinery

In the medium to long term, there is a huge potential
market especially overseas. The future of British coal
mining depends to a large extent on the relative price of

494

coal and oil, but there is a world wide thirst for minerals and with it, a demand for heavy earth moving equipment, such as dragline or bucket wheel excavators, large tracked vehicles somewhat similar to a tank. Management hair goes grey at the mention of heavy earthmoving tractors. But, as we have seen, the problem with the Vickers tractors was not, in the end, a problem of quality. The need for improved health and safety standards in the mines could also provide a challenge to the designers of mining equipment.

3. Inland Waterways

There are a number of proposals for making better use of Britain's canal system as an alternative form of transport. This would require a firm Government policy to resuscitate the canal system. These include:

(a) Pumps. As a result of the suggestions made by draughtsmen at Scotswood, the management asked British Waterways if there was any possibility of work for Vickers but the response was rather negative. The Manchester Shipping Canal want barges to clean the canal and pump houses. The canal gets silted up and the pumps are used to separate the silt into water and gravel. There is stiff competition from the Dutch who normally do this kind of work.

(b) Canal gates, particularly power-driven lock gates which would speed up transport through the canals.

(c) Earth moving equipment for widening the canals.

(d) Barges. A specific suggestion is floating containers hitched to a single power unit like a kind of water going cargo train.

4. Energy Equipment

(a) Fluidized bed boilers. Currently, boilers provide about 20 per cent of energy sources in Britain. There has been very little technical change in boilers and there is considerable potential for energy conservation in their design. Fluidized bed boilers pulverize coal and burn it over limestone like a liquid so that the limestone catches the sulphur. It is possible to improve the pressure control over oxygen and to minimize heat loss by burying the boiler tubes in the bed of limestone. The Americans are very interested because of the high sulphur content in American coal. Over the next ten years,

there is widely considered to be a real scope for this kind
of boiler.

(b) Heat pumps for industrial or office use. Heat pumps
function like refrigerators in reverse. They extract
heat from the environment and reproduce it in a purer
form. The Lucas Aerospace Workers are designing a
heat pump to be used in old peoples' homes. But heat
pumps on a larger scale for heating office blocks has
already been proved to be not only feasible but also
economic. The Manweb building in Manchester, the office
of Manchester's electricity board already uses a heat
pump and the Board is advertising for new designs.
Heat pumps are also useful in industry and could be
ideal for bottle washing machinery, like that which
Vickers manufactures at Crayford and Thetford. The
heat pumps would act as kind of heat recyclers extracting
heat from used water at the end of the bottle washing
process for reuse in the newly sterilized water at the
beginning of the process.

5.   Environmental Equipment

(a) Recycling plants. Currently, Vickers make the Logeman
domestic refuse recycling plant at Scotswood and the Seerdrum
pulverizing plant at Southampton. The Logeman equipment
bails (squashes) scrap metal and shreds and separates tin
from steel. Currently, there are snags in the separating
process which Vickers might try to solve. The Seerdrum
plant, which used to be made at Elswick, separates metal
from domestic refuses and turns the residual into fertilizers.
Seerdrum was taken over by Vickers and Vickers could consider
a development of the plant. Another suggestion is recycling
plants for other materials. An obvious candidate is brass in
order to service the non-ferrous metals division. Several
people think that the non-ferrous metals division should
have gone into plastics. A plastics recycling plant would
involve chemical engineering problems which Vickers is not
equipped to solve but it might be possible to collaborate
with workers in a chemical plant. Vickers could make the
heat exchangers and the pressure vessels.

(b) Tree transplantation equipment. Currently, tree and timber
sources are running out. The Dutch Elm disease has made the
situation especially serious. It is possible to use an
hydraulic cylinder system for the rapid up-rooting and transplantation of trees. The system is very simple and tree
transplantation would have to be done on a very large scale
to make production worthwhile.

(c) Oil spillage pumps. These are pumps which would reclaim
spilled oil through a kind of sweep. They would save marine
life and stop pollution on the beaches. They are very

expensive and difficult to make, however, and would
require a heavy social commitment to the environment.

6. Agricultural Equipment

The main ideas centre around agricultural systems for the
Third World. Vickers has helped India to establish a
factory for tanks; this experience might be extended to
tractors or other equipment. The very simplicity
required might, as one person said, be 'right up Vickers'
street'. A specific suggestion is the design of
irrigation systems to replace primitive systems in the
Third World. This would involve a combination of pumps
and earth moving equipment, such as tractors. The pumps
would need to be operated by a local source of power –
perhaps people, oxen, or camels. And there would need to
be repair shops for local maintenance and future construction. Ideally, a team of workers could spend time in a
village making their skills available to the local people
to help them build their own systems.
    The major impediment here is not so much the social
organization in Britain but the social structure of the
villages in which landlords with little interest in
improving the land and live off interest or rent paid by
dependent tenants or smallholders who cannot finance
their own investment.

7. Brewing Equipment

There is a general feeling that the variety and wholesomeness
of beer in the North East could be improved. Currently the
pubs are dominated by two large breweries and the beer
contains artificial chemicals which are designed to preserve
a standard quality in pubs, and to remove the instabilities
that result from lack of control over temperature and humidity.
Elswick could build small scale brewery plant which would
boost brewery employment and improve the beer. It could also
build pumps and vessels, with built-in temperature and
humidity control, for the pubs. The main problem is how to
wrest control of the pubs from the main breweries. It might
be necessary to think about alternative pubs.

8. The Reconstruction of Britain's Piers

Britain's piers are falling down. Rebuilding them is
essentially a nineteenth century engineering job for which the
Elswick plant was built. One could start with the Saltburn
pier which gives pleasure to many people in Newcastle.
    In addition to the suggestions on this list, there were
numerous other ideas. Some, like solar panels, mopeds, pedal

cars, windmills, hydrofoils for rapid service along the
Tyne, were ruled out because they were suitable for light
and medium engineering involving assembly lines or small
plants. Others, like hospital equipment or electro-
mechanical devices for controlling polluting emissions,
required very specialized skills in chemical or electronic
engineering although they could perhaps be developed in
collaboration with other workers in Vickers as well as other
companies. Many of the suggestions made by workers in
other plants might also be adopted. These include the
proposals for wave and tidal power and offshore engineering
made by workers at the Barrow Shipbuilding works (17) or
the proposals for firefighting equipment or for a hybrid
road/rail vehicle made by Lucas Aerospace workers. (18)
There are plenty of ideas around. It would be surprising
if workers who spend their lives in a factory environment
had not thought about different ways in which they might
use their skills and talents, about what they could be
making and how. Most of the ideas are apparently
technically feasible. The problem is political and
social. The problem is the organization of Vickers, the
current structure of the world market, and the priorities
of the Government. Yet once the possibilities for alternative products are explored and the political context
provided, the mobilization of human resources to replace the
current wastage offers extraordinary opportunities. People
who can build a Chieftain tank can and should build a whole
range of engineering products that society needs.

Notes

1. National Economic Development Office, <u>Mechanical
    Engineering: Summary of Findings and Recommendations
    of the Industrial Review to 1977</u> (London, April 1974).

2. Based on figures provided by the Stockholm International
    Peace Research Institute, the value per ton of Main
    Battle Tanks is as follows: AMX-30: $10,000 (?),
    Chieftain: $6,000, Leopard: $5,000, M-60: $4,000,
    T-54: $4,000,

3. For a description of the decline of West Newcastle,
    see Benwell Community Project, <u>West Newcastle Its
    Growth & Decline</u> (July 1976) and <u>Social Change in
    Benwell</u> (Oct. 1977).

4. Ministry of Defence News Release, 17 June 1976.

5. <u>Vickers Annual Report</u> (1976)

6. Vickers tanks sold to the Soviet Union under the
    picturesque name of 'the English Workman' are said
    to have played a key role in the subsequent evolution
    of Soviet tanks.

7. M.M. Postan, <u>British War Production</u> (HMSO and Longmans
    Green, London, 1952).

8. See <u>Jane's Weapons Systems</u> (1974-5).

9. J.D. Scott, <u>Vickers: A History</u> (Weidenfeld and Nicolson,
    London, 1962), p.284.

10. Whitesmiths are a Vickers' speciality. They work in cold
    metal as opposed to blacksmiths who work in hot metal.
    They were introduced in the 1920s to counter the militancy
    of the boiler-makers, who do roughly the same job. They
    belong to a separate union, the National Union of Sheet
    Metal Workers and Heating and Domestic Engineers.

11. See Vickers National Combine Committee of Shop Stewards.
    <u>A Workers Report</u> (Pluto Press, forthcoming).

12. Interview with Tyne-Tees Television.

13. <u>Vickers Annual Report</u> (1960).

14. J.D. Scott, <u>Vickers: A History</u>, p.145.

15. Clive Trebilcock,'"Spin-off" in British Economic History:
    Armaments and Industry, 1860-1914', <u>Economic History
    Review</u>, XXII (1969).

16.   'Vickers: A Proposal for Worker Involvement in Management'.

17.   See Part II-A 3 of this volume.

18.   Lucas Aerospace Combine Shop Stewards Committee,
      <u>Corporate Plan</u> (1976).

Table 1 : ELSWICK PRODUCTS, 1950-78

CENTURION and variants

CHIEFTAIN and variants

ABBOTT

VICKERS MAIN BATTLE TANK

NAVAL ORDNANCE (moved to Barrow)

CIVIL PRODUCTS

CLEARING PRESSES (transferred to Scotswood)
PRINTING MACHINERY (transferred to Vickers Crabtree)
'VSG' (transferred to Weymouth)
NON-FERROUS METALS
FORGE (closed)
FOUNDRY (Sdd)
HEAVY STEELWORK PLANT
DIES (discontinued)
RUBBER PROCESSING EQUIPMENT
EXPERIMENTAL TUNNELLING EQUIPMENT
STRIPWOUND PRESSURE VESSELS
PLASTIC INJECTION MOULDING
CONTAINER HANDLING CRANES
SEEDRUM PULVERISING PLANT
PRESSINGS
DECK MACHINERY (transferred to Scotswood)

1950 1952 1954 1956 1958 1960 1962 1964 1966 1968 1970 1972 1974 1976 1978

Table 2 Vickers Armoured Fighting Vehicles Since 1940

Valentine:
Specifications drawn up 1934.
Abandoned 1937.
Revived in 1939 with order for 100.
Total of 2,515 were delivered.
Mid 1943: Elswick were producing sixty a month.
Had eleven marks. There was a bridge laying
version and a seagoing version.

Archer:
Valentine chassis with 17 pounder gun. 800
ordered.
Deliveries began in 1944.

Centurion:
Design began in 1943. Designed at Chobham but
Vickers were responsible for coordinating design
and production. Production began at end of war
and was Elswick staple until 1959.
About 1,000 were produced. Stockholm
International Peace Research Institute (SIPRI)
value is about $100,000. Went through 13 marks
with numerous variants, including ARV, bridge
layer, Dozer tank, Engineer vehicle, artillery
observation tank, Beach Armoured Recovery
Vehicle. "Most versatile and successful tank
design ever developed" (Jane's Weapon Systems
74-5). Sold to India, Egypt, Iraq, Israel, Jordon,
Libya, S. Africa, Switzerland, Sweden,
Netherlands, Kuwait.

Chieftain:
Designed during 1950s. Production began in 1963,
by Vickers and ROF Leeds. 800 for British Army
SIPRI cost: $300,000. Bridge layer and ARV
versions. Seven marks. 1,500 sold to Iran.
165 sold to Kuwait.

Vickers Main
Battle Tank:
Developed between 1958-63 as a private venture.
Sold to India, Kuwait and Kenya. Produced under
licence in India as Vijayanta. 50 for Kuwait;
Cost £7m. Prototype production began in 1961.

105mm SP Gun
ABBOTT:
Vickers is sole manufacturer. Developed a
simplified version for export. Export orders
1970-72.

SP 30mm AA System
FALCON:
Uses Abbott chassis and has Hispano-S iza gun.

Table 3:   Vickers Exports and Sales from Overseas Subsidiaries

UK Exports

	Africa	N. America	S. America	Asia	Australasia	E. Europe	W. Europe
1965	4.7	1.6	0.8	2.5	1.1	4.3	2.6
1966	1.1	2.1	0.6	3.4	3.0	2.2	3.7
1967	1.4	3.4	0.7	5.4	1.5	0.1	4.1
1968	4.0	3.4	0.9	2.3	2.0	0.3	5.2
1969	2.5	2.6	1.2	3.2	1.5	2.6	5.7
1970	2.3	4.9	0.9	7.0	0.9	7.6	9.0
1971	3.7	4.9	2.1	16.9	1.2	1.2	7.5
1972	4.7	2.0	1.0	12.8	1.5	1.2	6.0
1973	2.6	2.9	9.1	13.0	0.9	1.8	7.7
1974	6.6	2.0	4.2	16.1	1.1	1.3	10.9
1975	3.9	2.9	5.2	13.3	1.8	2.5	9.7
1976	4.2	2.3	26.6	21.4	1.4	1.1	11.5
1977							

Sales of Overseas Companies

	Africa	N. America	S. America	Asia	Australasia	E. Europe	W. Europe
1965	0.2	24.6	1.6	0.4	7.1	3.9	1.5
1966	0.5	28.8	1.7	1.8	8.8	0.7	2.0
1967	0.4	22.5	1.1	2.9	9.5	1.0	1.4
1968	-	20.6	2.0	5.7	13.5	0.4	1.4
1969	1.0	19.3	1.8	1.3	14.4	0.1	14.8
1970	3.4	16.3	1.8	8.0	15.1	0.1	18.1
1971	1.8	10.9	0.2	2.2	18.0	0.1	17.0
1972	2.5	14.1	0.5	0.7	25.4	-	21.7
1973	2.3	16.3	0.4	0.8	25.7	0.8	28.5
1974	4.7	15.8	0.7	0.5	33.8	0.1	36.3
1975	4.0	28.4	1.3	1.2	45.0	0.1	50.7
1976	5.1	36.6	0.8	1.2	55.3	-	71.5

Source: Vickers Annual Reports

PART THREE

THE MINISTERIAL RESPONSE

STUDY INTO DEFENCE SPENDING - SUMMARY OF CONCLUSIONS

by John Gilbert, John Tomlinson and James Wellbeloved

(This paper by three government ministers who were
members of the Study Group is not actually a Study
Group paper as it was not presented until after
the report was completed. Its arguments were
nevertheless familiar to the Study Group as they
had regularly been presented and discussed during
the group's meetings. The main thrust of the
paper is to reject the policy of reducing military
spending which was the basis of the study group's
remit.)

Just as it would be wrong to endanger national security
in our concern for social justice, so it is no good having
a defence policy which could bankrupt the society it is
designed to defend.' Rt Hon Fred Mulley MP, Secretary of
State for Defence

In determining an appropriate level of defence expenditure,
many factors, often conflicting, have to be taken into account.
Not only do we need to consider the extent of our commitments
and the capabilities that we and our allies need to possess in
the light of our assessment of the existing and likely future
threats to our national security, we must also consider wider
economic and social factors. The difficulty about this type
of assessment is that very little of it can be based on hard,
indisputable facts.

There is no doubt that Soviet forces have a very powerful
capability at sea, on land and in the air. There is, however,
considerable room for argument about the likely intention of
the Soviet Pact leadership. These intentions must in turn
depend on their assessment of any threat to them and on their
perception of where their own short and long term interests
lie.

The economic arguments against defence expenditure are much
less clear-cut than those who would have us make drastic cuts
in defence expenditure often imagine. While no one would
argue for defence expenditure on the grounds of the employment
in generates, it is a fact that Defence is a large employer
both directly in the Armed Forces and their supporting
civilian personnel and through the defence industries. It
has, however, yet to be proved that a substantial diversion
of such manpower, and the resources they consume, to the
civil sector would necessarily increase the total national
product. Indeed, in the short term because of the inevitable
upheavals which an abrupt change in policy would cause, it is
quite likely that there would be a fall in total national
resources - a penalty that would be that much more difficult
to accept in a time of high unemployment like the present.
Equally, the manufacture of weapons and weapons systems
provides benefits in terms of technological spin-off for a
significant proportion of civil industry and, through arms
sales, is of positive benefit to the balance of payments. If
this manufacturing capacity and research and development
expenditure were diverted to the civil sector, it is possible
that alternative exports might develop, but it is - and will
remain - difficult enough for us to expand our exports of
these other types of goods, until such time as many of the
arms importers can be persuaded that they need to spend less
on their own defence.

In the following sections, we consider:

    the economic argument;
    the political and strategic argument; and
    the Arms Trade.

506

The industrial and employment implications of changing
from defence to civil production are considered separately.

The justification for defence spending rests with the
benefits it brings in terms of our own security and the
fulfilment of our international obligations. These have
to be major social objectives for any Government. The
proportion of the nation's resources to be devoted to them
can, however, only be determined by reference to the
overall economic position of the country and the competing
claims of other social needs.

High levels of arms expenditure and sales to third parties
raise a number of sensitive questions of morality and
expediency. This government has been in the frontline of
those wishing to improve détente and through this to achieve
balanced force reductions between East and West. In the
current situation, however, there is no reason to suppose
that unilateral disarmament on our part would cause others
to follow suit; in fact it could be argued that such a
course could only have dangerous destabilizing effects.
On the question of arms sales, this government always
takes account of the purposes for which arms might be used
and has frequently imposed total or partial embargoes on
sales to individual countries. It is, however, no part of
our policy to deny other countries who may be at risk the
means of defending themselves from potential aggressors.

No level of defence capability and expenditure is self-
evidently correct when measured against all the political,
military and economic factors which are constantly changing.

In the last resort, then, decisions as to the appropriate
level of defence expenditure must be matters of careful
balance. It would be folly to bankrupt the country by
spending more than our economy could afford. It would also
be folly - to put it no higher - to shirk the sacrifices
necessary to protect the way of life that we all hold dear.
For this reason it is important that defence expenditure
should be kept under constant review to ensure that sufficient
resources are devoted to it to meet our security needs, but
no more. It is because we think that our present level of
expenditure is as close to being correct as we can assess,
given the external and internal circumstances within which
Government has to operate, that we are unable to subscribe
to the conclusions of the majority of the Working Party.

Defence Expenditure and the Economy

The level of defence expenditure is affected not only by the vital strategic and political factors involved (see p.519), but also by decisions about the social allocation of resources, about how to mediate competing claims on public expenditure and how to reconcile strategic requirements with other important social needs. In this chapter we concentrate on the relation between our defence effort and the national economy. We would emphasize however that there is an intimate and reciprocal relationship between the arguments in this chapter and the strategic and political determinants discussed in the next two chapters. On the one hand the formation of our foreign and defence policies cannot be carried out in isolation from our national and international economic position; and on the other hand our economic situation is affected by the direction of foreign and defence policies.

The relationship between defence spending and the economy can for convenience be analysed in two ways. At one level there is the allocation of resources to defence, which takes place as part of the process of planning public expenditure as a whole and where one is looking at the opportunity cost of the expenditure in terms of other government priorities and at its impact on the general level of employment, prices, and taxation. At a lower level there is the allocation of resources within defence to various programmes where economic factors have an important influence on defence decision making. Although convenient to consider these separately, in practice they are part of a single 'closed-loop' process since programme decisions are influenced by the funds expected to be available and, in turn, the overall allocation of resources is affected by what capabilities can or cannot be furnished at a given budget level. (1) We start therefore with economic factors as they influence decision-making within defence circles.

Defence Decision Making and the Economy; Choice of Projects

The proportion of the British defence budget devoted to equipment has shown a slight tendency to fall over the last fifteen years, varying from a high of 42 per cent in 1965/6 to a low of 31 per cent in 1971/2. The latest estimates indicate a rise to 37 per cent in 1977/8. (2) Within that total, however, the increase in cost of new generations of military equipment has been remarkable. For example, it has been calculated (3) that the cost of producing 385 Tornados will be slightly greater than the entire production costs of Spitfire before and during World War Two. Taking account of inflation and other cost increases it has been argued that the cost per ton of warships has increased by anything from a factor of ten (the difference between an early post war 'Bay' class frigate and the last of a long series

of 'Leander' class frigates) to a factor of fifteen (the
difference between an early post war 'A' class submarine and
a modern 'Swiftsure' class submarine, or between the Vanguard
battleship and the new Anti Submarine Warfare Cruiser).

If anything these factors may be understated because they
refer only to production costs. Comparisons of total life
cycle costs will be even more striking since maintenance and
support costs tend to increase faster than production costs
owing to the complex relationship between the different
equipments which form part of the total system, particularly
where advanced electronics are involved.

The increases in costs are directly related to increases
in the sophistication and complexity of military equipment,
increases which are necessary if performance targets are to
be met. Many involve major quantum jumps in technology.
There are also continual developments in existing technologies - such as increases in the thrust of given types of jet
engines, and improvements in the accuracy of particular
kinds of guidance systems. These improvements at the margin
of performance tend however to involve more than proportionate
increases in costs. It has been calculated for example
that a 15 per cent increase in the speed of an aircraft might
involve a ten-fold increase in cost. (4)

The justification for increasing performance specifications,
and hence for having such high technology levels, comes from the
military planners' assessments of the environment in which the
weapons would have to operate. In the majority of cases there
is little choice but to adopt these high requirements. To
equip the Armed Forces with weapons that would be outclassed
by those of their likely opponents would be a clear act of
folly, to put it no more strongly. However with a relatively
fixed proportion of the defence budget available for equipment,
ever-greater sophistication in weaponry may be bought at the
expense of numbers, or of other parts of the equipment programme, to an extent that may invalidate the original strategic
assumptions. In these circumstances the advantages of
introducing ever higher levels of technology need to be
examined critically and in the chapter on the Strategic and
Political implications of Defence Cuts we look at these
problems as well as the applicability of some of the suggested
alternative solutions.

From the point of view of the individual defence contractor,
survival in business will depend upon keeping up with the
general level of technical advance and in some cases maintaining
technical leadership. It has been argued that this fact may
encourage an attitude within the firms in favour of technology
for its own sake. The history of the variable geometry
fighter in the 1950s has been quoted (5) as a case where 'it
became imperative that some military application can be offered
for all the theories and tests'. It is difficult to assess
the magnitude of this effect but it will probably become less
significant in the future as defence contractors increasingly

feel the effects of constraints on the defence budget, and as the increasing sophistication of modern weaponry leads to a decline in private venture R and D. The great bulk of R and D defence expenditure is now financed by the Ministry of Defence and is therefore closely tied to specific current defence requirements. It is through these operational requirements that the shape and size of defence R and D is determined and hence the rate of technical advance.

There will always be pressures on defence planners to support domestic industry even where this may not make the best economic sense or produce the best military solution.   It is evident that the state of the defence industry in terms of size, and of its managerial and technical ability to produce at internationally competitive prices, will be a major influence on procurement decisions.   Following the analysis of J.R. Kurth of a particular procurement decision in the USA (6) we may identify four main types of consideration that are likely to be important:

1. Strategic or geo-political considerations.

2. The particular models and assumptions currently held by both national and alliance defence planners.

3. Political considerations.

4. The needs and interests of defence contractors.

The Structure of the Defence Industry

A capacity to develop and produce military hardware depends not only on the amount of plant, machinery and labour available for R and D and production, but also on the vital, although unquantifiable, infrastructure of skills and techniques and on a network of experienced management in the services, and their contractors and sub-contractors.

In Britain the state has traditionally played an important part in developing and producing defence equipment through the Royal Dockyards and Arsenals and later the Royal Ordnance Factories.   With the nationalization of Rolls Royce, the largest single defence contractor, and of the shipbuilding and aerospace industries, public ownership has now been extended to much of defence industry and this should produce a more socially efficient use of resources by eliminating unnecessary competition and duplication of facilities.

With the increase in sophistication of military equipment which we have already noted, industry has become increasingly specialized.   Since the government is the main customer this means that the government must organize its defence procurement policy as to ensure the viability of its main suppliers. As Vice-Admiral Clayton has explained: (7)

We have to give the shipbuilders a regular rolling programme of orders. We depend upon the specialised warship builders. They have a very carefully balanced selection of trades — drawing office, steel workers and outfit trades — which are required specially for warship building and not for commercial shipbuilding, and of necessity we have to keep a flow of orders going to them.

With the increased unit costs of defence equipment and the limited funds available it is no longer sensible in many major systems areas to maintain competition between domestic defence contractors. In the last 15 years for example Hawker Siddeley has produced or developed 7 military aircraft, including 2 that were cancelled, and BAC have developed or produced 6 military aircraft including 2 that were cancelled. In the preceding 15 years, the companies that amalgamated to form Hawker Siddeley and BAC developed or produced 28 and 18 military aircraft respectively as well as several research aircraft. The number of aircraft produced of any given type has declined substantially along with the increase in unit costs. The same phenomenon is to be found in shipbuilding. In the period 1965 to 1974 about half as much warship tonnage was launched compared with the period 1945 to 1954, yet costs have increased by factors ranging from 10 to 15. (8) Successive Governments have therefore encouraged a reduction in the number of major contractors to a number consistent with the size of the defence programme.

If defence contractors are to be able to respond to government requirements they must keep up with wider international developments in military technology and they must be able to show that they can innovate. Firms facing the prospect of smaller production runs due to high unit costs will have an additional incentive to develop cheaper solutions to military problems. Such innovations benefit not only the defence industry but the defence forces themselves.

Attempted Solutions

The governments of all arms-producing countries where the defence sector plays a major economic role are well aware of the problem of rising unit costs caused by a combination of high R and D expenditure and reduced production numbers. Current attempts to rectify the problem identify three possible solutions: the first is to cultivate expanding export markets; the second as we have indicated above is the reorganization of the defence industries; and the third is international collaboration.

Arms exports accounted last year for 27 per cent of British arms production: this proportion has been relatively stable

511

over the last few years and could be expected to increase
significantly only in response to a major change in world
demands.   A full examination of the economic implications
of arms exports together with a discussion of political
issues is given in the section on Arms Sales (see p. 539).

The second solution, the reorganization of the defence
industry, cannot be said to be fully tested.   It is
difficult to draw lessons from the past experience of the
amalgamation of firms engaged in defence work in this
country since up until recently only limited rationalization
has taken place.   However with the major industries of
shipbuilding, aero-engines and aerospace now under public
control, significant rationalization can at last be carried
out.   Moreover, with the progressive reduction in the
number of different types of weapon delivery system the
process of rationalization must have a logical limit, at
least on a national basis.

International collaboration in the manufacture of weapons
on the other hand can also produce excess capacity on an
international scale and thus lead to demands for international
rationalization of resources.   However, so long as individual
nations protect their military manufacturing capacity there is
bound to be some wasteful duplication.   The arguments in
favour of international collaboration stem from the greater
economies which come from spreading very high R and D costs
over long production runs and from the economies which can be
made in Alliance logistics through standardization.   Collaboration can take several forms:

   a.   Agreements for reciprocal purchase of different
      national weapons systems.

   b.   Agreed specialization in given areas.

   c.   Full collaborative development and production of
      complete systems either multilaterally or bilaterally.

From the point of view of defence planners, full collaborative
projects have the disadvantage of greater administrative
complexity and project inflexibility, it is more difficult to
change the specification of the project, or indeed to cancel
it, in the light of changing priorities.   Considerable
experience has been built up in the last few years with
projects such as Tornado, Jaguar and the Anglo/German/Italian
FH70 and SP70 Howitzers and there is now much greater knowledge
about the problems of amalgamating design teams from several
different countries and reconciling the requirements of different
Armed Forces.   Nevertheless the success of these projects has
shown that collaboration is possible and that savings can be
made compared with single country development even if such
projects do not provide the complete answer which it was once
thought they would.

It can be said with a reasonable degree of certainty that

512

none of the three attempted solutions described above has on
its own reversed the trend that we have identified of rising
unit costs of defence equipment; but that is not to say that
in the long run unit costs will not be lower as a result of
these measures than they would have been.

The Economic Effects of Defence Expenditure

We are spending in Britain this year £6,329m on defence, or
5.1 per cent of GDP at market prices. It is self evident
that this money is not being spent in the hopes of a quanti-
fiable economic return; it is being spent because, in the
world in which we live, security and national independence
can be purchased in no other way. Jobs and arms exports
are not justifications for such a level of defence spending
although they must be numbered amongst the results.

When deciding upon the allocation of resources for defence
the government must consider both the strategic objectives
which are its primary justification together with the net
economic benefits or disbenefits which result. The true
cost of defence spending may be viewed in terms of opportunities
foregone in the production and provision of goods and services
in the private sector and in other parts of the public sector.
This definition is a 'reminder that spending entails allocation
and choice among competing opportunities'. (9)

These trade-offs do not apply only to defence spending:
they apply to all claims on the gross national product between
the main spending components - private consumption, private
investment, exports and public expenditure - which result from
society's decisions in the market place and from choices
expressed via its democratically elected representatives. In
these terms the true costs of high levels of public expenditure
may be thought of in terms of lower levels of private consumption, exports and private investment. This is true whether
higher public provision takes the form of spending in defence,
education, health and social security or any other public
sector programme. At another level there is the allocation
of resources between individual public expenditure programmes,
and in what follows emphasis is placed on the resource costs
of defence as against other forms of public expenditure.

Any assessment of resource costs must take into account the
following considerations: the extent to which factors of
production are highly specialized, the time period over which
the transfer or re-allocation of resources is envisaged - in
the short-run resource costs will tend to be high for this
reason; the demand for and supply of, the resources in
question, and the ease with which they can be absorbed elsewhere. (10)

We have applied this analysis to an examination of some of
the key resources involved, labour, capital expenditure and
research and development expenditure. We have also looked
at the effect of defence spending on exports and the balance
of payments.

(<u>A</u>) <u>Labour</u>. Current levels of defence expenditure provide
for the employment of 337,100 servicemen, about 300,000
civilians — industrial and non-industrial civil servants —
and about 200,000 workers directly employed in the defence
industries excluding those on export work.

Defence spending thus has a direct impact on jobs, being
responsible for the employment of over 3 per cent of the
economically active population. It has been an overriding
concern of the labour movement that recent reductions in
defence spending should not entail unacceptable increases
in unemployment, although it has always been recognized that
some loss of job opportunities is inevitable. For this
reason it is important that the existing machinery for
consultation between government, unions and firms should be
used to the full so that redundancy may be avoided wherever
possible.

Defence is not only a user of labour resources, it is also
a producer, particularly of skilled labour. Most servicemen
enlist shortly after school leaving age. They receive
education and technical training, often of an advanced kind,
from the Services and for the most part return to civilian
life while still young men, typically after three, six or nine
years service, better equipped to contribute to the civilian
economy.

Defence has been at the forefront of the development of
personnel management and selection methods and in the provision
of training, educational and medical services for its service and
civilian employees. The current defence budget includes
provision of some £633m for such 'social' expenditure. Defence
has also been instrumental in the development of modern techniques
for the more efficient use of resources such as operational
analysis, programming and other management techniques. The indirect role of the armed forces in pioneering new ideas and
introducing new techniques to society has a long history, for
example as Marx wrote to Engels in 1857:

> In general, the Army is important for economic development.
> For instance, it was in the Army that the ancients first
> developed a complete wages system. Similarly, among the
> Romans the Peculium Castrense was the first legal form in
> which the right of others than the fathers of families to
> movable property was recognised. So also the Guild
> system among the corporation of Fabri. Here too the
> first use of machinery on a larger scale ... The division
> of labour within one branch was also first carried out in
> the Armies.

(<u>B</u>) <u>Investment</u>. Defence spending can be broadly split into
direct expenditure on personnel (45 per cent), expenditure on
equipment (37 per cent), and miscellaneous works and other
expenditure (17 per cent). Taken all together the total of
individual public sector budgets reflects the government's
priorities about public as against private consumption, about

514

the level of demand in the economy, the requirements of
public sector borrowing and so on. It is of course open
to a government to restrict the growth of public spending
overall, in the interests of allowing greater growth in
the other components of GNP such as private investment or
private consumption. In such circumstances all expenditure
by local and central governments must be scrutinized in the
same light. Insofar as resources are employed on defence
tasks as opposed to other public programmes, this reflects
the priorities of the government of the day.

It has however been claimed, (11) that cuts in defence
spending would by themselves generate investment since
resources used in the production of defence equipment could
then be diverted to the production of more investment goods
in the private sector. Such an outcome could not however
be expected to occur automatically. Other public sector
programmes would have to be increased, or compensating tax
reductions made, in order to maintain the same level of
domestic demand. In part the release of resources would be
taken up by increased consumption and imports, particularly
since defence expenditure can be regarded as being financed
largely from taxation and not from private or corporate
savings. It is not obvious that private consumption as a
share of GNP has been stable in the period 1955-69; there
is some evidence to show that the share of private consumption
has fallen, while the shares of public consumption and private
investment have risen. For the UK the share of private
investment has, however, tended to rise more quickly than that
of public consumption, and more quickly than in the majority
of OECD countries. It cannot be inferred therefore that
historically defence spending and investment have been in
competition with each other, and in particular that lower
levels of defence spending would necessarily mean higher
private investment.

The relatively poor economic performance of this country
since the war has been ascribed in part to the deficiency in
domestic investment and the table on p.516 shows figures for
gross fixed investment as a percentage of GNP for the major
OECD countries. It will be noted that the United Kingdom's
low investment record is influenced by relatively low levels
of residential construction, it will also be seen that the
proportion of GNP spent by the defence industries is much
smaller than the percentage by which we have under-invested
compared to other countries. The explanation therefore for
our lower levels of gross fixed investment is more likely to
be related to the decline in the long run rate of return to
capital and to the falling share of profits in national
income, rather than to expenditure on the defence industries.

(C) Research and Development. The increasing sophistication
of weapons systems, in response to increased performance
specifications, is reflected in the high expenditure by the
Ministry of Defence on Research and Development. In the

TABLE

	Total Investment	Gross Fixed Investment as % of GNP (1968-72)			Defence Spending as % of GNP (1968-72)
		Machinery and Equipment	Residential Construction	Other Construction	
Belgium	20.8	8.8	4.7	7.3	3.3
Canada	21.4	7.9	4.4	9.1	2.7
Denmark	23.6	12.1	4.6	6.9	3.0
France	26.3	12.0	6.5	7.8	4.8
W Germany	25.9	12.4	5.2	8.3	3.9
Italy	19.8	8.4	5.9	5.5	3.2
Japan	38.7	32.1*	6.6	- *	-
Netherlands	25.3	10.5	5.7	9.1	3.9
UK	19.6	9.5	3.5	6.6	5.8
USA	17.0	7.0Ø	3.5	6.5	8.5

* Includes other construction

Ø Excludes Government expenditures on machinery and equipment

Source: OECD

current year (1977/78) it is planned to spend £826m on research and on the development of approved systems. This puts the defence industries in the forefront of technology in this country and these industries employ a greater number of scientists and engineers than civil industries producing less advanced products. In areas such as shipbuilding, miniaturization, guidance and control systems defence R and D is contributing to advances in the techniques used in civil work.

The great bulk of research and development for commercial purposes in this country is, however, financed privately. Like investment in capital goods, investment in new ideas is undertaken in the expectation of future profit. Attempts by the government to encourage an increase in such expenditure will be discounted by whatever view firms take about the future; and the informed instincts of businessmen are probably the principal factor determining the size of this expenditure. We have already noted the secular decline in the number of major defence contractors since the war, and the steps taken by successive governments to tailor capacity in the defence industries to requirements. It is likely that had those sectors of civilian industry which were expanding over that period needed additional R and D then the necessary resources would have been attracted from the defence sector or elsewhere. A NEDO study (12) into the mechanical engineering industry has shown on the one hand that employment in that industry of qualified scientists and engineers increased by 50 per cent between 1962 and 1971; and that on the other hand employment of such staff on R and D work increased not at all up to 1968 and by only 4 per cent over the whole period. This suggests that industry has been seeking to enlist the services of an increasing proportion of the most highly qualified staff within its total employment, but that it has apparently not chosen to employ them on R and D. As we have argued above, resource costs are principally a short-term phenomenon and there is no real evidence available to indicate that the use of highly-skilled manpower by the arms sector would have pre-empted its use elsewhere in industrial R and D had the real demand for it been there.

(D) Exports and the Balance of Payments. The chapter of this paper devoted to arms sales takes an overall look at the implications of this trade; the purpose of this brief section therefore is to summarize the results as they relate to the resource costs of this expenditure.

This year (1977/78) overseas sales of defence equipment are expected to reach £850m, representing 27 per cent of total military output, that is, equipment for HM Forces and for export.

Industrial resources devoted to defence work generate much the same order of export performance as the same quantity of resources would when put to comparable civil use. This is

not an argument for military arms expenditure, the
justification for which must rest on other grounds, but it
does represent an undoubted indirect benefit to be taken
into consideration when defence is compared with other
areas of public expenditure.

The net effects on the balance of payments of military
expenditure are hard to estimate since available figures
for Government invisible transactions, for example, local
expenditure by forces abroad, cannot be taken to reflect
the budgetary effect of the items concerned.   The published
figures show a small net debit in the immediate impact of
the expenditure.

	1976–77 £m	1977–78 £m
Cost of stationing troops abroad	690	696
Other military services *	160	147
Purchase of military equipment	167	258
Total debits	1017	1101
Receipts from US forces in Britain	30	42
Other receipts **	40	32
Sales of military equipment	720	854
Private expenditure by US forces in Britain	(80)	95
Total credits	870	1023

Source: Statement on defence estimates 1977, Cmnd 6735;
figures in brackets is an own estimate.

*   Includes contributions to infrastructure projects, R and D
    levies, and contributions to international defence
    organizations.

** Includes R and D levies

The Future Equipment Programme

The equipment programme outlined in the 1977 Statement on
the Defence Estimates absorbs a growing share of the total
defence budget.   On present plans, and assuming past trends
of productivity improvement, equipment expenditure over the
next 2 or 3 years is likely to ensure that broadly the
current level of 200,000 employed in the defence industries
remain engaged on work for the Ministry of Defence.

Within this overall figure, sectors of industry and
individual firms may experience expansion or contraction
as the effort on major programmes such as Tornado or Jaguar
builds up or declines.   It is part of the task of defence

industrial planning to anticipate such fluctuations in loading and so far as possible to iron them out or to mitigate any adverse consequences, including the high costs of disruption which any major transfer of resources is likely to cause in the short and medium term, especially in a period of recession and high general unemployment.

Although no changes in the defence programme are advocated in this report, the Study Group has examined the problems of conversion and diversification. Whether existing planning machinery is adequate for diverting released resources into a selection of socially useful industrial activities is a question which goes beyond the immediate scope of this study.

Conclusion

We have argued that the primary justification for defence spending must rest with the benefits that it brings in terms of domestic security, the fulfilment of overseas obligations and the protection of national interests overseas. How much of the nation's resources should be devoted to these social objectives can only be determined in the light of the national and international economic position of the country, and the assessment given to competing claims of other social needs. We have shown that there are indirect economic benefits to this country from defence, as well as indirect resource costs particularly in the short run. There is currently no conclusive evidence to show that the return from defence is any lower than that from other competing public expenditure programmes. Equally there is no reliable evidence of a link between military spending and the relatively poor economic performance of this country since the war, whether in terms of growth, investment, exports or inflation.

The Strategic and Political Implications of Defence Cuts

The strategic and political implications of either increasing or decreasing defence expenditure are questions which are central to any discussion of what is the most appropriate level for the defence budget. There is a tendency, particularly in certain circles of opinion, to view defence as a burden on the economy fulfilling no useful social objective. For those who take such a view, it is all too easy to see defence as a natural source of budgetary savings at a time of pressure on national finances and to ignore, or at best to play down, the consequences of such arbitrary reductions. For example arguments are advanced which suggest that, even if the UK were unilaterally to make massive reductions in its defence spending, the consequence would either be that our allies would willingly and without hesitation fill the gap left or even that such unilateral reductions would in some way induce the Warsaw Pact to follow suit and thereby further the cause

of disarmament. Such views ignore the political and
military realities which must influence defence decisions.

The labour movement's policy on defence must take account
of these realities. It is the view of the present government, as of all previous governments, that the security of
the country and of our people must remain one of the fundamental responsibilities of the party in power. Indeed as
a party with a particular concern for the democratic rights
and freedoms of all sections of our population, we quite
naturally wish to ensure that the benefits deriving from
our particular system of government are not put at risk by
any pressures that may be exerted from outside. We can only
be assured that our democratic rights will not be eroded if
adequate resources are devoted to defence   Expenditure on
defence should thus be seen as an essential complement to
the resources devoted to the other elements of our social
programme rather than merely as a drain on the exchequer
which acts in direct competition to our other objectives.

As is the case with all areas of government activity, it
is never possible to say that a particular level of defence
expenditure is absolutely right and therefore outside the
realms of debate and argument. Even less can one claim
that, as circumstances change, a chosen level of defence
expenditure will necessarily continue to remain valid.
For example, as a result of the continuing examination of our
defence programme the Government have been able to announce
since the Defence Review further savings in planned
expenditure amounting to several hundreds of millions of
pounds: most recently with cuts announced last year of
£200m in the 1977/1978 budget and a further £230m cut in
the planned 1978/9 budget. While, therefore, it may not
be possible to speak in these absolute terms about the
'right' level of defence expenditure, it is possible to
claim that, on the basis of a rigorous examination of all the
factors involved, including military, political and strategic
considerations as well as the constraints imposed by the
other priorities that the Government has set itself, a
particular programme for defence (involving a certain level of
expenditure) is the one most likely to meet the requirements
of the nation. It was just such an examination that this
government initiated on assuming office in 1974. The results
of that Defence Review continue to form the basis for our
defence planning into the 1980s. It is obviously always
possible to claim that the defence budget required by the
programme that resulted from the Defence Review is too large
by some arbitrary amount, say £1,000 or £1,500m. What is
unacceptable, whatever one's view of the Defence Review, is to
argue that the defence budget reduced by such an amount would
meet the nation's security needs. Applying such a principle
to other spheres of government concern, it would no doubt be
possible to cut the education or social security budgets by
an arbitrary £1,000m and to draw up a list of cuts on a rising
scale of pain. No one would suggest that, at the end of the

day, this way of going about matters would produce a balanced
and adequate programme for the country's health and social
security needs. There is no reason why defence should be
treated any differently.

The Nature of the Soviet Threat

One of the essential factors which must have an important
influence on the level of the defence budget is the threat
posed by any likely aggressor, which in our case means the
Soviet Union and its immediate allies. It is as well
therefore to consider the aspect of defence planning at
the outset.

The nature of the challenge that we face from the Soviet
Union is two fold. In the first place they have made it
clear that, while wishing to pursue the process of détente,
they would not wish to allow any let-up in the ideological
struggle. If this merely meant a peaceful battle of words
and ideologies, it is not something from which the West
should shrink and indeed, given the bankrupt nature of
Soviet ideology, it is a struggle in which the West should
whole-heartedly engage in the full expectation that the
principles on which the Western democracies are founded are
far more likely to stand up to rigorous scrutiny than the
barren ideology and inhumane practices adopted by Soviet
leaders. Unfortunately however it is clear that the
Soviet Union is not prepared to allow this clash of ideologies
to develop purely on this peaceful plain and we have seen that,
in various parts of the world and particularly in Eastern
Europe for example, they have not hesitated to resort to overt
military action in support of régimes embracing their type of
totalitarianism. It is indeed this continuing ambiguity in
Soviet claims to be seeking 'peaceful co-existence' on the one
hand yet promoting the 'ideological struggle' by military
means on the other that leads the West to seek other criteria by
which to judge Soviet intent. It is therefore clear that
Western nations must ensure that the Soviet Union is never
tempted to take such steps in areas of more direct concern to
our security interests.

The second way in which Soviet actions pose a direct challenge
to the Western nations is in the continuing growth in their
defence expenditure. In assessing the threat that these forces
pose to the West it is of course necessary to consider the
military capability that they offer to the Warsaw Pact nations.
At the same time it would be folly not to try to assess the
intentions of the Soviet leaders that give rise to this massive
build up. If it could be demonstrated beyond any reasonable
doubt that, despite this build up in arms, the intentions of
the Warsaw Pact countries were perfectly peaceable and that this
state of affairs would continue for many years into the future
there could well be reason for regarding the increase in the
Eastern defence budgets with less concern. Unfortunately it is
just not possible to view Soviet intentions with this degree of

confidence, particularly since the closed nature of that society conceals the process of decision making from their own public and from the Western nations. Even if it were true that the armaments build-up was brought about by a combination of internal and external factors none of which need necessarily derive from hostile intentions towards the Western democracies, it would be quite easy for these intentions to change as the options which an expanding military capability offers to the Soviet leadership increase. Recent history is all too full of examples of improvements in military capability leading to inflated foreign policy objectives. Such a change in intentions, even if it came about gradually rather than as a result of a sudden change in policy, would almost certainly occur in a shorter timespan than the West would need to achieve any substantial change in the balance of military strength.

Those who take an optimistic view of Soviet intentions point to their interest in consolidating their present spheres of influence and creating a limited land-based 'cordon sanitaire' on her Western front. This may or may not give an accurate indication of current Soviet aims, but the consolidation of existing spheres of control and the extension of Soviet hegemony are not mutually exclusive policy options. It can be argued that the first is a necessary prerequisite for the second. This view, moreover, ignores the growth of the Soviet Navy. It is fair to say that the Soviet Union is interested in the status quo insofar as it affects Eastern Europe, whereas in Western Europe and in the rest of the world they remain firmly attached to revolutionary change. In any case such an interpretation is quite inconsistent with the size and shape of Soviet forces both in Eastern Europe and elsewhere.

The fact of the matter is that, while there is no reason to suppose that the Soviet Union and its allies are considering launching a direct attack on the West, it is not possible to conclude that the Soviet Union would be unwilling to use its military strength to back up political and economic pressure aimed, to put it no higher, at extending the Soviet sphere of influence. That the Soviet Union has so far not chosen, or not been able, to use its military power to put pressure on the West in this way can be attributed, at least in part, to the credibility of NATO's policies of deterrence. Nonetheless we must continue to bear in mind that this indirect and insidious threat is in many ways more difficult to respond to than any direct military challenge. In the last resort we must be alive to the possibility that Soviet intentions could change to embrace the overt use of military force.

When considering Warsaw Pact capabilities in the context of the threat to our own security, it is as well to confine our examination to the situation in Europe and the Atlantic, since it is here that the great preponderance of forces challenging the security of Western Europe are deployed and where the confrontation is the most direct. Consideration

522

of the global balance is clearly complicated by the various interests of the superpowers outside Europe – for example, the Soviet Union is clearly concerned by recent developments in its relations with China and hence deploys forces along the Sino/Soviet border as a military safeguard. These forces can, at least for present purposes, be discounted as offering no direct threat to Western security, but, by the same token, it would be hard to claim that the main role of the Warsaw Pact forces stationed in Europe is not in Europe itself.

The most striking fact about the build up of forces in Eastern Europe is that the size of these forces is beyond anything that the Warsaw Pact could conceivably need for its own defence. There can be no question here that, as is often argued, the Warsaw Pact is merely attempting to make good a military advantage favouring the West. On the contrary the Eastern nations have since the last war always maintained a very sizeable advantage in conventional capability which they have been at pains to preserve and in some cases to improve. Even when the Western nations made a conscious effort to close the gap at a time when we were moving from the trip-wire philosophy to the more realistic flexible response strategy the response of the Warsaw Pact was to build up its conventional forces yet further, thereby fuelling the arms build-up and making impossible the establishment of relatively stable balance of forces in Europe. It is not safe to argue that, even though such an imbalance exists, it need be of no great concern to the West since the forces of the Soviet Union's allies are not particularly effective and could not be relied on to follow the Soviet lead if hostilities were initiated against the West. On the contrary, the indications show that the non-Soviet Warsaw Pact forces are well trained and equipped and capable of effectively carrying out an offensive role. As to the argument that they might in the event of hostilities not be mobilized, or that they might be used to hold down the populations of their own countries, it would be irresponsible of any government or Alliance to plan its defensive measures on the assumption that all the ready forces of the aggressor would not be used.

The basic facts about the current balance of forces in Europe are fairly well known, but, since the dangers inherent in the present situation are often forgotten, it is as well to go over the ground once again. Briefly, the Warsaw Pact enjoys a predominance over NATO forces in nearly all types of armament, and particularly in conventional weapons. In some sectors the predominance is very significant indeed and this is perhaps particularly true of the forces that exist to fulfil primarily offensive roles. As a broad indication of this imbalance the last assessment available to us shows that the Warsaw Pact have over 950,000 ground force personnel in the central region of Europe as against a NATO force of about 790,000, including the French. This gives the East an advantage in ground force personnel of over 150,000. In Main Battle Tanks, the foremost weapon for seizing and holding territory, the advantage is even more pronounced with the Warsaw Pact having over 16,500 tanks in active units in the central region against about 6,000 in NATO. This is by no means an isolated example of the superiority the East enjoys — similar superiority exists in, for example, the number of field guns — but it is perhaps a particularly significant example, given the role of their tank forces. Lest it should be felt that this preponderance in tank forces need not be a cause for concern it is perhaps salutary to quote from an interview that the Chief Marshal of Soviet Tank Troops, A. Babadzhayan, gave to Pravda last September:

> The excellent cross-country performance of our tanks enables them to overcome water obstacles and all kinds of barriers and encumbrances. In combination with high speeds and a substantial operating range, this enables the tank troops to undertake rapid marches, manoeuvre flexibly and to develop an offensive at great speed and to a great depth. The view has been expressed in the capitalist countries that in view of the consequence of guided anti-tank missiles, helicopters etc. (i.e. PGMs) tanks are beginning to lose their former importance. We do not think that this is so. The Soviet art of war recognises the tank troops' role as the main striking and manoeuvrable force of the ground troops.

The Soviet ground forces are therefore structured for Blitz-krieg. It would be open to the Warsaw Pact to adopt a more obviously defensive posture on the lines of the NATO forces facing them. They have failed to do so.

The Warsaw Pact naval forces in the Atlantic and air forces in Europe enjoy numerical advantages similar to those that they have in ground forces.

The geographical circumstances of NATO make us reliant on the maintenance of open sea routes in marked contrast to the Soviet Union and its Allies who can supply their own require-ments without the need for sea lines of communication. Thus, the build up of Warsaw Pact naval forces in the Atlantic. which outstrip the forces that NATO can deploy by about 50 per cent, is particularly disturbing. The threat from the Soviet submarine forces is especially significant; since 1968 the number of nuclear powered submarines in the Soviet Northern Fleet has more than doubled. In addition, the Soviet Union operates naval aircraft with a wide radius of action — its naval air forces are now being re-equipped with the Backfire long range bomber — and large, modern cruisers and destroyers equipped with up to date surface-to-surface missiles. More-over, the first aircraft carrier of the Kiev class has now been brought into service.

In combat aircraft, the Warsaw Pact enjoys an overall numerical superiority in central Europe over NATO of more than 2:1. It has been argued that, since much of this superiority is accounted for by the large numbers of fighter aircraft, the imbalance in offensive capability after reinforcement would not be particularly marked. It is important to realize, however, that the Warsaw Pact countries could, if they chose, augment their offensive capability by using their fighter aircraft in the attack role as well as in support of their ground forces. It is noteworthy that in recent years the re-equipment programme of the Soviet air forces has concentrated on ensuring that the Tactical Air Force is far better able to engage in conventional offensive air operations. Particularly significant amongst the improvements that have been made is the introduction of three new types of swing-wing tactical aircraft with both a conventional and nuclear capability.

If the East enjoys such advantages as a result of the deployment of their forces in Europe and the Atlantic, these advantages are compounded by the geographical factors favouring the Warsaw Pact. In particular the proximity of the Soviet Union to the central region would clearly favour the East should a period of heightened tension necessitate the reinforcement of those forces already on the ground. The Soviet Union has a considerable number of combat divisions in the Western Military districts immediately available to reinforce the central region; others are available at slightly lower levels of readiness. The West does of course have the means to counter any such reinforcement by the East. The US dual based forces could be moved very rapidly to the area and likewise rapid action could be taken by European nations, including the UK, to reinforce the central region forces from outside the area and to call up the reserve forces within. However, quite apart from the natural advantage enjoyed by the aggressor, it is quite clear that, as a consequence of the very long lines of communication between North America and Europe, NATO could not, in the early stages of reinforcement, rely on the main contributor to the Alliance for a rapid reinforcement of our forces to the same extent that the Warsaw Pact would be able to rely on the Soviet Union.

In view of the present state of the military balance as just described, it is particularly worrying that the Soviet Union is continuing to enhance the effectiveness of its forces by means of both qualitative and quantitative improvements. Its military expenditure is estimated to have grown in recent years by about 5 per cent a year in real terms and is now running at about 11-12 per cent of its GNP on military programmes. This is a clear indication of the priority given by the Eastern bloc to its defence programme. The more significant force improvements that this increase in expenditure represent are well documented in this year's Defence White Paper as well as in other sources. These improvements include the introduction of the new Backfire bomber, the deployment of the mobile SS 20 in the Western Military districts and the introduction of new types of inter-continental ballistic missiles such as the SS 17, SS 18 and SS 19. The Soviet Navy has introduced new nuclear-powered submarines and the first of a class of aircraft carriers equipped with helicopters and vertical take-off

aircraft.   In addition to these well publicized improvements
there have been others at a lower level which collectively
are no less significant.   In recent years the Soviet forces
have been re-equipped with new armoured personnel carriers,
self-propelled guns, surface-to-air missiles, tactical
rockets, guided weapons and bridging and chemical warfare
equipment.   Their logistic support and supply capability has
also been much enhanced.   These improvements have particularly
taken effect in central Europe and have been followed most
recently by the deployment in considerable numbers of the new
T-72 tank to the Soviet forces in East Germany.   Together
these improvements represent a significant erosion of the
NATO superiority in equipment quality which we have hitherto
relied on to offset the East's massive quantitative advantages.
While we are satisfied that the NATO forces are still sufficient
to carry out their deterrent and defensive roles, there can be
no doubt that the quantitative and qualitative improvements that
have been described have increased the military preponderance
of the Soviet Union and have notably reduced the amount of
preparation the Russians would have to make before they could
conduct sustained operations and consequently the amount of
warning NATO might receive of an impending attack.

It is sometimes argued that NATO's superiority in theatre
nuclear forces compensates for its conventional weaknesses.
This is to misunderstand the role of NATO's theatre nuclear
forces.   They form the second element in the Alliance's triad
of forces and are designed to provide NATO with the widest
possible range of options in the event of aggression by the
Warsaw Pact, thereby contributing to the credibility of
deterrence and to the maintenance of peace.   If NATO appeared
unable to sustain a stalwart conventional defence, a potential
aggressor might well conclude that the Allies were so neglect-
ful of their security that they would be unlikely to have the
collective determination to initiate the use of nuclear
weapons, with the appalling destruction that would ensue from
the widespread employment of these   As a consequence, an
aggressor might be prepared to take the risk.   Powerful
conventional forces are thus an essential element in the
deterrence of aggression, and theatre nuclear forces cannot
be seen as a substitute for them   The role of NATO's theatre
forces is to deter an aggressor from using such forces himself.
The Soviet Union is devoting immense efforts to augmenting
and strengthening its already formidable theatre nuclear
forces - and to give the Allies the option of signalling to
an aggressor, in the event of an attack, that he had under-
estimated the resolution of the Allies to defend themselves.
It is worth adding that the West has offered substantially
to reduce the number of US tactical nuclear weapons stationed
in Europe in the context of a satisfactory MBFR agreement.
This offer has so far met with only a negative response.

Alliance Response

In responding to the threat that the Warsaw Pact forces pose
to the West it is clear that the UK is bound to co-operate with
those countries that have a similar interest in defending their
democratic institutions. To attempt defensive measures in
isolation from our Allies would involve expenditure far beyond
anything that this country could afford. It is for this
reason that we must continue to make our effective contribution
through the Alliance in terms of our military preparedness and
do everything in our power to maintain the cohesion and
credibility of NATO.

In recent years NATO has become as much an instrument for
the pursuit of détente as a defence alliance designed to
ensure the maintenance of an effective deterrent. As the
atmosphere in international relations has become less cool,
disarmament negotiations have been opened with the East on a
number of fronts, and there have been attempts to extend
economic co-operation and cultural exchanges with the nations
of the Warsaw Pact. The British Government is concerned to
see that everything possible is done to make a success of
these efforts. Our long term objective is a state of complete
disarmament in which the formation of nations into military
alliances such as NATO and the Warsaw Pact will no longer be
necessary. However, it must be recognized that whilst
this is being pursued it is vital that neither side weakens its
defensive arrangements to such an extent that the other is able
to take advantage of this weakness to force through agreements
which, far from enhancing the stability of inter-state relations,
might actually harm them in the long term. Thus, from the
point of view both of the Alliance's interest in the pursuit of
détente and of the maintenance of an effective deterrent force,
it is important that Alliance members do not reduce their
contributions and, where possible, actually improve the
effectiveness of their forces so as to stem the widening gap
between the Warsaw Pact and NATO.

The importance that the UK attaches to its Alliance membership cannot be questioned. As a result of the Defence Review
we now devote 95 per cent of our defence effort to NATO. We
are the only European member to make a contribution to each
element of the triad of forces. However, the Defence Review
demonstrated that our commitment to the Alliance could be most
effectively fulfilled if we concentrated on certain areas of
Alliance defence rather than spread our efforts across all
areas and roles. Four general areas were selected as being
vital to our national security. These are the Central
Region of Europe where BAOR holds an important section of the
crucial Central Front and RAF Germany plays a major part in
the forward defence of the Region; the Eastern Atlantic and
Channel areas where the growth of Soviet maritime power is
particularly evident and where NATO depends critically on
Britain to provide the main element of the maritime forces
immediately available to the Alliance; the United Kingdom
and its immediate approaches whose security must be maintained if we are to play our full part in the Alliance; and

the NATO nuclear deterrent to which, with the Polaris force, we make a unique European contribution.

Britain's own security interests are fundamental in each of these areas. In them our interests and those of the Alliance coincide. By providing the capabilities to meet our interests in these areas, we ensure our own security and that of our allies. These capabilities are complementary and together make up a balance of forces on which our contributions to the Alliance rests. If one or more of the capabilities is reduced the overall level of the security of the UK and of NATO will suffer correspondingly.

Nonetheless, as a result of the Defence Review and subsequent decisions, UK defence spending should systematically fall from its present level of about $5\frac{1}{2}$ per cent of GNP to a level much more closely in line with those of our European allies.

The proportion of GNP is not, however, the final yardstick for determining what a country's defence effort should be, any more than defence spending per head should be. The success of NATO's policy of deterrence rests, in part, on whether the military capability of its members provides a credible defence; and the nature of the contribution which each partner makes and its cost are likely to vary from member to member because no two countries' circumstances are the same.

In the last resort each country will decide for itself what its contribution to meeting the perceived threat should be in the light of the sum of the contributions its allies have made to meeting the same threat.

Options for Reducing Defence Expenditure

In the end then, the choice of how much expenditure we should devote to our security needs is a matter of judgment. With all the pressures for resources to improve investment in industry and to help the balance of payments, it is natural that further substantial cuts in defence expenditure should be considered.

If defence expenditure is to be reduced by the quite arbitrary figure of £1,000m a year at 1974 prices, four main options have been identified:

   a. Reduce the Navy General Purpose Forces by paying off large surface ships, changing the new ship construction programme, and abandoning the second and third ASW cruisers.

   b. Reduce BAOR to an eventual strength of about 30,000.

   c. Reduce the RAF General Purpose Combat Forces by equipping the RAF to cover less than the full

spectrum of tactical air capabilities, with a concentration
of 'battlefield' roles and the cancellation of the MRCA
programme.

    d.   Abandon the Polaris force.

    To find £1,000m a year at 1974 prices would require
taking any two of options a-c plus option d together with
large proportionate cuts in support programmes.   In considering whether it is feasible to look for savings of this size
we need to look first of all at the strategic implications.
    We have noted earlier in this chapter the increasing quality
of the Soviet Navy with its growing emphasis on nuclear
powered submarines.   Much of it would be deployed in the
Eastern Atlantic and Channel areas where the Royal Navy makes
a particularly important contribution to the Alliance.   To
deter a large 'quality' fleet such as the Soviet Navy requires
sophisticated capabilities of which the large ship - such as
the new ASW cruisers - with its major anti-submarine capability
is an important constituent.   The matter is discussed at
greater length in Annex A.   Saclant has made it clear that
in his view his resources are not adequate to meet all the
contingencies of his major tasks and that accordingly the
defence of the Eastern Atlantic area will fall in substantial
part to the UK and our European allies, particularly in the
early days of any conflict.   We cannot expect others to
take the place of the RN in the Eastern Atlantic in view of
their other commitments elsewhere.   The Soviet Union would
therefore be left in a dominant position in the seas
immediately around this country through which the vital sea
routes to Europe all pass.
    On the Central Front the Warsaw Pact has superiority in
front-line manpower and equipment and is organized for
offensive operations at little warning, though the scale of
the attack would depend on the extent that the Warsaw Pact
traded preparation against surprise.   The essence of NATO's
deterrent posture must be to make it clear that an aggressor
cannot obtain an easy victory whatever type of attack it
chooses to adopt.   Without the flexibility of a substantial
level of conventional forward defence permanently stationed
in the most likely area of hostilities, the chances of an un-
reinforced attack by the Warsaw Pact are increased and the
Alliance's deterrent would revert to the trip-wire strategy,
postulating the early use of nuclear weapons.   To cut BAOR
by 45 per cent unilaterally would entail either accepting
that in the British area of responsibility there was no
longer a credible conventional defence or assuming that our
other NATO allies would fill the gap we created, though there
is no evidence at all to show that they would.   (These
arguments are set out at greater length in Annex B.)
    Quite apart from the strategic consequences of such cuts
it would clearly be extremely damaging to make such large
scale reductions before the negotiation of a satisfactory

agreement on mutual and balanced force reductions.   It is
only by demonstrating that NATO is prepared to deploy
effective deterrent forces and to match any improvements
made by the Warsaw Pact that we will be able to keep the
pressure on the Soviet Union for an agreement based on a
parity of conventional forces.   Any signs that NATO
countries individually are not prepared to sustain this
common front will merely demonstrate to the Soviet Union
that they can achieve much of what they seek in the
negotiations, namely the permanent establishment of
Eastern superiority in central Europe, without even the
need to attempt to reach an agreement with the West.
It is unrealistic to suppose that unilateral cuts in
the West will strengthen the hands of the 'doves' in
Moscow particularly since the decision makers in the East
are not, unfortunately, subject to the same pressures of
public opinion as those in the West.   If they had any
effect at all on the domestic pressures within the Soviet
Union unilateral cuts in the West would be more likely to
play into the hands of those that might argue that
greater political or military use should be made of the
increasingly favourable Warsaw Pact military posture.

In short, the process of multilateral negotiation is a
lengthy one and quick solutions cannot be expected.   If
successful, it offers a much better chance of establishing
more stable relations between East and West at a lower
level of forces and on a permanent basis.   The alternative
of unilateral cuts is essentially destabilizing;   it offers
no guarantee that the other side will follow suit;   and it
involves the significant risk that one side will attain an
overwhelming military advantage.

On the question of air forces NATO's strategy requires
the possession of a full range of capabilities, both
offensive and defensive.   It would not be realistic for the
RAF to opt out of some of the roles in its present spectrum
of capabilities.   To do so would require specialization
in roles by different NATO countries, since for all Alliance
air forces to concentrate on 'battlefield' tanks would leave
a number of vital roles unprovided for.   Thus the cancellation
of the MRCA would, for example, leave the air defence of
United Kingdom air space beyond the 1980s without adequate
cover and there would be absolutely no reason why our Allies,
seeing us completely unprepared to make adequate provision
for our fundamental national security requirements, should
choose to step in and fill the gap.   In any case, in the
light of the increasing Warsaw Pact threat, it is likely
that, if members of NATO agreed to specialize, they would
expect any resulting financial savings, if they material-
ized, to be used to improve the effective level of overall
defence rather than to reduce the efforts in individual
countries.   (These arguments are dealt with in greater
detail in Annex C.)

Abandonment of Polaris would lead to total reliance on the

US and France to provide the essential third element in the triad of conventional, tactical nuclear and strategic nuclear forces on which NATO's strategy of deterrence is based, and it would leave the French as the only European power with a strategic nuclear capability. It is unrealistic, not least because of the provisions of the Nuclear Non-Proliferation Treaty to expect other European powers, individually or collectively, to acquire a similar capability. Polaris can remain an effective part of NATO's strategic nuclear capability at comparatively little cost and contribute to the deterrent effect of NATO's strategic forces by providing an additional centre of nuclear decision within the Alliance which would compound the risks faced by an aggressor.

Thus to take any of these options would have a serious effect on the security of Britain. First there would be the direct military consequences of the reduction in military capability. Secondly, and more far-reaching, there would be the effect on the Alliance as a whole. There is no doubt that if Britain, as one of the major members of NATO, were to reduce its effort substantially in two or three of the four main areas in which it is now contributing to the defence of the West, the impact on the cohesion and solidarity of the Alliance would be severe and perhaps critical. Other European nations would be unlikely to have the means or the inclination to fill the gap left by the withdrawal of UK forces and some of them could well come under increasing pressure to make similar reductions in their defence efforts. In these circumstances it would not be surprising if the United States re-assessed its attitude to the defence of Europe with the possibility that this might mark the beginning of the unravelling of NATO. It is therefore misleading to argue that, because a cut of £1,000m at 1974 prices would represent gross reduction of only $1\frac{1}{2}$ per cent in total NATO spending, the consequences of such an action might not be serious. In European terms our contribution to the Alliance is vital and a reduction of this order cannot be dismissed lightly. We therefore do not believe that we can run the risks that such massive reductions in defence expenditure would involve: our security which depends on the Alliance is too high in our order of priorities to permit us to do so.

On the other hand we cannot take comfort from the application of new technology to weapon development through such innovations as Precision Guided Munitions (PGMs). There is no evidence yet that this new technology is necessarily cheaper than the old or that it will give a permanent advantage to the defender as against the aggressor. A number of the weapons and techniques now entering, or about to enter, service is of direct advantage to the attacker. Where new techniques can be said to be marginally more advantageous to the defender, such as certain of the PGMs, it would be foolish indeed not to expect, as has always happened in the past, that adequate offensive counter-measures will be found, probably sooner rather than later, or even that new aggressive techniques will eventually be developed requiring the development of further defensive weapons.

Conclusion

It would therefore be most unwise on political, military and economic grounds for us to make substantial cuts in our future defence expenditure.   Defence spending has now been established at a broad level that balances the needs of Britain's security and what, as a nation, we can afford to devote to it. We believe that our defence policy should continue to be based on that assessment of our priorities.

Naval General Purpose Forces

Against the naval threat described in this section, NATO's forces have to carry out a mission which includes deterrence of maritime harassment and the safeguarding of seaborne supply and reinforcement routes.   The question is whether this mission needs a naval presence of high quality including the big ship.

This task requires defences against submarines which represent the major threat to NATO at sea.   ASW is today carried out by a team made up of frigates and destroyers, submarines, big helicopters and long-range maritime reconnaissance aircraft operating from shore bases.   But at the same time the requirements of deterrence in all its aspects call for the defence of ships against Soviet tankers and surface ships if we are to achieve our ultimate purpose: use of the sea.   Defence against Soviet bombers requires fighters, missile defences and shipborne radars to detect incoming raids.   Lastly, defence against long-range missile attack by Soviet carriers and destroyers is provided by nuclear submarines or shore based strike aircraft carrying out pre-emptive attacks, or closer in, by ship and helicopter borne missile systems. At one time the Soviet threat was largely founded on numbers. But in recent years the quality of their ships, submarine and aircraft, and the weapons they carry, has grown quickly.   To deter such a threat requires defences which themselves are of a high quality.

The 1977 Defence White Paper states clearly that NATO depends critically on Britain to provide the main weight of the maritime forces immediately available to the Alliance in the Eastern Atlantic and Channel.   Apart from minesweepers, the only naval forces immediately available to back up those of Britain are small.   German, Norwegian and Danish naval forces have missions in the Northern Region of Allied Command Europe, and those of France are not integrated with the rest of NATO and may not be made available until after the start of hostilities.

It has been argued that the ASW cruiser, which will deploy the Sea King helicopter and now the Sea Harrier MVSTOL aircraft, is an unnecessary component of our maritime forces. The Sea King, which on task has the ASW capability of a frigate, has proved to be a very effective and flexible part of our anti-submarine forces.   The new cruisers will be capable of deploying them in significant numbers: for example a typical complement might be 9 Sea Kings as well as 5 Sea Harriers.   The new cruisers will also carry the Sea Dart air defence missile system and command and control facilities, including those required for shore-based aircraft operating on maritime missions. The next largest ship in the Navy's inventory of new ships is the Type 42 destroyer which has the Sea Dart missile system, but only carries 1 Lynx helicopter with considerably less ASW capability than the Sea King. It is possible to envisage a destroyer being designed to carry a Sea King, on the lines of the earlier County class destroyer/Wessex helicopter combination, but a very large force of destroyers would be required before the ASW capability of the cruisers could be matched.

The task of the Sea Harrier is complementary to that of shore based aircraft. It is required for shooting down long-range aircraft like the Bear which the Soviet Navy use for maritime reconnaissance and mid-course guidance of certain types of anti-ship missile. The Sea Harrier can also carry out over-the-horizon tectical reconnaissance and rapid-reaction attack against missile-armed surface ships.

The decision to build the ASW cruiser was a consequence of the decision made by the Labour Government in 1966 to phase out fixed-wing aircraft carriers and for the RAF to take on the task of maritime fighter defence and air strike; the new cruisers were seen as the most effective way of getting the Sea King helicopter to sea in the numbers required. The need to get the Sea King to sea in significant numbers as soon as possible before the arrival of the ASW cruisers led to the running-on of <u>Blake</u> and <u>Tiger</u> after conversion to carry the Sea King and the conversion of <u>Hermes</u> from the amphibious to the ASW role. <u>Ark Royal</u> also deploys a squadron of Sea Kings as well as providing a strike and air defence capability.

Because of the way in which the Alliance has developed over the last 25 years, only Britain provides NATO's maritime forces in the Eastern Atlantic and Channel with the essential elements of quality. If these were withdrawn only the USN could fill the gap, but at the expense of its own tasks in the rest of the Atlantic and elsewhere. Moreover, NATO's maritime affairs would become totally dominated by the United States. There would be no effective partnership to make a reality of Alliance cohesion at sea and maritime deterrence would be conducted entirely in super-power terms, at a time when some observers believe that the Soviet Union might choose the maritime environment to carry out limited-scale and localized conventional operations to probe NATO's political intentions and solidarity.

European Theatre Ground Forces

The front line units of BAOR are deployed in a vulnerable sector, vital to the defence of North-West Europe, against which the Warsaw Pact can launch the main weight of its attack. The Warsaw Pact forces enjoy the initiative, have superiority in front-line manpower, fire power and equipment and are organized for a short 'blitzkreig' attack in order to defeat NATO forces before reinforcements can become effective. The essence of NATO's deterrent policy must be to make it clear that an aggressor can obtain no easy victory. Without the flexibility afforded by a substantial level of conventional defence, NATO's deterrence would revert to the 'trip-wire' strategy, heavily reliant on the early use of nuclear weapons. Furthermore, if we accept Steven Canby's view that 'a strategy of defence by denial based on substituting nuclear weapons for conventional number is illusory', any case for a reduction of the manpower in BAOR from 55,000 to 30,000 must demonstrate convincingly that the smaller forces involved are capable of no less effective defence.

It has been suggested that, if the terms were right, lower force levels could represent a more realistic response to the politico-military threat. Certainly a scenario, based on the active and mutual development of détente, could include the maintenance of security at a lower level of forces; but it would be jeopardizing Western security to assume that scenario today. It will be sometime before we can properly evaluate Soviet commitment to the ideals of the CSCE and in Vienna the USSR seem to reject the concept of a common ceiling on ground forces as a sensible objective in the negotiation on MBFR. Neither should we necessarily count on a breakthrough in technological opportunities favouring the defences, or place greater reliance on rein-forcements to justify a lower level of forces in Germany.

There are a number of areas of technology which, it is argued, redress the attacker's present advantage over defence. Those which have the most bearing on operations in North-West Europe are surveillance and target acquisition, air defence, anti-tank defence and precision-guided weapons. The possession of improved weapons of these types would, it is sometimes suggested, allow the defender to spread his forces more thinly over the ground without any reduction in war fighting ability. But such weapons are by no means cheap and in any case require men to operate and support them. Nor can they yet be said to be a substitute for existing weapons. In particular nothing has happened to show that the tank is not as essential as it was for successful defensive as well as offensive operations. However as Hunt has pointed out, 'It is this particular point,

that technology is available to both sides and that developments by one may be expected to be matched shortly by the
other, that sets some limit to the extent to which new
weapons will help.' The benefits of new technology will
also be exploited by the Warsaw Pact forces, possibly
obliging the defender to disperse and leaving the tactical
advantage with the side that enjoys a mass superiority.
To the extent that NATO can exploit new technology imaginatively the better will it be able to implement its strategy
of forward defence and flexible response. It does not
however provide any justification for reducing front line
units in BAOR by 45 per cent.

Reinforcements can be found from regular units, and by
mobilizing the individuals and units of the TAVR. BAOR
would be more than doubled in size on mobilization, the
major contribution coming from reservists. It is often
suggested that reductions in numbers should be restored by
the greater use of reserve forces, but although this approach
acknowledges the need for maintaining a militarily credible
number of troops in BAOR it does not take sufficient account
of the extent to which BAOR is already dependent on reinforce—
ment. There must be a balance between regular forces and
reserves. It is unrealistic to expect a fighting force,
in which reserves are predominant, to bear the brunt of the
initial assault of heavily armed Soviet troops. Given the
restrictions on training and equipment, implicit in a further
£1,000 million cut in the Defence budget, reservists would
be operating over ground on which they had not trained and
would lack the tactical and technical expertise to enable
them to make full use of new advances in weapons and weapons
systems.

Against the background of these considerations, the military
planners have been more concerned with determining the critical
ratio between regular and reserve forces for operations in
BAOR, and the minimum strength of stationed forces in peacetime
for credible deterrence, than they might have been in satisfying the provisions of the Brussels Treaty. In the face of
Warsaw Pact superiority in numbers and equipment, it is unrealistic to believe that a radical restructuring of Nato, even
coupled with greater reliance on new technology and on reinforcement forces, would restore the capability lost by a reduction of
25,000 men in BAOR. Unless other members of NATO were able
and willing to take over that part of the British sector denuded
by such reductions, deterrence could only be maintained by a
reversion to a 'trip—wire' strategy, heavily reliant on the
early use of nuclear weapons.

Air Force General Purpose Forces

NATO strategy requires its air forces to possess a full range
of capabilities, both offensive and defensive.   To concentrate
on defence alone would be totally unconvincing.   An enemy with
strong offensive capability is not likely to be discouraged
from contemplating aggression if opposed by a force equipped
exclusively with defensive weapons systems: such a force
offers no threat to his initial capacity to launch an attack,
and could not guarantee immunity from air attack in all
circumstances.   NATO's offensive capability forces the enemy
to commit to air defence a substantial part of his resources,
which otherwise would augment even further his offensive
capability.

The air defence of the United Kingdom is inseparably
integrated with the NATO air defence network and the European
NATO countries have assigned all their air defence forces to
SACEUR in both peace and war.   Air defence of the United
Kingdom also includes the protection of our naval forces in
the Eastern Atlantic and around the coasts of the UK against
air attack, and adequate provision of tanker and AEW aircraft
will be needed to support maritime air defence conducted from
shore bases.

The MRCA ADV is a vital component of the UK and NATO air
defence structure of the 1980s and 1990s.   We do not possess
sufficient numbers of our current air defence aircraft, the
Phantom and Lightning, to consider extending their service
life.   Even if this were possible it would result in aircraft designed in the early 1950s facing an air threat that
has already progressed to the next generation of aircraft.
Nor can the United States offer an adequate and cheaper
fighter, even if it were acceptable to place a further burden
on our dollar reserves and to deal a severe blow to the
prospects of a future European aircraft industry.   Purchase
of the F16 for example (an aircraft that is inadequate for air
defence duties in the UK air defence region) would provide
Europe with work in assembling aircraft but would leave all
technological development to the United States, eventually
leaving Western Europe heavily dependent on the US for future
supplies of aircraft and spares.

The necessity for offensive air support for ground troops
and for local air superiority is accepted as a vital component
of the defensive land/air action that NATO would be forced to
fight on the Central Front.   The lessons of World War II,
Vietnam and the Arab/Israeli conflict show that attacks on
bases, headquarters, force concentrations and communications
from just beyond the battle lines to the deeper rear areas is
the most effective way of preventing an enemy from applying
the full weight of his offensive power against our own forces.
The Warsaw Pact armies are known to rely heavily on replacement

537

of their forward combat units by large formations held in reserve. There is thus a wide band of territory on the enemy side of the battle area which contains vital targets suitable for a combination of both the relatively short-range offensive aircraft, capable also of close air support, and the deeper penetration strike/attack/reconnaissance aircraft. The Jaguars and Harriers are designed to undertake the offensive support role and it is the MRCA which is designed to meet the deeper requirement, by day and night and in all weather. In addition, the MRCA IDS has the ability to deliver nuclear weapons on important WP targets should the strategic/escalatory situation demand it. This combination of capabilities greatly increases the resources and effort the enemy is forced to deploy to attempt to meet the threat.

It would indeed be a bonus to have cheap lightweight air superiority fighters operating particularly over the battle area in Europe, but we could only purchase these aircraft of limited utility at the expense of those needed for the much more essential offensive support and strike/attack roles. It is far better to develop the successor to the Jaguar and Harrier to include an effective air-to-air combat capability.

Nor would a cheap, simple, rugged fighter-bomber fit the offensive support requirement: a whole range of targets, behind the immediate front line but essential in the land and air battle, would be out of its reach; with unsophisticated navigation/attack equipment it could only operate in day time in good weather and still might not locate and destroy its target. Thus, the concept of a cheap and lightweight aircraft has to be considered in cost effective terms, and all the evidence points to the need for a fast, highly manoeuvrable battle support aircraft, equipped with first-class avionics.

The cheap and simple solution for the provision of aircraft is certainly at first sight an attractive one. While there are evident advantages in either the economy achieved or the greater numbers possible, the tasks that must be faced, which range from the detecting of nuclear submarines in the Atlantic to combat with some of the most advanced modern fighters and attack aircraft in the world, simply cannot be achieved without cost. The search all the time is for a compromise between the two requirements, cost and capability.

Nor does it seem possible for the RAF at this stage to 'opt out' of some of the roles in its spectrum of capabilities. This would require specialization in roles by different NATO member countries. However, in the light of the increasing Warsaw Pact threat, we cannot assume that the members of NATO would agree to specialize in this way unless the resulting financial savings were used to increase the effective level of air defence rather than simply to reduce the overall defence efforts of individual member nations. Air support is necessarily a combination of capabilities. The decision on how and where it will be used will depend on the strategic, technical and tactical situation at the time.

538

The Arms Trade

Introduction

The primary role of the British defence industries is to
produce the equipment needed by the UK Armed Forces to meet
their national and NATO commitments. Defence exports assist in
maintaining a viable defence industry, from which the British
Services' procurement needs can be met, and make a valuable
contribution to the UK's interests both in economic and foreign
policy terms. Successive British Governments have therefore
accepted the importance of supporting firms in their exporting
activities.

The present organization for Defence Sales was set up in
1966 following the Government's acceptance of recommendations
by Lord Stokes, who had been asked to advise on the system
which could best help to maximize, within recognized political
and security constraints, the sale of British equipment abroad.
The primary functions of the Organization are to help to
promote the sale abroad of defence equipment produced by
British industry and to sell direct the products of the
Ministry of Defence's Royal Ordnance Factories.

About a quarter of British arms production goes to the
export trade. This country is thus one of the 'Big Four'
arms exporters, probably slightly behind France, but considerably behind the US and the USSR. Any major programme for
cutting defence expenditure could therefore have implications
for the nature and extent of our defence exports.

The Political Arguments

It is often and understandably held that supplying arms to
the poorer countries of the world is in itself an act which
aggravates their economic problems, because spending on
defence diverts resources from those activities which could
contribute towards the elimination of poverty. For this
reason, it has frequently been suggested that the UK and other
arms exporters should take unilateral measures to control or
limit their sales to those countries. It is, however,
necessary to keep the size of the arms trade in perspective.

Leslie Gelb, now Director of the Bureau of Politico-
Military Affairs in the State Department recently wrote (13)
about the Arms Trade as follows:

> The volume of arms being transferred in the world today
> is not much larger than it was 10 years ago, except for
> sales to the Middle East, and even when it doubles in the
> next few years, US sales will be about the same as US
> grant military aid 25 years ago. What is different is
> the method of transferring arms, the kinds of arms being
> transferred, the recipients, and the relationship between
> supplier and recipient.

He continued:

Nor is there much evidence to show that arms imports are
more of a burden to non-oil-producing developing nations
than they were years ago. Of all the recipients of all
forms of American aid, the State Department has identified
only 14 states where the ratio of military expenditures
to GNP exceeded the average for less-developed countries.
In 1975, they were: the Middle East rivals (Egypt, Israel,
Syria); African states facing internal difficulties
(Ethiopia, Tanzania, Zaire, Zambia); Latin-American states
finally modernizing their defense establishments (Bolivia,
Ecuador, Nicaragua, Peru, Uruguay); and Cambodia and
Thailand. Put another way, military expenditures as a
percentage of GNP in all developing nations rose from 4.3
per cent in 1963 to 5.3 per cent in 1974.

Mr Gelb went on to say that none of this data was cited as an
excuse for arms transfers. It had been noted − 'simply to
establish that if there has been a real change in the arms
picture or a special problem resulting from US and worldwide
exports of arms through 1974, it was not a problem of quantity
except for the Middle East area.'

On 19 May, President Carter outlined (14) his administration's
new policy towards the control US exports of conventional arms.
In concluding his statement, he said:

In the immediate future, the United States will meet with
other arms suppliers, including the Soviet Union, to begin
discussions of possible measures for multilateral action.
In addition, we will do whatever we can to encourage
regional agreements among purchasers to limit arms imports.

There have been several attempts in the past to reach some
form of international agreement on arms transfers. These
have been resisted by many developing countries, who point out
that, as is recognized in Article 51 of the UN Charter, they
have a right to meet their legitimate external defence and
internal security needs. Restrictions imposed by supplier
countries are regarded as constituting neo-colonialist inter-
ference with the developing nations' right to determine national
priorities in the allocation of resources.

A further complication is the attitude of the Soviet Union.
They have consistently opposed initiatives in the UN relating
to a possible Register of Arms Sales, see below. As recently
as 31 March 1977 Mr Gromyko (15) questioned whether it was
possible to solve the problem of 'stopping the arms trade,
while leaving aside the solution of problems that create
centres of tension', in particular in places where people were
'waging a legitimate struggle for the liberation of their
territory, as in the case of the Middle East'. For international action to have any real chance of success in limiting

540

conventional arms transfers, the Soviet Union must be fully
involved.

It has been argued that the arms trade is dependent on a
high level of military hostilities and that defence exports
have an effect on the intensity, duration and, at times
initiation of hostilities. Little objective evidence has
been adduced, however, to support this claim. Most arms
are in fact acquired in order to deter aggression, not to
make war. The possession of arms is often a stabilizing
factor. The existence of NATO has contributed to peace in
Europe for many years, but it still remains essential to
maintain adequate defence power backed by national armaments
production industries.

No discussion about the arms trade would be complete
if it did not touch on the moral question involved. This
question is one which merits more debate than it is possible
to deploy in a paper of this kind. (16) Briefly, however,
the moral argument against defence exports rests on the
contribution to human suffering made by such exports and 'he
indifference of exporters to the effects of the trade,
particularly when exports to repressive régimes are involved.
The assumption that often lies behind this view is that the
possession and use of arms is automatically wrong. While
any serious discussion must respect the sincerity and strength
with which that view is held by those of a pacifist persuasion, it is not a view held by the majority of people in this
country, who believe that the UK is entitled to equip and
maintain armed forces for its own defence. Just as we from
time to time need to buy arms from our allies, so countries
which do not have their own arms industries look to the UK
and other industrialized nations to supply them. However,
before any export licence is granted, such factors as the
character of the régime to which the export is being made,
and the possibility that the arms might disturb the balance
of security in sensitive parts of the world or might be used
by certain régimes for internal repression, are carefully
considered by the Labour Government. That is why we operate
an embargo on arms sales of any sort to Chile and South Africa.

The Economic Argument

Work on overseas orders provides direct employment for
70,000-80,000 (17) workers and indirectly for many more. The
clearest example of financial benefits to Britain from defence
sales is the contribution which export contracts make to the
balance of payments; this is expected to reach £850 million
in 1977-8.

Assumptions made about the level of domestic procurement
are crucial in assessing the full economic value of arms
exports. If the assumption is that our security needs could
be met with a greatly reduced equipment budget, some of the
advantages of exporting arms would be diminished. If, on the

541

other hand, it is assumed that our own procurement must continue at something like its present level for the foreseeable future, there are many points to be made in favour of defence sales. Previous chapters have shown why it is on the latter assumption that analysis must be based.

It has been argued that it is misleading to look at the contribution of arms exports to the balance of payments because overall military expenditure produces a deficit in foreign exchange, and that the military sector is poor at producing exports when compared with other sections of the economy. The figures in the section on Defence Expenditure and the Economy p.518 show the whole picture, and illustrate that, when arms exports are included, there is only a very small overall deficit in foreign exchange. Moreover, it is unrealistic to compare the export performance of the military sector with the economy as a whole. UK arms exports are constrained by non-commercial considerations and as defence expenditure serves a public good, its value cannot therefore be wholly assessed in commercial terms. The usefulness of expenditure on social services, health, and schools is never assessed by their contribution to exports.

It has been claimed that expenditure on Research and Development by the military sector seriously pre-empts resources that would otherwise be available for the development of British Industry. It has however been shown earlier that no link can be proved between military expenditure and the relatively poor performance, by international standards, of the British economy. Low investment, low productivity, poor managements and restrictive practices, are not afflictions that can be blamed on the use of resources for defence purposes.

US experience suggests that the most rapid increase in civilian R & D since 1945 have tended to coincide with the periods of most rapid growth in military R & D; and the slowest growth in civilian R & D has occurred during periods of military retrenchment. There are many valuable technological spin-offs from weapons development. Moreover, in practice, private enterprise can rarely support the lengthy and complex research commonly required by military projects.

The use of industrial resources in weapons production is sometimes said to reduce export opportunities in the civil sphere. As has already been argued, however, we need to develop a range of highly sophisticated weapons in order to meet our defence commitments. This gives Britain a comparative advantage in world markets, which should not lightly be thrown away. Alternative products would have to be sold in fiercely competitive world markets in which there is no guarantee that the UK would be any more successful than at present.

It has been pointed out that many more countries will be producing their own defence equipment in years to come and that our share of the market will consequently reduce. This is merely to acknowledge, that the demand for arms, as for

542

any other export, is subject to change. There is no evidence to suggest that our market share will decline in the near future and therefore no reason, on this account alone, to search for alternative exports.

Within these wider economic considerations, there are a number of specific points to be made about the relationship between arms exports and domestic procurement. First of all, exports lead to longer production runs and help to reduce unit costs. At the very least, if a defence manufacturer exports a large number of items this will help to spread overheads, and enable the British Services to buy more cheaply.

Secondly, exports help to offset the very high R & D costs through the direct levies which are paid by manufacturers on export contracts designed and developed at Government expense, as well as through profits. Indeed, without defence exports we might not be able to support the level of R & D which we maintain at the present time. We should then be forced to buy from abroad more of the sophisticated equipment required for our own forces. The consequences both for employment and the balance of payments could be severe.

It is sometimes said that the growth in the value of arms exports in recent years reflects 'surplus capacity' in the defence industries, and that the use of capacity in this way distorts defence procurement decisions. In support of this view it is pointed out that the setting up of the Defence Sales Organisation in 1966 coincided with a major reduction in defence expenditure in the mid-1960s. It is argued that certain equipments, for example the Sea Harrier, are of little use to our own forces and that we only order them in order to assist exports.

This concept of 'surplus capacity' is misleading. Those industries involved in defence production are also, for the most part, engaged in civil production and in many areas it is simply not possible to identify separately the amount of capacity which is available for defence production. While the concept of 'spare capacity' implies that identifiable resources are lying idle or are being misemployed, it is only in a few areas that surplus resources suitable only for defence production exist. Indeed, in many more areas, export requirements have to be balanced carefully against our own needs, because the available capacity is not able to meet all the demands being placed on it.

Defence exports absorb a larger proportion of our military production than they once did, but this reflects not only the reduction in the quantity of equipment ordered for our own forces, but the great increase since the war in demand from abroad. This increase in overseas demand has stemmed both from the emergence of many new independent nations and from the growing prosperity of many developing countries. The formation of the Defence Sales Organisation, although mainly a rationalization of existing functions, reflected this

543

increasing demand.   It was certainly not set up to assist
in filling 'surplus capacity' in the armaments industry.
There are several items of military equipment which have
been sold abroad but not bought by the UK Armed Forces, and
it is simply not the case that the British Services are
forced to buy equipment so that it can be sold abroad.   The
decision to order Sea Harrier for the Royal Navy was fully
justified on operational grounds, in no way determined by its
export potential.

Conclusion

Arms exports assist us greatly in meeting our own defence
procurement needs.   Furthermore, there is no evidence to
suggest that the use of resources to manufacture arms exports
acts as a hindrance to the economy in general.   The benefits
to the balance of payments and to employment are easy to see.
Those who argue for a reduction in defence exports on economic
grounds have therefore to demonstrate:

    a.   that we could without selling arms abroad meet our
         own defence equipment needs just as economically as
         we do at present, and

    b.   that the plant and labour currently employed in
         manufacturing arms could, without excessive cost
         or loss of jobs, be redeployed to manufacturing
         alternative products which would be successfully
         sold abroad with as great a profit to our GNP and
         as great a benefit to our balance of payments.

These points have nowhere been conclusively demonstrated.

UK Controls on Defence Sales

The British Government has never considered that arms should
be exported from this country indiscriminately.   Under the
Export of Goods Control Order all items falling within the
category 'Aircraft, Arms and Military Stores and Appliances',
are strictly controlled.   The Order also controls a number
of other goods, mainly of an industrial nature.   If any items
falls within the description of these industrial goods and
also within a description of the arms, military stores and
appliances list, it would be treated as 'military'.   Before
an export licence for such an item is issued, each application is carefully examined in the light of all the relevant
criteria.   Regard is paid, among other things, to the
proposed end-use of the item in question, to our international
undertakings, to the needs and security of our own Services,
and to human rights in the country concerned, as well as to
security and financial factors.   The UK's economic interests
are also a factor — though by no means necessarily the most

important - in the decision. Particular study is given to
proposals to sell to any area, such as the Middle East, where
an imbalance of arms might promote a conflict, or where there
are indications that conflict is likely to arise.

The policy is one of responsible restraint, and the penalty
accepted is a lower level of sales than would be achieved by
treating the export of arms solely as a commercial transaction.

How can Arms Sales be further controlled?

Notwithstanding the considerable benefits to the UK of
maintaining the present level of defence exports, and the
inevitable difficulties of deploying resources at present
involved in this area, Labour Governments have long been
concerned lest the uncontrolled spread of conventional
weapons throughout the world should lead to instability,
and have been actively involved in seeking international
agreement to control arms transfers, while recognizing
that the subject cannot be regarded in isolation from the
general question of controlling military expenditure.
When Lord Chalfont was appointed Minister of State at the
Foreign and Commonwealth Office in 1964 he was known as
'Minister for Disarmament'. Considerable attention has
been paid to this subject by his successors, notably by
Lord Goronwy-Roberts, who singled out conventional arms
transfers as a priority subject for control in speeches
to the Conference of the Committee on Disarmament in July
1976 and to the First Committee of the UN General Assembly
in November 1976. When he addressed the UN in September
1975 as Foreign Secretary Jim Callaghan also emphasized
our concern at the rapid growth of global spending on arms,
and on 15 March this year he told the House of Commons that
in his talks with President Carter 'the major part of our
discussion on arms was on arms control and the prospect of
reducing arms expenditure rather than increasing it'. We
have actively supported the UN Secretary General's study on
how the military budgets of countries with different
economic systems could be realistically compared, as a
preliminary to discussing agreements on limitation. We
have also supported various attempts to study the specific
question of arms transfers.

The UK has already taken measures on a unilateral basis
to control arms sales. To be really effective, however,
measures aimed at reducing the world-wide trade in arms
must be universally recognized and accepted both by supplier
and by customer nations. As a first step towards the
achievement of world wide limitations it has been suggested
that a Register of Arms Sales should be created, perhaps
under UN auspices.

There have already been several abortive attempts in the
UN to introduce such a Register. In November 1965 Malta

tabled a draft resolution in the UN First Committee
inviting the 18 Nation Disarmament Committee (now the CCD) to
submit proposals for establishing a system for publishing arms
transfers through the UN.   The draft was rejected in the
First Committee by 19 votes, including the USSR, to 18, among
whom were the UK, with 39 abstentions, including the USA.   In
November 1968 Denmark tabled in the First Committee a draft
resolution requesting the Secretary-General to ascertain the
views of member governments on, inter alia, the registration
with the Secretary General of 'all imports and exports of
conventional arms, ammunition ... and on the publication of
information on these transfers'.   The Draft was strongly
opposed by developing countries, including India, Argentina,
Saudi Arabia and Egypt, and was not put to the vote.   Most
recently, at the 1976 session of the UN General Assembly the
Japanese tabled a resolution, which the UK supported,
calling for a study of conventional arms transfers.   This
was prevented from coming to a substantive vote by a procedural
motion introduced by India and strongly supported by non-aligned
and Soviet and East European members.
     The main reasons for the rejection of these motions are clear.
In many cases, both the customer and the supplier are unwilling
to disclose details of their transactions on the grounds of
commercial confidentiality.   In addition, many customer countries
consider that the publication of information relating to their
military capability would be prejudicial to their national
security.   It is in deference to these views that HMG has
consistently throughout the years declined to publish details of
individual British defence sales, although the Government has
never sought to prevent publication if firms and customers have
no objection.
     The developing countries' objections to initiatives designed
at controlling the arms trade are fundamental.   Unless they
and, amongst the supplier countries, the Soviet Union, will
support multilateral restrictions, attempts to limit the world
wide trade in arms are unlikely to bear fruit.   Mr Brezhnev
recently told (18) the Soviet Trades Council that: 'In general,
the problem of the international arms trade seems to merit an
exchange of views.'   It is to be hoped, therefore, that
President Carter's recent attempts to initiate international
discussions on this subject will meet with a favourable response from the USSR.   The British Government has said in the
meantime that it looks forward to hearing President Carter's
detailed proposals, and that it will support realistic multilateral measures aimed at achieving a world wide limitation
on arms transfers.

Conclusion

Until such time as there is substantial progress towards
such international agreement it would be inconsistent with
the UK's security and economic interests, and ineffective in
reducing the world wide trade in arms, for the UK to take
unilateral steps to reduce its defence exports. The
Government will, however, continue to exercise stringent
controls over the export of defence equipment in conformity
with our arms control policies and will continue to work for
international agreement aimed at curbing the world wide
build-up in conventional arms.

Notes

1. David Greenwood, _Budgeting for Defence_ (RUSI, 1972).

2. Statements on the Defence Estimates, 1965–77.

3. M. Kaldor, 'Defence Cuts and the Defence Industry', SGP8. The calculations are based on estimates made by the Stockholm International for Peace Research Institute (SIPRI) of the real resource cost in 1973 US dollars of individual aircraft. Warship costs are based on sources provided by the Annual Appropriation Accounts: _Jane's Fighting Ships Annual._

4. See Sir Michael Cary, 'Military Procurement', _RUSI Journal_ (March 1974).

5. Derek Wood, _Project Cancelled_ (Macdonald and Jane's, London, 1975).

6. J.R. Kurth, 'Why we Buy the Weapons we do', _Foreign Policy_, No.11 (Summer, 1973).

7. Evidence of Vice-Admiral Clayton, _Fifth Report from the Commission on Public Accounts,_ Session 1975–6, HCP, 556.

8. M. Kaldor, 'Defence Cuts'.

9. Greenwood, _Budgeting for Defence_, p.63.

10. Ron Smith, 'The Resource Cost of Military Expenditure', SGP7.

11. Ibid.

12. 'Mechanical Engineering', NEDO Industrial Review to 1977, National Economic Development Office, February 1974.

13. Leslie Gelb 'Arms Sales', _Foreign Policy_ No.25 (Winter, 1976–7).

14. President Carter's announcement of 19 May.

15. Press Conference in Moscow, 31 March.

16. For a discussion on the ethical aspects of arms sales, see Arthur Hockaday, 'The Transfer of Arms', _Crucible, The Journal of the General Synod Board for Social Responsibility_ (Jan–March, 1977).

17. _Hansard_, 19.10.76. col.421–2.

18. Statement to the Soviet TUC, 21 March 1977.

APPENDIX ONE: A LIST OF PAPERS SUBMITTED TO THE NEC STUDY
GROUP ON DEFENCE EXPENDITURE, THE ARMS TRADE AND ALTERNATIVE
EMPLOYMENT

All papers were prepared for the Study Group unless stated.
Papers are listed in order of presentation.

* Past member of the Defence Study Group.

+ Paper published elsewhere.

The Extent of Defence Cuts

The Financial Implications of the Party's Commitment on
Defence Expenditure — Office
Military Expenditure: Gross Domestic Product in NATO
Europe — F. Blackaby
* Observations on the Proposal to Align UK Defence
Expenditure with the Average Percentage Spent on Defence
by the FRG, Italy and France — P. Cockle
Note on Military Expenditure and National Product: UK and
Certain Other Countries — F. Blackaby
* Impact of Recent Changes in Italian Defence Expenditure
on the Study Group's Assumption of the Average European
Defence Burden — P. Cockle
* The Equi-Burden Calculation: 1975 Figures — P. Cockle

How Cuts Could Be Made

A Note on Approaches to Cutting Defence Spending — M. Kaldor
and D. Smith
Defence Programme Options to 1980-81—D. Greenwood
* Note on the Budget Approach and a Programme Option
Which Retains the UK Nuclear Strategic Force — P. Cockle

Defence Expenditure and the Economy

Defence, Industrial Capacity and the Economy — M. Kaldor
The Employment and Other Economic Consequences of Reduced
Defence Spending — D. Greenwood
The Resource Cost of Military Expenditure — R.P. Smith
+ Military Expenditure, Exports and Growth — K.W. Rothschild,
Kyklos, vol.xxvi, 1973
Defence Cuts and the Defence Industry — M. Kaldor
The Opportunity Costs of Defence — M. Kaldor

The Strategic and Political Implications of Defence Cuts

+ New Weapons Technologies and European Security — R. Burt,
Orbis, Summer 1975.
+ Defence Review: An Anti-White Paper — R. Cook, D. Holloway,
M. Kaldor and D. Smith, Fabian Research Series, November 1975.
* Defence Programme Options to 1981: Political — Strategic and
Economic Implications — A.L. Williams

Strategic and Political Implications of Reduced Defence
Programmes — D. Smith
The Political and Strategic Consequences of a Cut in British
Military Expenditure — F. Blackaby
A Political Approach to Defence Cuts — D. Holloway and M.
Kaldor
Is There a Russian Threat? — R. Neild (talk presented to the
group)

Creating New Jobs

+ The Clyde Submarine Base and the Local Economy — Timothy
Stone and David Greenwood, paper presented to the Scottish
CND Conference, 15 February 1975.
The Multi—Role Combat Aircraft (MRCA) — D. Smith
The Anti—Submarine Warfare Cruiser — M. Kaldor
Note on the Employment Consequences of a £1,000 Million Cut
(at 1974 prices) in Military Expenditure Over Five Years — F.
Blackaby ˌ
+ Arms, Jobs and the Crisis — CND, July 1975
* The Crisis in the Capital Goods Industry — A. Doll—Steinberg
Military Expenditure Cuts: Note on the Transfer of Resources —
F. Blackaby
Aspects of Conversion of Arms Industries — D. Smith
+ Corporate Plan — Lucas Aerospace Combine Shop Stewards
Committee
Alternative Employment for Naval Shipbuilding Workers: A
Case—study of the Resources Devoted to the Production of the
ASW Cruiser — M. Kaldor and A. Booth
TORNADO — Cancellation, Conversion and Diversification in
the Aerospace Industry — D. Smith
Community Planning and Base Conversion — D. Smith
Correspondence Relating to Vickers Oceanics — A. Booth and
E. Varley

The Arms Trade

* Defence Exports: Some of the Benefits to the British
Defence Effort and Economy — A.L.Williams
Defence and the Arms Trade — R.F. Cook
First Steps Towards Limiting Arms Transfers — S. Merritt
British Armaments Sales Abroad — F. Allaun
US Disclosure of Information on Defence Exports — J. Gilbert

APPENDIX TWO

A LIST OF PRODUCTS IDENTIFIED AS BEING SUITABLE FOR
MANUFACTURE IN THE CONVERTED DEFENCE INDUSTRIES

In the following appendices, based on the material collected
by the NEC's Defence Study Group, a large number of alternative
products have been suggested.   This list draws together the
major proposals.   It is not an exhaustive list, but it shows
the kind of practical possibilities which could be realized if
a planned conversion from military production were to take
place.   Some of the products listed are obviously of less
importance than others, and their inclusion in this list does
not suggest that there is automatically a possibility of
direct conversion from military production.   The purpose of
this list, therefore, is to show the scope and volume of
opportunities which exist.

AEROSPACE AND RELATED MILITARY INDUSTRIES

Aircraft

Short to medium range civil aircraft seating up to 200
Civil helicopters to service North Sea oil installations
Short take-off and landing passenger and freight aircraft
Helium airships for air freight
Robot helicopter for crop spraying

Marine Vessels

Jet propulsion of ships
Submerged production systems
Micro-processors for submersibles
Marine mineral exploitation and marine agriculture

Transport

Retarder brake system for trains and coaches
Development of other brake systems for all vehicles
Speed/distance related warning systems
Battery cars
New rolling stock
Monorail development
Hybrid engines containing internal combustion engine,
generator, batteries, electric motor.

Energy

Nuclear material disposal
Integrated energy systems

Components for low energy heating (e.g. solar heating)
Fuel cell power-plants
Standby power units for the computer industry
Power packs for oil pumping
Processing plants (e.g. sewage, proteins from fine
  chemicals, ethylene recovery from oil, etc.)
Barrage schemes (Tidal power)
Extended application of gas turbine systems.

Medical

Pacemakers and renal dialysis machines
Medical electronic equipment, including hospital communica-
  tions, computers in hospitals, etc.
Personalized machinery for the disabled
Telechiric machines.

Building

Industrial soundproofing
Prefabricated parts for building

Mechanical Engineering

Ball-screws and machine tools to produce them
Computer controlled servo-hydraulically operated machine tools
Other digitally controlled machine tools

Motors

Linear motors operating pumps and compressors
High-speed motors

Other

Electronic libraries
Self-teaching devices
Mechanized agricultural equipment
Test facilities for manufactured products

SHIPBUILDING AND RELATED MILITARY INDUSTRIES

Fabricated Structural Metals

Metal doors, sash, frames, etc.
Fabricated plate work
Sheet metal work
Architectural metal work
Miscellaneous metal work

Railroad Equipment

Locomotives
Railroad cars

Construction and Mining Machinery

Construction machinery and equipment
Mining machinery and equipment
Oilfield machinery and equipment

Special Industrial Machinery

Food products machinery
Textile machinery
Woodworking machinery
Paper industries machinery
Printing trades machinery
Other special industry machinery
Marine agriculture machinery

Materials Handling Equipment

Elevators
Conveyors
Hoists, cranes, monorails
Industrial trucks, etc.

Trailers and Miscellaneous Transportation Equipment

Truck trailers
Trailer coaches
Miscellaneous transportation
Pipe-laying barges
Semi-submersible oil-rigs

General Industrial Machinery and Equipment

Pumps, compressors and equipment
Blowers and fans
Industrial patterns
Mechanical power transmission equipment
Industrial process furnaces and ovens
Other general industrial machinery

Service Industry Machines

Automatic merchandising machines
Commercial laundry machines
Refrigerators and air conditioners
Automated stockholding and issuing systems
Other service industry machines

Farm Machinery

Farm machinery and equipment

Metalworking Machinery

Metal cutting machinery
Metal forming machinery
Special dies, tools, jigs
Machine tool accessories
Miscellaneous metalworking machinery
Telecheiric devices for metal bearing nodules

Furniture and Fixtures

Public building furniture
Wood office furniture
Metal office furniture

Others

Machine shops, jobbing and repair
Iron and steel forgings
Castings and engravings
Turbines and steam engines
Internal combustion engines
Electrical measuring instruments
Non-electrical heating equipment
Steel springs
Other fabricated metal products
Screw machine products
Bolts, nuts, screws

TANK PRODUCTION

Mining Machinery

Heavy earth moving equipment
Large tracked equipment

Inland Waterways

Pumps for cleaning
Canal gates
Earth moving equipment for widening canals
Barges

Energy Equipment

Fluidized bed boilers
Heat pumps

Nodding ducks for wave power
Tidal barrage

Environmental Products

Recycling plants for various materials
Tree transplantation equipment
Oil spillage pumps

Agricultural Machinery

Tractors and similar vehicles
Irrigation systems

Transport

Locomotives
Hybrid road/rail vehicle

Other

Car presses
Brewing equipment
Reconstruction of piers
Fire fighting equipment
Medical mass screening systems
Electronic blood analysis machinery

Frank Blackaby   Deputy Director of the National Institute
for Economic and Social Research, formerly a Senior Research
Fellow at the Stockholm International Peace Research Institute
and editor of their Year Book on Armaments and Disarmaments.

Paul Cockle   Formerly the Economist at the International
Institute for Strategic Studies (1972-6) and a member of the
Study Group in his personal capacity.

John Gilbert   Minister of State at the Ministry of Defence.

David Greenwood   Director of the Centre for Defence Studies
at Aberdeen University and author of numerous studies
including Budgeting for Defence, he was a consultant to the
Study Group.

Ron Hayward   General Secretary of the Labour Party.

Mary Kaldor   Fellow of the Science Policy Research Unit at the
University of Sussex, principal co-author of The Arms Trade
with the Third World (Stockholm International Peace Research
Institute, 1971) and author of The Disintegrating West (Allen
Lane, 1978).

Ian Mikardo   Member of Parliament for Bethnal Green and Bow,
formerly Chairman of the Labour Party's International Committee
and of the Study Group and one of the longest serving members
of Labour's National Executive Committee.

Olof Palme   Former Prime Minister of Sweden and Chairman of
the Swedish Social Democratic Workers Party.

Dan Smith   Research Officer in the Department of Economics
at Birkbeck College, London, formerly General Secretary of the
Campaign for Nuclear Disarmament (1974-5) and Research Fellow
at the Richardson Institute for Conflict and Peace Research
(1976-8), he was the editor of Alternative Work for Military
Industries (the Richardson Institute, 1977) and is the author
of numerous articles and pamphlets on defence and disarmament.

Ron Smith   Lecturer in Economics at Birkbeck College, London,
his published work includes Consumer Demand for Cars in the USA.

John Tomlinson   Under Secretary of State at the Foreign and
Commonwealth Office.

Steve Vines   Research Officer in the International Department
of the Labour Party and Secretary to the Defence Study Group,
he was previously a freelance journalist and researcher,
specializing in the Middle East.

<u>James Wellbeloved</u>   Under Secretary of Defence for the
Royal Air Force.

<u>Alan Lee Williams</u> Member of Parliament for Hornchurch,
Chairman of the Parliamentary Labour Party Defence Group and
author of several defence publications including <u>Crisis in
European Defence.</u>

INDEX

Compiled by Dennis J Nisbet JP,
Member of the Society of Indexers

A-7D Aircraft 283
ACE (Allied Command Europe) 148-50,
533
ADV see Air Defence Variant of MRCA
AFCENT (Allied Forces Central Europe)
133-4, 150, 153, 161, 187, 527, 529
AFGPCF see Air Force General Purpose
Combat Forces
ASM (Anti-Ship Missiles) 219, 222
AST 403 Aircraft 189, 293, 424
ASW Cruisers see Anti-submarine
Warfare Cruisers
ATGW see Anti-tank Guided Weapons
Aberdeen University Defence Studies
Unit 111, 370
Air Defence Variant (ADV) of MRCA
161, 418-20, 422-3, 441, 537
Air Force General Purpose Combat
Forces (AFGPCF) and 'balanced
forces' 153-4, 187-8; comparison
of, with Warsaw Pact forces 78;
cost of 35, 40-2; cuts in 86-7,
198, implications of 234-7, 537-8,
options for 143-6, 161-5, 168-70,
206-7, 528-9; see also Tornado
Aircraft
Air Staff Target (AST) 403 Aircraft
189, 293, 424
Allaun, Frank, MP 550
Allied Command Europe (ACE) 148-50,
533
Allied Forces Central Europe
(AFCENT) 133-4, 150, 153, 161,
187, 527, 529
Amazon frigate 291-2
Anglo-French Helicopters 56, 293
Anti-ship Missiles (ASM) 219, 222
Anti-submarine Warfare (ASW)
Cruisers: background to 292,
394-6; cancellation of 159, 210,
528; case study of resources
devoted to 393-415; cost of 50,
55, 396-8, 400; effect on defence
budget of 8; employment aspects
of 334, 399; options for 188,
225-8, 533-4; ordering of second
180; savings on 39, 84, 85
anti-tank guided weapons (ATGW)
86, 215, 219-21, 335
Arab/Israeli War, 1973 195, 220,
221, 230, 481, 537
Argentina 546

Ariadne Frigate 292
Ark Royal 39, 83, 158, 210, 225,
228, 534
Arms Control Agreements 5, 28, 81-3
arms race 1,4,5,28,75,81-2
arms sales/trade 9, 27-8, 115-24,
539-47; see also exports
Armstrongs 480, 484, 490
army combat forces (other than
ETGF) 35, 40, 160-1, 206
Aspaturian, Vernon 90
Aspin, Les 79, 91, 215-18, 245-6
Atomic Energy Authority 446
atomic weapons 70, 81
Attwater and Sons 446
Augustine, Norman R. 219-20, 246,
312, 440
Australia 476

BAC: aircraft produced by 54, 511;
conversion at 101, 427-33, 447;
employment at 105, 425, 427, 446;
and guided weapons 335; and
Panavia 417; preservation of
286-7, 300; Preston's dependence
on 98, 446; Shop Stewards Combine
Committee 101, 105-6, 427; see
also Tornado Aircraft and Ward,
Peter
BAOR (British Army of the Rhine :
Rhine Army): air transport and
234; cuts proposed in 40, 84, 210,
528; effect of cuts in 85-86,
230-1, 233, 529, 535-6; importance
of 527; possible savings in 160;
present strength of 32, 145;
restructuring of 180, 229; tanks
in 231-2
Babadzhayan, Marshal A. 524
Backfire bomber 74, 524, 525
balance of payments 27, 50, 62-3,
517-18, 541
Barnett, Richard J. 387, 391-2
Barrow Shipyard see Vickers
Bear submarine 227
Belgium 140, 334, 417
Belgrade Review Conference 83
Belize 116, 133
Benn, Tony, MP 459
Berger, Rear Admiral 211, 213
Berliner, J.S. 91

Blackaby, Frank: author of papers
252-61, 355-9, 361-5, 549, 550;
biography 556; quoted 95, 448
Blake cruiser 225, 227-8, 534
Blechman, Barry M. 213, 245
Blue Water Missile 93
Bolivia 540
Booth, Albert, MP: author of
papers 393-415, 550; quoted
114, 370
Booz-Allen and Hamilton Report
400
Boyd-Carpenter, John 263
Brandon, Henry 124
Brazil 118, 288
Brezhnev, President L. 204, 546
British Aircraft Corporation
see BAC
British Army of the Rhine see
BAOR
British Leyland 446, 486
British Nuclear Fuels Ltd 446
British Rail 97, 234
British Waterways 495
Brown, David 335
Brown, Robert, MP 390
Brown, Ronald, MP 423
Buccaneer Aircraft 234, 290
Burt, Richard 194, 222, 247, 549

CIA (Central Intelligence Agency)
74, 79, 217
CLDC see Central Lancashire
Development Corporation
CSB see Clyde Submarine Base
CSCE see Conference on Security
and Co-operation in Europe,
Helsinki
CSEU see Confederation of Ship-
building and Engineering Unions
Callaghan, Jim, MP 545
Camborn, J.R. 113, 391
Cammell Laird 207, 334
Campaign Against Arms Trade
123, 124
Canada 417
Canby, Steven 92, 192-3, 221,
229, 235, 247-9, 535
Cardiff Destroyer 292
Carey, Sir Michael 286, 312, 548
Carrington, Lord 396
Carter, President 116, 194, 540,
545, 546
Central Intelligence Agency see
CIA
Central Lancashire Development
Corporation (CLDC) 98, 447
Centre for Alternative Industrial
and Technological Systems 465, 473

Centurion Tank 481-4
Chalfont, Lord 250, 545
Checinski, Dr. Michael 65
Chieftain Tank 479-87
Chile 118, 123, 541
China 10, 74, 77, 215, 523
'Christmas cuts', 1975 327, 331
Churchill, Winston, MP 68
Clayton, Vice-Admiral 53, 65, 286,
288, 312, 510, 548
Clyde Submarine Base (CSB) 110-11,
451-2; see also Polaris Submarines
Coates, Ken 462, 478
Cockle, Paul: author of papers
130-6, 196-202, 549; biography
556; quoted 92, 211
Coffey, J.I. 250
'Common MRCA' (IDS) 162, 419-20,
422-3. 441, 538
Community planning (USA) 97-8, 112,
445-57
Concorde 56, 61, 263-4, 461, 463,
464
Condor ASM 219
Confederation of Shipbuilding and
Engineering Unions (CSEU) 475, 476
Conference on Security and Co-
operation in Europe, Helsinki
(CSCE) 3, 70, 83, 152, 535
conscription costs 131-2, 136, 257-9
Conservative Party: and defence
budget/programme 10, 98-9, 210,
321, 322, 331, 357; and small
carriers 395; and taxation 88;
and the USSR 52, 68, 226
conversion: and aerospace industry
422, 425-38; and arms industry
366-92; British experience of
96-7; defined 94, 422; detailed
studies for 388-9; occupational
102-4, 373-8; planning required
for 99-101, 385-8; production
101-2, 369-73, 551-5; target
industries for 383-5; US experience
of 97-8, 103, 370, 372, 387, 448-51;
at Vickers 404-5
Cook, Robin F. 30, 549, 550
Cooley, Mike 449, 457, 459, 462-70,
475-8
Cuba 432
Currie, Malcolm R. 247
Customs and Excise Exports of Goods
(Control) Order, 1970 124
Cuts in defence expenditure:
amounts of possible 33-7, 68;
economic consequences of 315-54,
355-9; ministerial response to
504-48; strategic and political
implications of 67-90, 95-6, 203-8,
209-50, 515, 519-38; summary of

arguments for 25–30; transfer of
resources arising from 361–5;
ways of securing 37–42, 83–90;
see also Defence Expenditure
and Employment
Cyprus 32, 87, 133, 161, 206
Czechoslovakia 72

Dalton, Hugh, MP 19
De Havilland 96
Decca 105
defence expenditure: aligned with
other countries 130–6; current
31, 82, 142–8, 263, 513–5; and
the defence industry 281–314;
and the economy 49–65, 203–8,
251–61, 315–59, 506–19, 541–4;
and investment 50, 58–60, 119;
resource cost of 262–80; see
also Cuts in defence expenditure
and employment
defence industry: conversion
aspects of 366–92; defence
costs and 281–314; reorganisation
of 55; structure of 52–4, 510–11;
surplus capacity in 53–6, 543–4;
see also BAC, Lucas Aerospace,
Vickers and other firms and
employment
Defence, Ministry of (MOD): and
air defence system 236, 424;
and arms trade/exports 120–1,
124, 288; attitude of, on
sharing sites 451–2; attitude
of, on Study Group 24;
conscription costs analysed by
131, 259; employees of 32, 95;
Expenditure Committee on military
equipment sales by 55; and naval
forces 291, 394; R & D expenditure
by 510; tank production and 481,
484, 486; technical change and
284–5; and variable geometry
aircraft 51, 285;
Defence Programme Options: to
1981 42–5, 137–95, 196–202,
203–8, 528–38; to 1984 45–6,
180–5
Defence Review, 1974:
institutional change and 11;
manpower aspects of 317, 321–6;
outcome of 138–9, 141, 176, 520,
527–8; withdrawals as a result
of 206
Defence Sales Organisation 539,
543
Defence White Papers see under
White Papers
Denmark 533, 546

detente: climate of 1, 30, 81–2;
defined 5; Labour Government and
507; NATO and 527; paradox of 4;
Soviet policy on 72–3, 82, 204,
521
Digby, James 219, 231, 246, 247
disarmament 1–3, 30, 371, 527,
545–6
diversification: and aerospace
industry 104–8, 422, 425–38;
defined 94, 422; and shipbuilding
industry 109–10; and submarine
base 110–12; at Vickers 405
Doll-Steinberg, A. 550
Duffy, Pat, MP 248
Dumas, Lloyd J. 374–6, 391
Dunlop 105

ECM (Electronic Counter Measures)
220
ENMOD Treaty 82
ETGF see European Theatre Ground
Forces
Eastern Atlantic (EASTLANT) and
Channel Areas 134–5, 149, 155,
156, 161, 529, 533–4
East-west relations 4, 73, 81–3,
136
Economic Survey, 1953 268
economy and defence expenditure
49–65, 203–8, 251–61, 315–59,
506–19, 541–4
Ecuador 121, 540
education 263
Edwards, Sir Geoffrey 293
Egypt 118, 121, 220, 221, 540, 546
Eisenhower, President 263
'Electronic Battlefield' 219
Electronic Counter Measures (ECM)
220
Elliott, Dave 459, 478
Elswick Works see Vickers
employment: alternative, for
shipbuilders 393–415; creation
of new 93–112; defence cuts and
315–59; defence expenditure and
58; in military service and
industry 32, 318–9, 506, 514, 518,
541; 1974 Defence Review and
321–6, 340; 1975 White Paper and
328–32, 340; 1976 PES and 350–2;
regional and local effects on 342–3;
at Vickers 480, 483, 488, 491
English Electric 97, 289
Enthoven, Alan C. 78, 91, 239, 241,
249–50
Erickson, J. 193
Ethiopia 540

European Theatre Ground Forces
(ETGF): comparison of with
Warsaw Pact forces 77-8; cost of
35, 40; cuts in 85-6, 187, 198,
implications of 229-33, 535-6,
options for 43-6, 148-52, 160,
168-70, 205-6; see also BAOR
Evans, Mike Parry 468
Export of Goods Control Order 544
exports: analysis of 288, 503;
and balance of payments 517-18;
and home market 27; as proportion
of production 55, 61-2, 511-12;
see also arms sales / Trade
Exxon 112

F16 Aircraft 537
F-111 Bombers 80
FRG see Germany, Federal Republic of
Fairchild A10 Aircraft 41, 162
Fairhall, David 248
Falkland Islands 133
Farrar-Hockley, Major-General A.H.
220-1, 246, 247
Fearless 39, 83, 158, 210, 225
Ferranti 105, 335
Fiat 417
Fiorello, Marco R. 312
Fleet submarines 225, 228, 334
Flynn, G.A. 91
Ford, President 450
Foxbat Aircraft 74, 424
France: and arms trade 115, 539;
and GDP comparisons 33, 34, 36,
253-61; and GNP comparisons 131,
139-40; military budget/
expenditure of 7, 82, 88, 92;
and NATO 77, 533; as a nuclear
power 32, 531; resource costs in
271
Frei, President 124
Frigates 50, 55, 508

GDP see Gross Domestic Product
GEC Marconi-Elliott 105, 123, 335
GNP see Gross National Product
Gallois, General 56, 293
Gan 206
gas turbine engines 51, 283
Gelb, Leslie 539-40, 548
Germany, Federal Republic of (FRG):
and arms trade 118, 122;
British troops in 50, 86, 334;
economic growth in 59, 61; and
GDP comparisons 33, 34, 36,
253-61; and GNP comparisons 131,
139-40; military budget/
expenditure of 7, 82, 88, 89, 92;
naval forces of 533; resource costs
in 271, 275; and tanks 232, 481;

and Tornado 104, 417-18; and USA
134; unemployment in 58, 460
Gibraltar 39, 159
Gilbert, John, MP: author of
papers 505-48, 550; biography 556
Gilmour, Sir Ian, MP 68, 240, 250
Gordon, G.G 349
Greece 273
Greenwood, David: author of papers
137-95, 315-54, 549, 550; biography
556; quoted 38, 56, 197-200, 205,
207, 210, 211, 225, 226, 235, 244,
263-4, 285, 370, 378, 422
Griffith, W.E. 91
Griffiths, David 391, 457
Gromyko, A. 540
Gross Domestic Product (GDP) 15, 33,
34, 47, 253-61, 513
Gross National Product (GNP):
defence burden on 138-40;
increase in 34; share of: spent
on defence 31, 59, 60, in
Defence Review 204, 528,
compared with other Western
countries 8, 15, 20, 23, 29, 131,
176, 515-16, 540, by USSR 74, 204,
525
Guatemala 116

HS 146 Aircraft 460, 478
Halzman, F.D. 91
Harland and Wolff 400
Harrier Aircraft: and ASW cruisers
394, 398; expenditure on 41, 84,
87, 210, 234; exports of 288;
replacement of 189, 424, 538;
role of 234-5, 237, 538; weakness
of 162
Hawker Siddeley 54, 97, 286-7, 299,
331, 335, 460, 511
Hayward, Ron 17, 556
Healey, Denis, MP 290, 395
Heath, B.O. 418
helicopters, Anglo-French 56, 293
Hendin, Jim 489
Hermes 39, 83, 158, 210, 225-8, 534
Highlands and Islands Development
Board 108, 470
Hitch, C.S. 264, 277
Hockaday, Arthur 548
Holland, Stuart 392
Holloway, Admiral James L. 214, 245
Holloway, D. 549, 550
Hong Kong 32, 133, 149, 206
Horn, R.G. 90
Howitzers 512
Huckfield, Les, MP 475
Hunt, Kenneth 192-3, 211, 231, 245,
248, 535
Hunters Point Shipyard 452

ICI 123

IDS see 'Common MRCA'
IISS (International Institute for
Strategic Studies) 15, 77, 421
Ikle, Dr 115, 124, 125
Imperial Chemical Industries (ICI)
123
India 484, 497, 546
Interceptor ADV see Air Defence
Variant of MRCA
inter-continental missiles 70
international collaboration 55-6,
417, 512
International Institute for
Strategic Studies (IISS) 15, 77,
421
International Research and
Development Co. Ltd 484
Intrepid 39, 83, 158, 210, 225
investment and defence expenditure
50, 58-60, 119
Invincible ASW cruiser 8, 159, 225,
227-8, 292
Iran 117, 121, 288, 481, 482, 487
Israel: Arab/Israeli war 195, 220,
221, 230, 481, 537; and arms trade
117, 121, 288, 540; and nuclear
weapons 82; and tank warfare 220,
230, 481-2
Italy: and GDP comparisons 33, 34,
36, 253-61; and GNP comparisons
131, 140; resource costs in 271;
and Tornado 104, 417-18

Jaffray, Mr (MOD) 225, 227-8,
239-40
Jaguar aircraft: expenditure on
41, 56, 84, 87, 145, 210, 234,
293; international collaboration
on 512; replacement of 189, 424,
538; role of 234-5, 237, 538;
weakness of 162
James, Mr (MOD) 240
Jane's All the World's Aircraft
295, 313
Jane's Fighting Ships 212, 247,
296-8, 308, 398
Jane's Weapons Systems 247, 499
Japan: and arms trade 118, 122,
546; military expenditure of
58, 59, 275; and motorcycle
industry 384; use of gas turbines
in 380, 432
John, Brynmoor, MP 392
Judd, Frank, MP 191

Kaiser, Robert 70, 90
Kaldor, Mary: author of papers 4-15,
281-314, 393-415, 549, 550;
biography 556; quoted 64, 267, 277,
370, 391, 432.

Kaufmann, Prof. William 91
Kennet, Lord 194
Kenya 484
Kfir Aircraft 121
Kissinger, Henry 80
Kitson, Brigadier F. 9, 15
Klare, Michael 76, 91, 214,
245-6
Korea 71, 118
Kurth, J.R. 52, 65, 263, 277,
510, 548
Kuwait 484

LACP see Lucas Aerospace
Corporate Plan
Labour Party: Annual Conference,
1931 19; Annual Conference, 1973
140; and arms trade 122; and
conversion 96; and defence
expenditure cuts 10, 58, 81, 88,
90; Defence Study Group Report
2, 3, 4, 14, 16-125, 316;
General Election Manifestos, 1974
23, 33, 110, 138, 224, 367; and
German military expenditure 89;
National Executive Committee 14,
24, 316; and Polaris 84, 224
Latimer, Clive 466
Lebanon 71
Leontief, W. 277
Levgold, R. 91
Lightning aircraft 537
Lockheed scandal 118
long-range cruise missiles 70
Lucas Aerospace 105, 460, 476;
Corporate Plan (LACP) 12-13,
458-78; Shop Stewards Combine
Committee 107-8, 380, 389, 433,
459, 496, 498, 550

MBB 417
MBFR see Mutual and Balanced Force
Reductions
MFR see Mutual Force Reductions Talks,
Vienna
MIRV (Multiple Independently
Targeted Re-entry Vehicles) 70
MOD see Defence, Ministry of
MRCA (Multi-role Combat Aircraft)
see Tornado Aircraft
MVEE (Military Vehicles and
Engineering Establishment) 484
McDonagh, James J. 106, 113, 371,
384, 391-2, 439
McGeogh, Vice-Admiral Sir Ian
396-7
McGuire, Michael 91
McKean, R.N. 264, 277
Maclean, Fitzroy 249
McNamara, Robert S. 74, 91

Malaysia 206
Malta 206, 545
manpower see employment
Manpower Services Commission 102
Marathon 111
Marconi-Elliott Avionics Systems
  (MEASL) 423-4
Marconi-Elliott, GEC 105, 123, 335
Maritime Harrier 121-2, 335
Marshall, Andrew 218
Marx, Karl 514
Masirah (Oman) 206
Mason, Roy, MP 192, 390, 439, 460
Mauritius 206
Medical Research Council 61
Melman, Seymour 113, 367, 369, 390
Merglen, General A. 230, 247
Meriden Co-operative 384
Merritt, S. 550
Middle East 9, 117, 118, 545; see
  also individual countries
Mikardo, Ian, MP 17, 19-21, 556
Milan ATGW 221, 335
Military Balance, The 15, 46, 79,
  90, 91, 92, 216, 246, 247
military expenditure see defence
  expenditure
Military Vehicles and Engineering
  Establishment (MVEE) 484
Milne, James 391
mission programmes 35, 38-42, 145
Monroe, Elizabeth 246, 247
Moran, Terry 475
Morris, William 462, 478
motorcycle industry 384-5
Mulley, Fred, MP (Secretary of
  State for Defence) 25, 194, 506
Multiple Independently Targeted
  Re-entry Vehicles (MIRV) 70
Multi-role Combat Aircraft (MRCA)
  see Tornado Aircraft
Mutual and Balanced Force Reductions
  (MBFR) 526, 530, 535
Mutual Force Reduction (MFR)
  Talks, Vienna 80, 152, 242

NATO: air forces 215, 524-5,
  537-8; arms programmes 5; and
  arms trade 121; balance of power
  within 67, 68; Britain's role in
  7, 10, 29, 32, 133-6, 147-9;
  cohesion of 204, 239-44, 522,
  527-8; defence expenditure 131;
  effect on, of defence cuts 88-9,
  529-38; GDP statistics 34;
  ground forces 77, 215-17, 229-33,
  523, 529, 535-6; homegeneity of
  74; military strength 81; naval
  strength 85, 212-13, 533-4; 1977
  White Paper on 68; nuclear strength

80, 84, 238-9, 526; and Panavia
  417; tanks 78, 215, 523-4
NEB see National Enterprise Board
NEDC/NEDO see National Economic
  Development Council/Office
NGPCF see Naval General Purpose
  Combat Forces
NHS see National Health Service
National Coal Board 364, 461,
  494-5
National Economic Development
  Council/Office (NEDC/NEDO) 480,
  499, 517, 518
National Enterprise Board (NEB)
  112, 368, 371, 376
National Health Service (NHS) 12,
  13, 99, 263, 467
nationalization: House of Lords
  and 426-7; and multinational
  companies 294; opportunities
  presented by 52-53, 65, 112,
  368, 405-6, 510; and Vickers
  110, 404
Naval General Purpose Combat
  Forces (NGPCF) and 'balanced
  forces' 155-6, 188; Comparison
  of, with Warsaw Pact forces
  76-7; cost of 35, 39; cuts in
  85, 198, implications of 225-9,
  533-4, options for 43-6, 158-9,
  168-70, 205, 528; see also
  Anti-submarine Warfare Cruisers
Naval (Royal) Dockyards 97, 207,
  286, 334, 370, 510
Neild, R. 550
Netherlands 140, 160, 417
Neutron bomb 21
Newhouse, John 237-8, 249
Newton, D. 113, 391
Nicaragua 540
North Atlantic Treaty Organisation
  see NATO
North Sea oil 134, 255, 394
Northern Army Group (NORTHAG) 160
Northern Ireland 9, 86, 230, 233,
  234
Northrop scandal 118
Norway 133, 533
nuclear balance 79-81, 135
nuclear energy 61
Nuclear Non-Proliferation Treaty
  531
nuclear strategic forces see
  Polaris Submarines
nuclear threshold 237-9
numbers game 211-19
numerically controlled (NC)
  machines 483, 486
Nunn, Sam 187, 194

Office of Economic Adjustment
(OEA) 97-8, 113, 448-50, 452,
455-6
Olding, Jack & Co. 489, 490
Oman (Masirah) 117, 133, 206
Oppenheimer, Dr J. Robert 21

PGM see Precision Guided
Munitions
Palme, Olof 1-3, 556
Panavia 417
Paris Agreements, 1954 150
Parsons 484, 485, 488
Pattie, Geoffrey 438
Perry, Robert 312
Peru 540
Phantom aircraft 41, 87, 161,
234-6, 537
Philippines 118
planning agreements 99, 368
Plessey 105
Poland 72, 82
Polaris submarines: abandonment of
84-5, 223-5; cost of 35, 145;
deterrent effects of 135, 186;
employment aspects of 110-12,
334; retention of 43,84, 148,
197-9, 205, 530-1; rundown of
158, 168, 210, 332, 334, 451-2;
savings on 38-9, 44-5, 158
Poseidon submarine 225
Postan, M.M. 499
Pratt and Whitney 107, 431
Precision Guided Munitions (PGM)
219-23, 531
Preston area 98, 446-7, 454-5
procurement determinants 52, 121,
221
Public Expenditure Survey, 1975
317, 326-31
Public Expenditure Survey, 1976
350-4
Purdy, D. 274, 277

RAF see Air Force General Purpose
(Combat) Forces
R and D see Research and Development
ROF see Royal Ordnance Factories
RPV see Remotely Piloted Vehicles
Racal 123
Rand Corporation 284, 310
Record, Jeffrey 215, 239, 246,
249
Reed, Arthur 65, 313
register of arms sales 540, 545-6
Remotely Piloted Vehicles (RPV)
222, 235
research and development (R and D):
costs of 60-1, 511, 515, 517; and
exports 121, 122, 543; growth of

542; manpower in 402
Rhine Army see BAOR
Robens, Lord 405, 482, 484
Roberts, Lord Goronwy 125, 545
Rodgers, William, MP 392, 438
Rolls-Royce: conversion and
diversification in 106, 427-32,
461; employment in 105, 331,
335; nationalization of 510;
and tanks 486; and Tornado 417,
425-32; and tractors 489
Rothschild, Professor Kurt W.
62, 66, 549
Royal Air Force see Air Force
General Purpose (Combat) Forces
Royal (Naval) Dockyards 97, 207,
286, 334, 370, 510
Royal Ordnance Factories (ROF)
286, 335, 370, 446, 451, 482,
485, 510
Rumsfeld, Donald H. 245
Russia see USSR

SACEUR (Supreme Allied Commander,
Europe) 150
SALT see Strategic Arms Limitation
Talks
SAM see Surface-to-air Missiles
SIPRI see Stockholm International
Peace Research Institute
SS-X-20 Missiles 80
Salter, Dr Stephen 403
Sandys, Duncan 289
Saudi Arabia 117, 288, 546
Scarbrow, E. 459
Schlesinger, James 217, 246
Schwartz, M. 90
Scott, J.D. 485, 490, 499
Scottish Campaign for Nuclear
Disarmament 110, 113, 370, 374,
382, 391-2, 451, 457
Sea Harrier: cancellation of 39,
84, 159, 210, 227-8; capabilities
of 227, 533-4; retention of 195,
544
Sea King helicopters 394, 398,
533-4
Senghas, Dieter 312
Sheffield Destroyer 291-2
Short Bros and Harland 96, 286,
335
Simpson, John 313
Singapore 206
Sino-Soviet Conflict see China
Smith, Dan: author of papers 4-15,
209-50, 366-92, 416-44, 445-57,
549, 550; biography 556; quoted
81, 285
Smith, Ron P.: author of papers
262-80, 549; biography 556;

quoted 57, 65–6
Smiths 336
Socialist International 3
South Africa 123, 125, 541
Soviet Union see USSR
Sparrow missiles 87, 161
Spitfire 50, 283, 508
Stanhope, Henry 91
Stockholm International Peace
  Research Institute (SIPRI):
  on arms trade 123, 124; on cost
  of equipment 65, 120, 282, 295,
  499; on GDP 15; on Soviet budget
  79
Stokes, Lord 539
Stone, Timothy 338, 348–9, 457,
  550
Strategic Arms Limitation Talks
  (SALT) 5, 82–83, 186
Striver, Herbert 375
Sudan 118
Suits, D.B. 348
support programmes 36, 42, 87,
  145–6, 165–6
Supreme Allied Commander, Europe
  (SACEUR) 150
Surface-to-air Missiles (SAM) 41,
  87, 162, 219–22, 234–7
Swan Hunter 121, 207, 292, 334
Sweden 7, 10, 72, 110, 476
Syria 118, 220, 540

TSR-2 11, 55, 285, 289–91
tanks 78, 215, 222, 231–3, 480,
  524–5; see also Chieftain Tank
Tanzania 13, 470, 540
taxation 50, 88
technological change 10, 12,
  51–2, 219–23, 282–5, 509–10
Terminal Homing Programme 220
Thailand 540
Thatcher, Mrs Margaret, MP 52
Third World 10, 71, 118, 214, 466,
  497
Thornley, Stephen 424
Tiger Cruiser 225, 227–8, 534
Tomlinson, John,.MP 505–48, 556
Topeka Air Base 98, 452, 456
Tornado aircraft (MRCA): case study
  on 416–44; contractors for 442–3;
  cost of 26, 50, 56, 104–5, 283,
  441, 508; effect of cancellation
  of: on aerospace industry 207,
  422–38, on air defence 41, 84,
  86–7, 161–3, 210, 234–7, 530,
  537–8; effect on defence budget
  of 8; international collaboration
  on 417, 512; mission of 188, 189;
  organisation 417–20; origin of
  52, 285; Preston and 447;

resources devoted to 105; second
  batch of 180; and TSR2 11, 285,
  291; see also BAC
Transport and General Workers'
  Union 19, 475
Trebilcock, Clive 274, 277, 499
Turkey 118, 273

UN see United Nations
USA: aircraft 537; arms control
  by 115–16, 118, 121, 123; and
  arms trade 539–40; community
  planning in 97–98, 112, 448–52,
  456; Congress and 'backsliders'
  133–4; conversion in 103, 109,
  369, 370–6, 387, 401, 455;
  GNP, defence share of 59, 131;
  ground forces 77; military expenditure
  79, 82, 204, 217–19, 257–8, 271,
  273; military procurement 52;
  NATO commitment 84; Navy 213–14,
  222; and nuclear option 237; and
  Polaris role 223–4; relative
  decline of 10; tanks 481; ties
  with UK and Germany 134; /USSR
  parity 70; unemployment in 58,
  460; see also NATO
USSR: arms industry in 53; and
  arms trade 539, 540–1, 546;
  concern over strength of 10, 52,
  68–82, 204, 521–6; and disarmament
  conferences 152; diplomatic
  pressure 133; effect on, of
  German military expenditure 89;
  and GNP 74, 204, 525; ground
  forces 77–8, 215, 535; and IRBM
  forces 186; intentions and
  interests of 69–73, 506, 521–6;
  military expenditure 78–9, 217–9;
  military manpower 215–17; naval
  forces 76–77, 85, 134, 212–4,
  225–7, 522, 529, 533–4; nuclear
  strength 79–81; tanks 78, 215,
  499; see also Warsaw Pact
Ulam, Adam B. 91
Ullman, John E. 379, 384, 387, 392
unemployment 2, 3, 12, 23, 58,
  207–8
unilateral government actions 6–7,
  10, 83, 116, 507, 519
United Nations Department of
  Economic and Social Affairs 371
United Nations General Assembly
  1, 2, 125, 545
United States see USA
Upper Clyde Shipbuilders 396
Uruguay 540

VTOL aircraft carriers 76, 314,
  394

Vance, Cyrus 116
Variable Geometry aircraft 51, 56, 283, 285, 291, 293, 509
Varley, Eric, MP 550
Vershbow, Alexander R. 70, 90
Vickers Ltd: and Chieftain Tank 480-7; commercial experience at 487-92; conversion at 404-15, 492-8; effect of defence cuts on 334-5; employment at 399; and exports 121, 207, 288; MOD and research at 52, 285; Main Battle Tank 484; National Combine Committee of Shop Stewards 407, 479-503; and nationalization 110; offshore engineering 109; orders for 286, 292, 303; workforce skills at 12, 103
Vietnam 9, 71, 219, 285, 537
Vincent, Dr R.J. 241, 250
Vines, Steve: author of papers 4-15, 458-78; biography 556
Vosper Thorneycroft Ltd 286, 288, 302

WEU (Western European Union) 150, 151
Walker, William B. 312, 420-1, 439
Ward, Peter 106, 113, 433, 439-40
Warsaw Pact: air force 78; arms programmes of 5; ground forces 77, 133, 529-30; homogeneity of 74; military strength 81, 151; naval strength 212-13; 1977 White Paper comment on 68; nuclear strength 80; unilateral action and 519; see also USSR
Wayne-Smith, K. 249
Wellbeloved, James, MP: author of paper 505-48; biography 557
West Germany see Germany, Federal Republic of
Western European Union (WEU) 150, 151
Westlands 331
Westmoreland, General William C. 219, 246
Weyland, General Fred C. 221, 247
White Paper, 1957 11, 55, 289
White Paper, 1966 11, 291, 394
White Paper, 1975 33, 131, 228, 326, 372
White Paper, 1976 77, 214, 238
White Paper, 1977 68, 77, 533
White, William D. 92, 193, 249
Williams, Alan Lee, MP: author of papers 203-8, 549, 550: biography 557; quoted 240, 249
Williams, Dr Geoffrey 312, 313
Wilson, Andrew 249

Wohlstetter, A. 91
Wolfe, Thomas W. 91
Wood, Derek 65, 304, 307, 312-13, 548

Yarmolinsky, A. 273, 277
Yarrow and Co Ltd. 207, 286, 288, 301
Yom Kippur War, 1973 195, 220, 221, 230, 481
Yugoslavia 72, 116

Zaire 540
Zambia 470, 540
Zimmerman, Steven M. 106, 113, 371, 384, 391, 392, 439